The Role of Pendrin in Health and Disease

Silvia Dossena • Markus Paulmichl
Editors

The Role of Pendrin in Health and Disease

Molecular and Functional Aspects of the SLC26A4 Anion Exchanger

Springer

Editors
Silvia Dossena
Paracelsus Medical University
Institute of Pharmacology
and Toxicology
Salzburg
Austria

Markus Paulmichl
Paracelsus Medical University
Institute of Pharmacology
and Toxicology
Salzburg
Austria

ISBN 978-3-319-43285-4 ISBN 978-3-319-43287-8 (eBook)
DOI 10.1007/978-3-319-43287-8

Library of Congress Control Number: 2017931637

Printed on acid-free paper

This Springer imprint is published by Springer Nature
The registered company is Springer International Publishing AG
The registered company address is: Gewerbestrasse 11, 6330 Cham, Switzerland

To Florian Lang, for the continuous support and his unconditional enthusiasm, which is an endless source of inspiration

Preface

Disease is the consequence of deranged molecular events affecting the function of cells, tissues, organs and eventually the whole body of living organisms. Therefore, understanding a pathophysiological process often requires a multidisciplinary approach and the joint efforts of scientists from the biochemical, biological, biomedical and clinical fields.

The Role of Pendrin in Health and Disease has the peculiarity of focusing on an individual molecule, of which a detailed description of the multifaceted function is provided from different points of view. Pendrin is a protein localized in the plasma membrane of cells in various tissues and organs and transports negatively charged ions. Malfunction of pendrin causes genetically inherited diseases of which hearing loss is the main feature, such as Pendred syndrome and non-syndromic deafness.

The main aim of this book is to summarize and contribute to the diffusion of the current knowledge on the genetics, molecular biology and physiology of pendrin as well as the derangements occurring as a consequence of alterations in its function. After an introductory historical background (Chap. 1), the book is organized into five parts, of which the first four illustrate the pathophysiology of pendrin in the inner ear, thyroid, kidney and airways, respectively. Specifically, Chaps. 2 and 3 describe mouse models that greatly contributed to the understanding of pendrin function and dysfunction. Chapters 4, 5 and 6 focus on clinical aspects of the inner ear and thyroid alterations in case of pendrin malfunction, and Chaps. 7, 8 and 9 illustrate how combined *in vitro* and *in vivo* studies integrate in the understanding of the molecular pathophysiology of pendrin. Part V includes two chapters focusing on the genetic, epigenetic, structural and molecular aspects of pendrin.

Although mainly focused on pendrin, some chapters of this book may be of great interest for scientists not directly involved in pendrin research. Specifically, Chap. 4 includes the anatomy of the inner ear, Chap. 5 is a precious guide for geneticists in the challenging effort of providing a genetic diagnosis of deafness and Chap. 6 illustrates aspects of thyroid pathology. This book, besides providing scientists with a detailed picture of the current knowledge on pendrin, may help otolaryngologists and endocrinologists in the diagnosis of Pendred syndrome and pendrin-related deafness and can be a useful tool for PhD and undergraduate students undertaking studies in physiology and pathology.

We have appreciated the contribution of scientists actively engaged since years in the research on various aspects of pendrin genetics, pathophysiology and pharmacology; to ensure a fresh perspective, some of the chapters are co-authored by young investigators.

We gratefully acknowledge all authors that, after working at the laboratory bench, patient's bedside or in the classroom, were able to find the time to contribute to *The Role of Pendrin in Health and Disease*.

Also, as science is an ever-evolving discipline and new knowledge is gathered quickly and continuously, we apologize for possible omission of information made accessible during the last steps of the editorial process of this book.

Salzburg, Austria — Silvia Dossena
Salzburg, Austria — Markus Paulmichl

Contents

Introduction 1

Silvia Dossena and Markus Paulmichl

1.1 Historical Background

Sometimes in science, it is important to address the right question. It was in 1896 that a young doctor, Vaughan Pendred (1869–1946), observed two cases of deaf-mutism and goiter in a large Irish family resident in Durham (Pearce 2007) and asked himself, as well as the readers of the prestigious medical journal *The Lancet*: "Why this association?" (Box 1.1) (Pendred 1896). In 1960, Fraser, Morgans, and Trotter proposed the eponymous of "Pendred syndrome" for defining the association of congenital deafness and sporadic goiter. The same authors attempted to explain the reason of this association with an interference between thyroxine synthesis and auditory function by toxic substances produced by an aberrant metabolic pathway (Fraser et al. 1960; Fraser 1965), a hypothesis that was never demonstrated. In the same period, Thould and Scowen postulated that deafness in Pendred syndrome could arise from an inadequate exposure to thyroid hormones *in utero* (Thould and Scowen 1964). This hypothesis was refuted by the fact that the nature of deafness caused by hypothyroidism – nerve (De Vos 1963) or conduction type – is different from that observed in Pendred syndrome (perceptive, i.e., sensorineural) (Woolf 1965), and the cause of the syndrome remained obscure.

Several years later, the advent of molecular genetics greatly accelerated the discovery of genes involved in disease. In 1996, Coyle and Sheffield independently showed that Pendred syndrome maps to chromosome 7q, in the region containing the gene of nonsyndromic deafness DFNB4 (Coyle et al. 1996; Sheffield et al. 1996). In 1997, Everett and collaborators, using a positional cloning strategy, identified the gene mutated in Pendred syndrome (formerly called *PDS*) and described three potentially deleterious sequence alterations segregating with the disease

S. Dossena (✉) • M. Paulmichl
Institute of Pharmacology and Toxicology, Paracelsus Medical University,
Strubergasse 21, Salzburg A-5020, Austria
e-mail: silvia.dossena@pmu.ac.at; paulmichl@me.com

S. Dossena, M. Paulmichl (eds.), *The Role of Pendrin in Health and Disease*,
DOI 10.1007/978-3-319-43287-8_1

(Everett et al. 1997). The same authors reported that the *PDS* gene produces a transcript of approximately 5 kb that was found to be expressed at significant levels in the thyroid and at lower levels in the kidney. The 780 amino acid (86 kD) protein predicted to be encoded by this transcript was named pendrin and was found to share significant sequence homology with known sulfate transporters. Therefore, these authors initially suggested that an altered sulfate transport may play a fundamental role in thyroid physiology and pathogenesis of congenital deafness. These findings readily boosted numerous studies aimed at the functional and molecular characterization of pendrin. In parallel, from genetic studies, the heterogeneity of pendrin sequence alterations associated with Pendred syndrome and nonsyndromic deafness became immediately obvious (Coyle et al. 1998). The importance of molecular analysis of the *PDS* gene in the assessment of families with congenital hearing loss and to obtain a definite diagnosis of Pendred syndrome was firmly established (Coyle et al. 1998; Kopp 1999).

A few years later, functional studies showed that the pendrin protein does not transport sulfate (Scott et al. 1999; Kraiem et al. 1999; Bogazzi et al. 2000), but rather shows affinity for monovalent anions such as iodide, chloride, and bicarbonate (Scott et al. 1999; Royaux et al. 2000, 2001; Soleimani et al. 2001). Being that pendrin is expressed in the thyroid (Royaux et al. 2000) and inner ear (Everett et al. 1999), an impairment in its function may cause pathological changes in both organs, leading to deafness and goiter. Accordingly, a mouse model lacking pendrin expression manifests inner ear malformations similar to those encountered in patients with Pendred syndrome. Pendred's question was finally answered.

1.2 Perspectives and Open Questions

Often in science, the answer to a question germinates a myriad of new questions. After its cloning, an enormous amount of work was devoted to define the physiological role of pendrin in the different organs in which it is expressed, and to understand the precise cellular events leading to disease following alteration of pendrin function by genetic mutation. In this scenario, several aspects still remain to be fully elucidated. For example, the striking intrafamilial variability in terms of deafness and goiter phenotype in Pendred syndrome has always intrigued scientists: What other factors, besides pendrin mutations, contribute to the severity of the disease? Why are some individuals with biallelic pendrin mutations affected by nonsyndromic deafness (DFNB4), while others develop the full spectrum of symptoms typical of Pendred syndrome? What are the precise molecular and cellular events culminating to inner ear malformations and deafness following alteration of pendrin activity?

Also, the precise functional role of pendrin in the thyroid is still a matter of debate. Is pendrin the true apical iodide transporter of the thyrocyte, responsible for the iodide flux into the follicular lumen, or does it rather control the cytosolic and/or follicular pH by acting as a transporter of bicarbonate? If pendrin is the apical iodide transporter of the thyrocyte, why do mice models lacking pendrin expression not develop goiter? On the other hand, if pendrin is not the apical iodide transporter, why do patients with Pendred syndrome develop thyroid dysfunction and goiter?

Besides being expressed in the inner ear and thyroid, pendrin is also found in the kidneys, airways, and several other tissues, although the expression levels are lower compared to those of the thyroid. Although kidney and lung diseases are not manifested in Pendred syndrome, emerging evidence points to an important role of pendrin in controlling the acid–base status of the organism, blood pressure, and airway surface liquid volume.

Therefore, pendrin seems to be an attractive drug target, and it remains to be seen whether or not the transporter can be effectively and safely targeted in the treatment of hearing loss, hypertensive states, or airway distresses.

Box 1.1

The original article of V. Pendred, describing the disorder which came to be known as Pendred syndrome. Taken from Pendred (1896), with permission.

Clinical Notes:

Medical, Surgical, Obstetrical, and Therapeutical.

Deaf-Mutism and Goitre.

By Vaughan Pendred, M.R.C.S. Eng., L.R.C.P. Lond., Late house surgeon Guy's Hospital

The curious association of deaf-mutism and goiter occurring in two members of a large family has induced me to record these cases. Why this association? Perhaps, some readers of THE LANCET may be able to throw some light on the causes of this combination of diseases: Absence of thyroid–cretinism; overgrowth of thyroid–deaf-mutism. I append the family history as recounted to me by the mother. The family is an Irish one, and the parents have been upwards of 40 years resident in Durham. The father, aged 66 years, and the mother, aged 67 years, are alive and healthy. They have had ten children, five sons, and five daughters. In an episode of smallpox 25 years ago, the whole family was attacked with the exception of the younger of the deaf-mutes, and four males and one female died, although all had been vaccinated, and this recently, as they were children. The remaining son and two of the daughters are healthy and vigorous. The first goiter case is the first-born of the family – a spare woman now aged 38 years. She is deaf and can only mumble indistinctly; little care has been taken to educate her and so she is imbecile. The goiter is a large multilobular hard tumor, the greater part on the right side of the neck; from time to time, she suffers from dyspneic attacks. The growth was first observed after the smallpox, i.e., at 13 years of age. The second surviving girl is now aged 28 years and is the fifth of the family; she is a small, spare, intelligent woman, her expression being in marked contrast to her sister's. She is not absolutely deaf and can mumble incoherently; her education has been attended to with so much success that she has been "in service." The tumor is larger than in the other case, but is of the same character; it has been growing for about 15 years, and during the last year has caused both dyspnea and dysphagia, which have become so urgent that I have sent her today to Newcastle Infirmary for operation.

Durham.

References

Bogazzi F, Bartalena L, Raggi F, Ultimieri F, Martino E (2000) Pendrin does not increase sulfate uptake in mammalian COS-7 cells. J Endocrinol Invest 23(3):170–172

Coyle B, Coffey R, Armour JA, Gausden E, Hochberg Z, Grossman A, Britton K, Pembrey M, Reardon W, Trembath R (1996) Pendred syndrome (goitre and sensorineural hearing loss) maps to chromosome 7 in the region containing the nonsyndromic deafness gene DFNB4. Nat Genet 12(4):421–423

Coyle B, Reardon W, Herbrick JA, Tsui LC, Gausden E, Lee J, Coffey R, Grueters A, Grossman A, Phelps PD, Luxon L, Kendall-Taylor P, Scherer SW, Trembath RC (1998) Molecular analysis of the *PDS* gene in Pendred syndrome. Hum Mol Genet 7(7):1105–1112

de Vos JA (1963) Deafness in hypothyroidism. J Laryngol Otol 77:390–414

Everett LA, Glaser B, Beck JC, Idol JR, Buchs A, Heyman M, Adawi F, Hazani E, Nassir E, Baxevanis AD, Sheffield VC, Green ED (1997) Pendred syndrome is caused by mutations in a putative sulphate transporter gene (*PDS*). Nat Genet 17(4):411–422

Everett LA, Morsli H, Wu DK, Green ED (1999) Expression pattern of the mouse ortholog of the Pendred's syndrome gene (*Pds*) suggests a key role for pendrin in the inner ear. Proc Natl Acad Sci U S A 96(17):9727–9732

Fraser GR (1965) Association of congenital deafness with goitre (Pendred's syndrome) A study of 207 families. Ann Hum Genet 28:201–249

Fraser GR, Morgans ME, Trotter WR (1960) The syndrome of sporadic goitre and congenital deafness. Q J Med 29:279–295

Kopp P (1999) Pendred's syndrome: identification of the genetic defect a century after its recognition. Thyroid 9(1):65–69

Kraiem Z, Heinrich R, Sadeh O, Shiloni E, Nassir E, Hazani E, Glaser B (1999) Sulfate transport is not impaired in pendred syndrome thyrocytes. J Clin Endocrinol Metab 84(7):2574–2576

Pearce JM (2007) Pendred's syndrome. Eur Neurol 58(3):189–190

Pendred V (1896) Deaf-mutism and goitre. Lancet 148(3808):532

Royaux IE, Suzuki K, Mori A, Katoh R, Everett LA, Kohn LD, Green ED (2000) Pendrin, the protein encoded by the Pendred syndrome gene (*PDS*), is an apical porter of iodide in the thyroid and is regulated by thyroglobulin in FRTL-5 cells. Endocrinology 141(2):839–845

Royaux IE, Wall SM, Karniski LP, Everett LA, Suzuki K, Knepper MA, Green ED (2001) Pendrin, encoded by the Pendred syndrome gene, resides in the apical region of renal intercalated cells and mediates bicarbonate secretion. Proc Natl Acad Sci U S A 98(7):4221–4226

Scott DA, Wang R, Kreman TM, Sheffield VC, Karniski LP (1999) The Pendred syndrome gene encodes a chloride-iodide transport protein. Nat Genet 21(4):440–443

Sheffield VC, Kraiem Z, Beck JC, Nishimura D, Stone EM, Salameh M, Sadeh O, Glaser B (1996) Pendred syndrome maps to chromosome 7q21-34 and is caused by an intrinsic defect in thyroid iodine organification. Nat Genet 12(4):424–426

Soleimani M, Greeley T, Petrovic S, Wang Z, Amlal H, Kopp P, Burnham CE (2001) Pendrin: an apical $Cl^-/OH^-/HCO_3^-$ exchanger in the kidney cortex. Am J Physiol Renal Physiol 280(2):F356–F364

Thould AK, Scowen EF (1964) Genetic studies of the syndrome of congenital deafness and simple goitre. Ann Hum Genet 27:283–293

Woolf DL (1965) Congenital deafness and sporadic goitre. Br Med J 1(5449):1511–1512

Part I

The Role of Pendrin in the Inner Ear

Mouse Models Reveal the Role of Pendrin in the Inner Ear

2

Philine Wangemann and Andrew J. Griffith

Abstract

In 1896, Vaughan Pendred, MD, wrote a case report about two siblings that presented with hearing loss and goiter. This initial report was followed over the next 100 years with additional cases, and the condition became known as Pendred syndrome. The underlying gene, *SLC26A4*, which codes for the protein pendrin, was discovered in 1997, and mutations of *SLC26A4* have since been recognized to underlie not only Pendred syndrome but also nonsyndromic hearing loss associated with an enlargement of the vestibular aqueduct (EVA) and variable deficits in vestibular function. In 2001, Dr. Lorraine Everett, in a team led by Dr. Eric Green, reported the first mouse model that recapitulates EVA, deafness, and vestibular dysfunction. This and other mouse models have proven to be tremendously valuable in the quest to understand the role of pendrin in hearing and vestibular function. This chapter summarizes work on these mouse models that are revealing the role of pendrin in the inner ear.

P. Wangemann (✉)
Anatomy & Physiology Department, Kansas State University, Manhattan, KS, USA
e-mail: wange@vet.ksu.edu

A.J. Griffith
Otolaryngology Branch, National Institute on Deafness and Other Communication Disorders, National Institutes of Health, Bethesda, MD 20892, USA

S. Dossena, M. Paulmichl (eds.), *The Role of Pendrin in Health and Disease*,
DOI 10.1007/978-3-319-43287-8_2

2.1 Introduction[1]

The gene *SLC26A4* (MIM #605646) codes for the protein pendrin, which is an electroneutral exchanger for anions such as HCO_3^-, Cl^-, and I^- (Scott et al. 1999; Scott and Karniski 2000; Soleimani et al. 2001). Pendrin is expressed in the apical membrane of epithelial cells in the inner ear, the thyroid, and the kidney (Royaux et al. 2000, 2001; Wangemann et al. 2004). Additional expression has been found in airways, mammary glands, uterus, placenta, liver, platelets, and adrenal glands (Bidart et al. 2000; Suzuki et al. 2002; Rillema and Hill 2003; Nakao et al. 2008; Alesutan et al. 2011; Pelzl et al. 2013; Lazo-Fernandez et al. 2015). The expression of *SLC26A4* in the inner ear and the thyroid is consistent with the observations that mutations of *SLC26A4* cause hearing loss associated with an enlarged vestibular aqueduct (EVA; MIM #600791), Mondini-like dysplasia of the cochlea, vestibular dysfunction, and enlargement of the thyroid (Pendred syndrome; MIM #274600). The prevalence and spectra of *SLC26A4* mutations vary among different populations (Park et al. 2003; Tsukada et al. 2015). The highest prevalence of *SLC26A4* mutations has been reported in East Asian populations in China, Taiwan, Japan, and Korea. Notably, in some Chinese populations, 13.7, 16.8, or 18.6 % of children attending schools for the deaf carry mutations of *SLC26A4* (Yuan et al. 2009; Zhu et al. 2015; Jiang et al. 2015). Between 82 and 98 % of East Asian patients with EVA carry mutations of *SLC26A4* (Choi et al. 2009; Tsukada et al. 2015). The prevalence of *SLC26A4* mutations in European and North American Caucasians is much lower. About 50 % of North American and European Caucasian patients with EVA carry mutations of *SLC26A4* (Campbell et al. 2001). Of these patients, about one-half carry one detectable mutation in the coding regions and splice sites of *SLC26A4*, whereas the other one-half of patients carry mutations on both alleles. Hearing loss is sometimes congenital, but typically fluctuating and progressive with an onset before or around the time of oral-auditory speech and language acquisition (Pryor et al. 2005; Choi et al. 2009; Miyagawa et al. 2014). Vestibular dysfunction is less overt and less common (Abe et al. 1999; Sugiura et al. 2005; Suzuki et al. 2007; Zhou and Gopen 2011; Zalewski et al. 2015; Jung et al. 2016), and the onset of goiter occurs typically after puberty (Fraser et al. 1960; Fraser 1965; Ajij et al. 2016). In contrast to mutations of other genes that cause hearing loss, there is no close correlation between mutation type and phenotype for *SLC26A4* (Choi et al. 2009; Tsukada et al. 2015). However, in north American and European Caucasian EVA patients, the number of mutant alleles of *SLC26A4* is correlated with both the thyroid and auditory phenotypes. Bilateral EVA and a biochemical defect in thyroid iodine organification are penetrant features of two mutant alleles of *SLC26A4*, whereas unilateral EVA and a normal thyroid iodination phenotype are observed in association with one mutant allele of *SLC26A4*. The incomplete penetrance of EVA and variability in the severity of hearing loss suggest that additional genetic or environmental factors play a

[1] With permission of S. Karger AG this chapter represents an update of an article published in 2013 (Wangemann 2013).

critical role, which merits further investigations into the physiology and pathobiology of pendrin-related deafness with the goal to ameliorate or prevent hearing loss in individuals carrying mutations of *SLC26A4*.

Most research into the physiology and pathobiology of pendrin-associated hearing loss has been carried out in mice. The goal of this review is to provide a brief summary of studies in mouse models that have been developed to delineate the role of pendrin in the physiology of hearing and balance with the ultimate goal to develop strategies to preserve hearing in individuals that carry mutations of *SLC26A4*. Clinical phenotypes and the genetics of hearing loss associated with EVA are reviewed elsewhere (Griffith and Wangemann 2011; Ito et al. 2011, 2013).

2.2 Synopsis of the Development of the Murine Inner Ear

The development of the inner ear begins in mice at embryonic day 9.5 (E9.5) with the formation of an otocyst (*reviewed in*: (Mansour and Schoenwolf 2005)). The otocyst encloses amniotic fluid, which is a plasma-like fluid containing ~140 mM Na^+ and ~10 mM K^+ (Cheung and Brace 2005). Between E10 and E10.5, two protrusions begin to extend from the otocyst: one forms the cochlea and the other forms the endolymphatic sac. While the protrusions elongate, the center of the otocyst reorganizes into the vestibular labyrinth. The lumen of the endolymphatic sac opens at E10.5, and the lumen of the cochlea opens at E14.5 (Kim and Wangemann 2010). Lumen formation during the growth phase of the inner ear is controlled by a balance of fluid secretion and absorption. Fluid secretion appears to occur in the vestibular labyrinth and fluid absorption in the endolymphatic sac (Kim and Wangemann 2010). The hypothesis that NaCl secretion and NaCl absorption control luminal volume during the growth phase of the inner ear is consistent with the finding that the luminal fluid in the cochlea is filled until E17.5 with a solution that contains ~140 mM Na^+, ~126 mM Cl^-, ~10 mM K^+, and ~25 mM HCO_3^- (Kim and Wangemann 2011, Li et al. 2013b).

At E17.5, closure of the utricular-saccular duct separates the interconnected fluid compartments of the inner ear into two fluid systems (Cantos et al. 2000). One system consists of the cochlea, the saccule, and the endolymphatic sac, and the other consists of the utricle, the three *ampullae*, and the three semicircular canals (Fig. 2.1). The separation of the vestibular labyrinth from the cochlea coincides with the acquisition of mechanosensitivity of vestibular hair cells at E17 (Geleoc and Holt 2003; Lelli et al. 2009). The onset of vestibular function occurs ~8 days later, at postnatal day 4 (P4), concurrent with the general maturation of the organ and the maturation of the innervation (Nordemar 1983; Desmadryl and Sans 1990). With the conclusion of the growth phase of the inner ear, the luminal fluid, endolymph, in the cochlea and the vestibular labyrinth changes from a NaCl-rich to a KCl-rich solution. The onset of K^+ secretion in the utricle is currently under investigation. The onset of K^+ secretion in the cochlea, however, has been found to occur at ~ E19.5. Two days later, at P0, when cochlear sensory cells

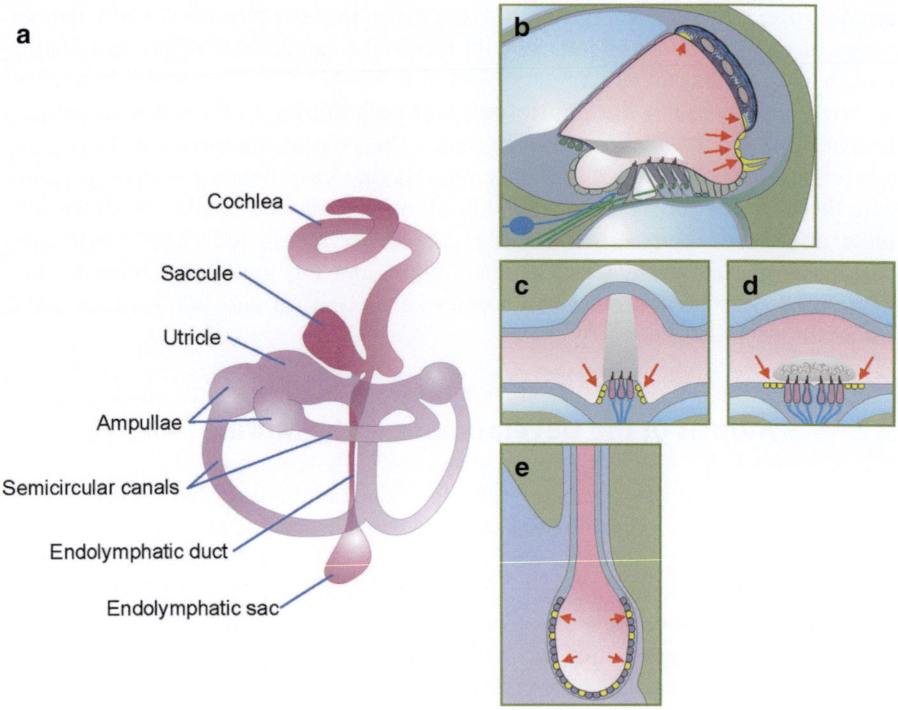

Fig. 2.1 Pendrin expression in the inner ear. (**a**) Schematic overview. The inner ear consists of the cochlea, saccule, and utricle and three *ampullae* with semicircular canals and the endolymphatic sac and duct. All compartments are lined with epithelial cells and filled with endolymph. (**b**) Cross section of one turn of the mature cochlea. Pendrin is expressed in epithelial cells of the spiral prominence (*long arrows*) and in spindle cells of *stria vascularis* (*short arrows*). (**c**–**d**) Cross sections of the saccule or utricle and an *ampulla*. Pendrin is expressed in transitional cells, which are epithelial cells surrounding the *maculae* in the saccule and utricle and the *cupulae* in the *ampullae*. (**e**) Cross sections of the endolymphatic duct and sac. The endolymphatic duct runs through a bony canal, which is called the vestibular aqueduct. Pendrin is expressed in mitochondria-rich cells in the endolymphatic sac (*arrows*). Similar diagrams were contributed to other papers (Li et al. 2013a, Wangemann 2013)

acquire mature mechanosensitivity, cochlear endolymph contains ~70 mM K^+ (Lelli et al. 2009, Li et al. 2013b). Five days later, at P3, endolymph contains ~100 mM K^+, which is close to the mature concentration of 150 mM K^+ (Li et al. 2013b). Concurrent with the rise of the K^+ concentration occurs a decline of the Na^+ concentration (Li et al. 2013b). The onset of hearing at ~P12 coincides with the maturation of the endocochlear potential, which rises between P5 and P15 from ~10 mV to the mature voltage of ~90 mV (Steel and Barkway 1989; Yamasaki et al. 2000; Wangemann et al. 2007). In the mature inner ear, the Na^+ concentration is ~1 mM and the Ca^{2+} concentration is 22 μM, both concentrations being unusually low for an extracellular fluid since most extracellular fluids contain ~150 mM Na^+ and 1.5 mM Ca^{2+}(*reviewed in*: Wangemann 2006).

2.3 Pendrin Expression in the Inner Ear

Pendrin in the inner ear functions as an exchanger of Cl^- and HCO_3^- and thereby may contribute to an elevated endolymphatic HCO_3^- concentration and an elevated endolymphatic pH (Nakaya et al. 2007; Wangemann et al. 2007; Kim and Wangemann 2011). In the mature inner ear, pendrin is expressed in the apical membrane of nonsensory epithelial cells in the cochlea, the vestibular labyrinth, and the endolymphatic sac (Royaux et al. 2003; Wangemann et al. 2004). During development, the earliest expression of pendrin in the inner ear occurs at E11.5 in the endolymphatic sac (Kim and Wangemann 2011). Between E13.5 and E14.5, expression in the endolymphatic sac increases dramatically. At E14.5, virtually all pendrin expression in the inner ear is located in the endolymphatic sac. In the cochlea, the earliest expression of pendrin is found in the hook region at E14.5. The hook region is the most basal part of the cochlea. Between E14.5 and E17.5, pendrin expression expands from the hook region to the lower and then to the upper turn of the cochlea. The onset of pendrin expression in the vestibular labyrinth occurs at E14.5 in the utricle and at E16.5 in the *ampullae*.

2.4 Mouse Models That Lack Pendrin Expression

The first mouse model, *Slc26a4*$^{\Delta/\Delta}$ (formerly called *Slc26a4*$^{-/-}$ or *Pds*$^{-/-}$), is a knockout mouse maintained in the 129S6 strain, in which exon 8 of *Slc26a4* is replaced with a neomycin cassette (Everett et al. 2001). The replacement introduces a frame shift, which prevents the generation of a functional protein. *Slc26a4*$^{\Delta/\Delta}$ mice develop an enlarged vestibular aqueduct and a Mondini-like dysplasia of the cochlea, fail to develop hearing and display vestibular deficits. The hearing and balance phenotype of *Slc26a4*$^{\Delta/\Delta}$ mice is more severe than the phenotype in most patients with DFNB4 or Pendred syndrome, who, in many cases, are born with residual hearing that often deteriorates within the first 3 years (Kim et al. 2015).

Consistent with the recessive inheritance pattern, *Slc26a4*$^{\Delta/+}$ mice develop normal sensory systems and normal hearing and balance. *Slc26a4*$^{\Delta/\Delta}$ and *Slc26a4*$^{\Delta/+}$ mice have been used extensively to investigate the consequence of a complete lack of pendrin for the development of the inner ear. Further, *Slc26a4*$^{\Delta/\Delta}$ and *Slc26a4*$^{\Delta/+}$ mice have been used as background for transgenic mouse models that feature temporally or spatially limited pendrin expression from a transgene (Choi et al. 2011, Li et al. 2013a).

Several of the phenotypic features of *Slc26a4*$^{\Delta/\Delta}$ mice have been observed in the pendrin-mutation knock-in mouse, *Slc26a4*$^{\mathrm{Tm1Dontuh/Tm1Dontuh}}$. *Slc26a4*$^{\mathrm{Tm1Dontuh/Tm1Dontuh}}$ mice, which are maintained in the C57BL/6 strain, contain a splice-site mutation at exon 8 of *Slc26a4*. This mutation introduces a frame shift and a new stop-codon designed to recapitulate the human mutation c.919-2A>G, which is the most prevalent mutation in China, Taiwan, and Mongolia (Lu et al. 2011; Tsukada et al. 2015). Similar to *Slc26a4*$^{\Delta/\Delta}$ mice, *Slc26a4*$^{\mathrm{Tm1Dontuh/Tm1Dontuh}}$ mice lack exon 8 of *Slc26a4*, which leads to a similar phenotype.

The phenotypes of *Slc26a4*$^{\Delta/\Delta}$ and *Slc26a4*$^{Tm1Dontuh/Tm1Dontuh}$ mice bear a clear resemblance to the human phenotype. In contrast, *Slc26a4*$^{Tm2Dontuh/Tm2Dontuh}$ mice, also maintained in the C57BL/6 strain, contain a single-nucleotide mutation designed to recapitulate the human mutation p. H723R, which is the most prevalent mutation in Japan and Korea (Lu et al. 2013; Tsukada et al. 2015). *Slc26a4*$^{Tm2Dontuh/Tm2Dontuh}$ mice, however, have normal hearing and balance (Lu et al. 2013). The basis for this latter observation is unknown, but illustrates the challenges in developing mouse models that closely approximate the human EVA phenotype.

2.4.1 Development of the Cochlea Without Pendrin

The first pathobiological alteration of the inner ear in *Slc26a4*$^{\Delta/\Delta}$ mice is the abnormal enlargement of the luminal volume that begins at E14.5. The enlargement coincides with cochlear lumen formation and persists throughout adulthood (Everett et al. 2001; Kim and Wangemann 2011). Formation of the enlargement during the growth phase of the inner ear appears to be the consequence of an imbalance between NaCl secretion and pendrin-dependent NaCl absorption. At E18.5, when the growth phase of the inner ear comes to an end, the enlargement amounts to a ~10-fold larger volume of scala media in the cochlea of *Slc26a4*$^{\Delta/\Delta}$ mice compared to that of *Slc26a4*$^{\Delta/+}$ mice.

The second pathobiological alteration in the inner ear in *Slc26a4*$^{\Delta/\Delta}$ mice is the acidification of cochlear endolymph that begins at E15.5 and coincides with the enlargement and the failed rise in pendrin expression at E14.5 (Kim and Wangemann 2011). Acidification of the luminal fluid also occurs in the endolymphatic sac and has been documented in the mature inner ear in the cochlea and the utricle of the vestibular labyrinth (Nakaya et al. 2007; Wangemann et al. 2007).

Luminal enlargement and acidification are the primary pathobiological alterations, which distribute the effect of pendrin deficiency from pendrin-expressing cells to the entire inner ear (Kim and Wangemann 2010). Luminal acidification alters pH-sensitive mechanisms, and luminal enlargement limits cell-to-cell communication mechanisms that rely on contact or diffusible factors transmitted via the luminal or abluminal compartment. A remarkable number of secondary consequences of the lack of pendrin expression have been observed. Impaired cell-to-cell communication may be responsible for the premature onset of connexin 26 expression in basal cells of *stria vascularis* at E18.5, for the retarded development of the layered structure of *stria vascularis* observed at P3, the retarded development of the organ of Corti between age P5 and P10 that includes a delayed opening of the tunnel, failure to express BK K^+ channels in inner hair cells, and a delayed innervation of the organ of Corti (Wangemann et al. 2009; Kim and Wangemann 2011).

One of the diffusible factors affected by the impaired cell-to-cell communication appears to be thyroid hormone (T3) that is generated from the pro-hormone thyroxine (T4) produced in the thyroid gland and delivered to the inner ear via the vasculature. In the cochlea, T3 is generated from T4 in fibrocytes located in the modiolus, the spiral limbus, and the spiral ligament. Receptors for T3, however, are located in the organ of Corti and the cochlear capsule. Enlargement and consequently delays in thyroid hormone signaling lead to signs of a local cochlear hypothyroidism that is evident between P5 and P10 from a delayed opening of the tunnel of Corti, delayed innervation, thickening of the tectorial membrane, failure to express BK K^+ channels in inner hair cells, and a delayed ossification of the cochlear capsule (Wangemann et al. 2009; Kim and Wangemann 2011; Dror et al. 2014).

Additional consequences of the enlargement include an increase in the rate of K^+ secretion by strial marginal cells (Li et al. 2013b), an increase in the rate of Na^+ absorption by Reissner's membrane epithelial cells (Kim et al. 2014), oxidative and nitrative stress in *stria vascularis* (Singh and Wangemann 2008), loss of the K^+ channel KCNJ10 in intermediate cells of *stria vascularis* and loss of the endocochlear potential (Wangemann et al. 2004), a rise in the endolymphatic Ca^{2+} concentration (Wangemann et al. 2007), and finally degeneration of sensory cells and *stria vascularis* (Jabba et al. 2006). The increase in the rate of K^+ secretion is evident from the finding that differences in endolymph K^+ concentrations between *Slc26a4*$^{\Delta/\Delta}$ and *Slc26a4*$^{\Delta/+}$ mice never exceed a factor of 2, while the volume of endolymph differs by a factor of 10 (Kim and Wangemann 2010, Li et al. 2013b). Whether the increase in the rate of K^+ secretion is a function of the enlarged luminal volume or a function of the lower pH is not known. Marginal cells sense K^+ concentrations at the apical membrane by an unknown mechanism, and low apical K^+ concentrations lead to an increase in the rate of K^+ secretion (Wangemann et al. 1995; Wangemann et al. 1996). A pH effect on K^+ secretion is also conceivable since the K^+ channel KCNQ1/KCNE1 (formerly known as KvLQT1/minK or IsK) in the apical membrane of marginal cells is activated by extracellular acidification (Unsold et al. 2000). In spite of the elevated rate of K^+ secretion, the K^+ concentration in endolymph remains slightly lower in *Slc26a4*$^{\Delta/\Delta}$ compared to *Slc26a4*$^{\Delta/+}$ or *Slc26a4*$^{+/+}$ mice (Royaux et al. 2003, Li et al. 2013b). The lower K^+ concentration may be balanced by a higher endolymphatic Na^+ concentration, which, in turn, may be responsible for the observed upregulation of Na^+ reabsorption and ENaC in Reissner's membrane epithelial cells (Kim et al. 2014).

A plausible consequence of higher metabolic rates necessary to maintain higher rates of K^+ secretion is oxidative and nitrative stress that results in elevated amounts of nitrated and oxidized proteins in *stria vascularis* of *Slc26a4*$^{\Delta/\Delta}$ mice (Singh and Wangemann 2008). Incompletely developed defense mechanisms may contribute to the free radical load. Expression levels of the K^+ channel KCNJ10 are reduced in the combined presence of oxidative and nitrative stress (Singh and Wangemann 2008), and loss of KCNJ10 is sufficient to abolish the endocochlear potential (Marcus et al. 2002). The endocochlear potential, however, depends not only on the expression of KCNJ10 but also on the integrity of epithelial and endothelial barriers that separate the intrastrial fluid space from endolymph, perilymph, and blood. Studies which used fluorescent beads (~20 nm diameter) as tracer have failed to demonstrate a compromise of the barrier between the intrastrial fluid space and blood; however, studies that used the smaller tracer biotin have demonstrated a compromise in the barrier between the intrastrial fluid space and perilymph in *Slc26a4*$^{\Delta/\Delta}$ mice (Ito et al. 2014). This compromise in the barrier integrity in conjunction with the loss of KCNJ10 appears to cause the observed loss of the endocochlear potential.

The loss of the endocochlear potential and the acidification of endolymph are likely the main factors that contribute to the elevation of the endolymphatic Ca^{2+} concentration in *Slc26a4*$^{\Delta/\Delta}$ mice. The concentration of Ca^{2+} in endolymph of normal mice, such as *Slc26a4*$^{\Delta/+}$ mice, is 22 μM, whereas the Ca^{2+} concentration in *Slc26a4*$^{\Delta/\Delta}$ mice is higher by a factor of ~100 (Wangemann et al. 2007). The reduction of the endocochlear potential could contribute to the observed elevation of the endolymphatic Ca^{2+} concentration since suppression of the endocochlear potential has been shown to coincide with a rise in the endolymphatic Ca^{2+} concentration (Ninoyu and Meyer zum Gottesberge 1986; Ikeda et al. 1987). The elevated luminal Ca^{2+} concentration in *Slc26a4*$^{\Delta/\Delta}$ mice is likely to contribute to the degeneration of sensory hair cells that becomes overt at P15 (Everett et al. 2001). Contributing factors may be the luminal acidification, local cochlear hypothyroidism, and the lack of the endocochlear potential. Similarly, marginal cells of *stria vascularis* degenerate after P15, and macrophages appear in stria vascularis of *Slc26a4*$^{\Delta/\Delta}$ mice at P30 (Jabba et al. 2006). Macrophages that accumulate in *stria vascularis* are strongly pigmented and give *stria vascularis* a dark appearance (Wangemann et al. 2004). It is unclear whether pigmentation is inherent to macrophages or whether pigmentation is acquired by phagocytosis of melanin-containing intermediate cells of *stria vascularis*.

2.4.2 Development of the Vestibular Labyrinth Without Pendrin

Secondary consequences of the lack of pendrin expression in the vestibular labyrinth include an increase in the endolymphatic Ca^{2+} concentration and the formation of giant otoconia which leads to vestibular dysfunction (Everett et al. 2001; Wangemann et al. 2004; Nakaya et al. 2007). Otoconia are protein-containing $CaCO_3$ crystals that enable the detection of linear acceleration in the utricle and saccule of the vestibular labyrinth. Normal otoconia vary in size but do not exceed ~20 μm, whereas giant otoconia reach sizes of ~200 μm. Giant otoconia have been reported in *Slc26a4*$^{\Delta/\Delta}$ mice as well as in *Slc26a4*$^{loop/loop}$ and *Slc26a4*$^{Tm1Dontuh/Tm1Dontuh}$ mice (Everett et al. 2001; Wangemann et al. 2004; Dror et al. 2010; Lu et al. 2011). The concentration of Ca^{2+} in vestibular endolymph of normal mice, such as *Slc26a4*$^{\Delta/+}$ mice, is 250 μM, whereas the Ca^{2+} concentration in *Slc26a4*$^{\Delta/\Delta}$ mice is higher by a factor of ~10 (Nakaya et al. 2007). The higher endolymphatic Ca^{2+} concentration may be a function of the larger transepithelial potential (−4 mV in *Slc26a4*$^{\Delta/\Delta}$ mice vs. −1.5 mV in *Slc26a4*$^{\Delta/+}$ mice), the more acidic pH (pH 7.1 in *Slc26a4*$^{\Delta/\Delta}$ mice vs. pH 7.4 in *Slc26a4*$^{\Delta/+}$ mice), and a failure to absorb Ca^{2+} (Nakaya et al. 2007; Yamauchi et al. 2010).

2.5 Mouse Models That Express Hypomorphic Pendrin

Slc26a4$^{loop/loop}$ mice were identified in a mutagenesis screen for neurosensory disorders and found to contain a point mutation, S408F, that reduces the anion exchange activity of pendrin without affecting protein expression (Dror et al. 2010). Interestingly, this hypomorphic mutant results in a phenotype that is similar to the phenotype of

Slc26a4$^{Δ/Δ}$ mice, which features a complete loss of pendrin expression. Similar to *Slc26a4*$^{Δ/Δ}$ mice, *Slc26a4*$^{loop/loop}$ mice do not acquire hearing or vestibular function, develop an enlargement of the cochlea, and form giant otoconia in the vestibular labyrinth. Further, *Slc26a4*$^{loop/loop}$ mice, similar to *Slc26a4*$^{Δ/Δ}$ mice, develop signs of cochlear hypothyroidism including a thickened tectorial membrane with reduced beta-tectorin expression, a lack of BK K^+ channels in inner hair cells, and a delayed ossification of the temporal bone (Dror et al. 2014). These findings underscore the importance of sufficient pendrin function during development of the inner ear.

2.6 Mouse Models with Spatially Limited Pendrin Expression

Slc26a4 transcription is controlled by various regulatory elements in the promoter region of *Slc26a4* that allow the expression to be cell-specific and responsive to specific situations (Yang et al. 2007; Adler et al. 2008; Rozenfeld et al. 2012; Vanoni et al. 2013). Notably, expression in mitochondria-rich cells of the endolymphatic sac of the inner ear is driven by the transcription factor FOXI1, whereas expression in the cochlea and the vestibular labyrinth is driven by other regulators (Yang et al. 2007). FOXI1 in mitochondria-rich cells of the endolymphatic sac drives not only the expression of pendrin but also the expression of ATP6V1B1, a subunit of the H^+ ATPase that is expressed in the inner ear only in these cells (Vidarsson et al. 2009). FOXI1 and ATP6V1B1 have therefore been used to generate mouse models that exclude the expression of pendrin from the endolymphatic sac or that limit the expression of pendrin in the inner ear to the endolymphatic sac.

2.6.1 Inner Ears Without Pendrin Expression in the Endolymphatic Sac

The mouse model *Foxi1*$^{-/-}$, which lacks expression of the transcription factor FOXI1, expresses pendrin in the cochlea and in the vestibular labyrinth but lacks pendrin expression in the endolymphatic sac (Hulander et al. 2003). The observations that *Foxi1*$^{-/-}$ mice are deaf, have vestibular dysfunction, and develop an enlargement of the inner ear point to the importance of the endolymphatic sac for the development of the cochlea and the vestibular labyrinth.

2.6.2 Inner Ear Without Pendrin Expression in the Cochlea and Vestibular Labyrinth

Slc26a4$^{Δ/Δ}$ and *Slc26a4*$^{Δ/+}$ mice have been used as genetic background for a model that features a spatially limited pendrin expression (Li et al. 2013a). Tg(*B1-hPDS*)*Slc26a4*$^{Δ/Δ}$ mice contain a transgene, which consists of the promoter for human *ATP6V1B1* that drives the expression of human *SLC26A4*, formerly called *hPDS*. *ATP6V1B1* codes for the B1-subunit of the H^+ ATPase, which is expressed in mitochondria-rich cells of the endolymphatic sac but not in the cochlea or in the vestibular labyrinth. Thus, Tg(*B1-hPDS*)*Slc26a4*$^{\Delta/\Delta}$ mice express human pendrin in mitochondria-rich cells of the endolymphatic sac but lack pendrin expression in the cochlea and the vestibular labyrinth of the inner ear (Li et al. 2013a). Tg(*B1-hPDS*)*Slc26a4*$^{\Delta/\Delta}$ mice do not develop the enlargement typically observed in *Slc26a4*$^{\Delta/\Delta}$ mice. This finding demonstrates that restoration of pendrin expression in the endolymphatic sac is sufficient to restore fluid absorption in the endolymphatic sac and to restore normal fluid homeostasis throughout the inner ear. Most interestingly, hearing and balance are restored in Tg(*B1-hPDS*)*Slc26a4*$^{\Delta/\Delta}$ mice. These findings raise the possibility that a spatially limited therapy focused on the endolymphatic sac (a structure that is relatively remote from the cochlea) might restore normal hearing and balance. Although this finding suggests that the expression of pendrin in the cochlea and the vestibular labyrinth is dispensable for hearing and balance, it remains to be determined whether a cochlea without pendrin has a robust hearing phenotype or whether pendrin expression has a protective or homeostatic role in response to common stressors such as noise and aging.

2.7 Mouse Models with Temporally Limited Pendrin Expression

Slc26a4$^{\Delta/\Delta}$ and *Slc26a4*$^{\Delta/+}$ mice have also been used in combination with a binary transgenic line that permits temporally limited pendrin expression. Tg[E],Tg[R]*Slc26a4*$^{\Delta/\Delta}$ mice contain two transgenes, an effector transgene, Tg[E], which consists of the murine promoter of *Slc26a4* in series with the coding sequence of the tet-on transactivator rtTA, and a responder transgene, Tg[R], which consists of a response-element for the doxycycline-bound rtTA-regulating expression of a murine *Slc26a4* cDNA (Choi et al. 2011). Since rtTA expression is driven by an *Slc26a4* promoter and regulatory elements, the temporal and spatial domains of expression of pendrin are limited to those in which pendrin is expressed in wild-type mice. Both transgenes, which were randomly inserted at unlinked locations in the genome, were crossed with the *Slc26a4*$^{\Delta/\Delta}$ line so that pendrin is expressed in the inner ear when doxycycline is present. Omission of doxycycline prohibits expression, and withdrawal of doxycycline leads to a rapid cessation of pendrin expression. Thus, pendrin expression can be controlled through the administration of doxycycline. When doxycycline was present from conception onward, mice developed a normal anatomy of the inner ear with no evidence of enlargement and normal hearing thresholds similar to *Slc26a4*$^{\Delta/+}$ mice, and when doxycycline was absent, mice developed an enlargement of the inner ear and failed to acquire hearing similar to *Slc26a4*$^{\Delta/\Delta}$ mice. Interestingly, cessation of pendrin expression in a fully functional inner ear did not affect hearing (Choi et al. 2011). These findings demonstrate that pendrin is required for the development but not for the maintenance of hearing in a normally developed ear.

2.7.1 Pendrin Expression Is Required During a Critical Time Period During Development

The mouse model Tg[E],Tg[R]*Slc26a4*$^{\Delta/\Delta}$ was used to determine the time period during which pendrin expression is required for normal cochlear development (Choi et al. 2011). Mice developed normal hearing, when doxycycline was administered from conception onward. Normal hearing thresholds developed even when the normal onset of pendrin expression at E11.5 was delayed by 5 days to E16.5 in the conditional mutant mice. Further, mice developed normal hearing thresholds, when pendrin expression was terminated by withdrawal of doxycycline as early as P2. These data define E16.5 to P2 as the most critical time period during which pendrin is needed for the development of normal hearing thresholds (Choi et al. 2011). The time period needed for the development of an uncompromised endocochlear potential and a normal endolymphatic pH begin earlier and last longer, which may imply that the hearing phenotype may be less robust in spite of normal thresholds when pendrin expression is restricted to the most critical time period (Choi et al. 2011). Taken together, the observations open the prospect that a temporally limited therapy focused on the prenatal phase of development may restore hearing in individuals bearing mutations of *SLC26A4*.

2.7.2 Pendrin Deficiency During Development Leads to Degeneration of *Stria Vascularis* and Causes Fluctuating and Progressive Hearing Loss

The mouse model Tg[E],Tg[R]*Slc26a4*$^{\Delta/\Delta}$ was used to determine the effect of pendrin deficiency on hearing thresholds and on the stability of the hearing phenotype (Ito et al. 2014). Doxycycline was administered from conception until E17.5 and thus, pendrin expression was terminated within the critical time period. Mice developed a normal anatomy of the inner ear with no evidence of an enlargement; however, the hearing phenotype was not stable. Mice developed fluctuating and progressive hearing loss, which recapitulates a hearing phenotype often observed in patients with mutations of *SLC26A4* (Griffith and Wangemann 2011; Ito et al. 2014, 2015).

Several observations point to *stria vascularis* as the major player in the development of fluctuating hearing loss (Ito et al. 2014; Ito et al. 2015). Hearing thresholds correlated with the magnitude of the endocochlear potential in pendrin-deficient mice. The expression level of KCNJ10 was reduced, and the normal complex morphological interdigitations between marginal and intermediate cells were diminished (Ito et al. 2014). Progressive hearing loss correlated with degeneration of *stria vascularis* including loss of KCNQ1 expression and enlargement of apical cell surfaces of marginal cells, hyperpigmentation, and expression of macrophage markers such as CD68 (Ito et al. 2015). Degeneration of *stria vascularis* in pendrin-deficient mice resembles, to some extent, the degeneration observed, at a faster time scale, in *Slc26a4*$^{\Delta/\Delta}$ mice that are devoid of pendrin expression (Jabba et al. 2006). The link between pendrin deficiency and degeneration of *stria vascularis* may involve the pendrin-expressing spindle-shaped cells of *stria vascularis* (Wangemann et al. 2004; Nishio et al. 2016).

Spindle-shaped cells provide the upper and lower limits of *stria vascularis*, face endolymph on their apical membrane, and connect marginal cells to basal cells (Katagiri et al. 1968; Luciano et al. 1995). Unlike marginal cells, spindle-shaped cells do not express KCNQ1 in their apical membrane but do express pendrin and ATP-gated cation channels P2RX2 (Wangemann et al. 2004; Housley et al. 2013).

2.7.3 Reinstatement of Pendrin Expression Alleviates Fluctuating Hearing Loss

The mouse model Tg[E],Tg[R]*Slc26a4*$^{\Delta/\Delta}$ was further used to determine whether restoration of pendrin expression has a beneficial effect on hearing (Nishio et al. 2016). Doxycycline was administered from conception until E17.5, which resulted in the development of fluctuating and progressive hearing loss. Doxycycline was then reinstated either at P6 or at P30, which is before or after the establishment of the endocochlear potential and the onset of hearing. Reinstatement of pendrin expression at P6 alleviated hearing fluctuations, but reinstatement at P30 had no beneficial effect. An inverse correlation was noted between hearing thresholds and the amount of pendrin expression in the apical membrane of spindle-shaped cells. These observations underscore the importance of spindle-shaped cells for the homeostasis of *stria vascularis*, link deficient spindle-shaped cells to fluctuating hearing loss, and indicate that restoration of pendrin expression may be beneficial to ameliorate progression and fluctuations of hearing in individuals bearing mutations of *SLC26A4*.

Conclusions

Studies in mouse models have provided tremendous insights into the role of pendrin in inner ear development and the etiology of an enlargement of the vestibular aqueduct. Studies of pendrin-related mouse models have revealed pathobiological mechanisms that may have broad implications beyond hearing loss associated with loss-of-function or hypo-functional mutations of *SLC26A4*. The concept that a temporally and spatially limited therapy may be sufficient to restore normal hearing or to ameliorate fluctuations and progression of hearing loss provides an imperative to develop interventions that secure a lifetime of normal hearing for individuals bearing mutations of *SLC26A4*.

Acknowledgment The authors thank Dr. Thomas Friedman and Dr. Michael Hoa for critical review of the manuscript. The work was supported by Kansas State University and The National Institute on Deafness and Other Communication Disorders grants NIH-R01-DC012151 to P.W. and NIH-Z01-DC000060 and NIH-Z01-DC000060 to A.J.G.

References

Abe S, Usami S, Hoover DM et al (1999) Fluctuating sensorineural hearing loss associated with enlarged vestibular aqueduct maps to 7q31, the region containing the Pendred gene. Am J Med Genet 82:322–328

Adler L, Efrati E, Zelikovic I (2008) Molecular mechanisms of epithelial cell-specific expression and regulation of the human anion exchanger (pendrin) gene. Am J Physiol Cell Physiol 294:C1261–C1276
Ajij M, Shambhavi, Patra B et al (2016) Pendred syndrome in a newborn with neck swelling: a case report. J Trop Pediatr pii: fmw002
Alesutan I, Daryadel A, Mohebbi N et al (2011) Impact of bicarbonate, ammonium chloride, and acetazolamide on hepatic and renal SLC26A4 expression. Cell Physiol Biochem 28:553–558
Bidart JM, Lacroix L, Evain-Brion D et al (2000) Expression of Na^{+}/I^{-} symporter and Pendred syndrome genes in trophoblast cells. J Clin Endocrinol Metab 85:4367–4372
Campbell C, Cucci RA, Prasad S et al (2001) Pendred syndrome, DFNB4, and PDS/SLC26A4 identification of eight novel mutations and possible genotype-phenotype correlations. Hum Mutat 17:403–411
Cantos R, Cole LK, Acampora D et al (2000) Patterning of the mammalian cochlea. Proc Natl Acad Sci U S A 97:11707–11713
Cheung CY, Brace RA (2005) Amniotic fluid volume and composition in mouse pregnancy. J Soc Gynecol Investig 12:558–562
Choi BY, Stewart AK, Madeo AC et al (2009) Hypo-Functional SLC26A4 variants associated with nonsyndromic hearing loss and enlargement of the vestibular aqueduct: Genotype-phenotype correlation or coincidental polymorphisms? Hum Mutat 30:599–608
Choi BY, Kim HM, Ito T et al (2011) Mouse model of enlarged vestibular aqueducts defines temporal requirement of Slc26a4 expression for hearing acquisition. J Clin Invest 121:4516–4525
Desmadryl G, Sans A (1990) Afferent innervation patterns in *crista ampullaris* of the mouse during ontogenesis. Brain Res Dev Brain Res 52:183–189
Dror AA, Politi Y, Shahin H et al (2010) Calcium oxalate stone formation in the inner ear as a result of an *Slc26a4* mutation. J Biol Chem 285:21724–21735
Dror AA, Lenz DR, Shivatzki S et al (2014) Atrophic thyroid follicles and inner ear defects reminiscent of cochlear hypothyroidism in *Slc26a4*-related deafness. Mamm Genome 25:304–316
Everett LA, Belyantseva IA, Noben-Trauth K et al (2001) Targeted disruption of mouse *Pds* provides insight about the inner-ear defects encountered in Pendred syndrome. Hum Mol Genet 10:153–161
Fraser GR (1965) Association of congenital deafness with goitre (Pendred's syndrome) a study of 207 families. Ann Hum Genet 28:201–249
Fraser GR, Morgans ME, Trotter WR (1960) The syndrome of sporadic goiter and congenital deafness. Quart J Med 29:279–295
Geleoc GS, Holt JR (2003) Developmental acquisition of sensory transduction in hair cells of the mouse inner ear. Nat Neurosci 6:1019–1020
Griffith AJ, Wangemann P (2011) Hearing loss associated with enlargement of the vestibular aqueduct: mechanistic insights from clinical phenotypes, genotypes, and mouse models. Hear Res 281:11–17
Housley GD, Morton-Jones R, Vlajkovic SM et al (2013) ATP-gated ion channels mediate adaptation to elevated sound levels. Proc Natl Acad Sci U S A 110:7494–7499
Hulander M, Kiernan AE, Blomqvist SR et al (2003) Lack of pendrin expression leads to deafness and expansion of the endolymphatic compartment in inner ears of *Foxi1* null mutant mice. Development 130:2013–2025
Ikeda K, Kusakari J, Takasaka T et al (1987) The Ca^{2+} activity of cochlear endolymph of the guinea pig and the effect of inhibitors. Hear Res 26:117–125
Ito T, Choi BY, King KA et al (2011) *SLC26A4* genotypes and phenotypes associated with enlargement of the vestibular aqueduct. Cell Physiol Biochem 28:545–552
Ito T, Muskett J, Chattaraj P et al (2013) *SLC26A4* mutation testing for hearing loss associated with enlargement of the vestibular aqueduct. World J Otorhinolaryngol 3:26–34
Ito T, Li X, Kurima K et al (2014) Slc26a4-insufficiency causes fluctuating hearing loss and *stria vascularis* dysfunction. Neurobiol Dis 66:53–65
Ito T, Nishio A, Wangemann P et al (2015) Progressive irreversible hearing loss is caused by *stria vascularis* degeneration in an Slc26a4-insufficient mouse model of large vestibular aqueduct syndrome. Neuroscience 310:188–197

Jabba SV, Oelke A, Singh R et al (2006) Macrophage invasion contributes to degeneration of *stria vascularis* in Pendred syndrome mouse model. BMC Med 4:37
Jiang Y, Huang S, Deng T et al (2015) Mutation spectrum of common deafness-causing genes in patients with non-syndromic deafness in the Xiamen area, China. PLoS One 10, e0135088
Jung J, Seo YW, Choi JY et al (2016) Vestibular function is associated with residual low-frequency hearing loss in patients with bi-allelic mutations in the *SLC26A4* gene. Hear Res 335:33–39
Katagiri S, Kawamoto K, Watanuki K (1968) Some surface views of the *stria vascularis* and its adjacent areas. Acta Otolaryngol 66:386–398
Kim HM, Wangemann P (2010) Failure of fluid absorption in the endolymphatic sac initiates cochlear enlargement that leads to deafness in mice lacking *pendrin* expression. PLoS One 5, e14041
Kim HM, Wangemann P (2011) Epithelial cell stretching and luminal acidification lead to a retarded development of *stria vascularis* and deafness in mice lacking pendrin. PLoS One 6, e17949
Kim BG, Kim JY, Kim HN et al (2014) Developmental changes of ENaC expression and function in the inner ear of pendrin knock-out mice as a perspective on the development of endolymphatic hydrops. PLoS One 9, e95730
Kim BG, Roh KJ, Park AY et al (2015) Early deterioration of residual hearing in patients with *SLC26A4* mutations. Laryngoscope. doi:10.1002/lary.25786
Lazo-Fernandez Y, Aguilera G, Pham TD et al (2015) Pendrin localizes to the adrenal medulla and modulates catecholamine release. Am J Physiol Endocrinol Metab 309:E534–E545
Lelli A, Asai Y, Forge A et al (2009) Tonotopic gradient in the developmental acquisition of sensory transduction in outer hair cells of the mouse cochlea. J Neurophysiol 101:2961–2973
Li X, Sanneman JD, Harbidge DG et al (2013a) SLC26A4 targeted to the endolymphatic sac rescues hearing and balance in *Slc26a4* mutant mice. PLoS Genet 9, e1003641
Li X, Zhou F, Marcus DC et al (2013b) Endolymphatic Na^+ and K^+ concentrations during cochlear growth and enlargement in mice lacking Slc26a4/pendrin. PLoS One 8, e65977
Lu YC, Wu CC, Shen WS et al (2011) Establishment of a knock-in mouse model with the *SLC26A4* c.919-2A>G mutation and characterization of its pathology. PLoS One 6, e22150
Lu YC, Wu CC, Yang TH et al (2013) Differences in the pathogenicity of the p.H723R mutation of the common deafness-associated *SLC26A4* gene in humans and mice. PLoS One 8:e64906
Luciano L, Reiss G, Reale E (1995) The junctions of the spindle-shaped cells of the *stria vascularis*: a link that completes the barrier between perilymph and endolymph. Hear Res 85:199–209
Mansour SL, Schoenwolf GC (2005) Morphogenesis of the inner ear. In: Kelley MW, Wu D, Popper AN, Fay RR (eds) Development of the inner ear Springer handbook of auditory research. Springer, New York, pp 43–84
Marcus DC, Wu T, Wangemann P et al (2002) KCNJ10 (Kir4.1) potassium channel knockout abolishes endocochlear potential. Am J Physiol Cell Physiol 282:C403–C407
Miyagawa M, Nishio SY, Usami S (2014) Mutation spectrum and genotype-phenotype correlation of hearing loss patients caused by *SLC26A4* mutations in the Japanese: a large cohort study. J Hum Genet 59:262–268
Nakao I, Kanaji S, Ohta S et al (2008) Identification of pendrin as a common mediator for mucus production in bronchial asthma and chronic obstructive pulmonary disease. J Immunol 180:6262–6269
Nakaya K, Harbidge DG, Wangemann P et al (2007) Lack of pendrin HCO_3^- transport elevates vestibular endolymphatic [Ca^{2+}] by inhibition of acid-sensitive TRPV5 and TRPV6. Am J Physiol Renal Physiol 292:1314–1321
Ninoyu O, Meyer zum Gottesberge AM (1986) Changes in Ca^{++} activity and DC potential in experimentally induced endolymphatic hydrops. Arch Otorhinolaryngol 243:106–107
Nishio A, Ito T, Cheng H et al (2016) Slc26a4 expression prevents fluctuation of hearing in a mouse model of large vestibular aqueduct syndrome. Neuroscience 329:74–82
Nordemar H (1983) Postnatal development of the vestibular sensory epithelium in the mouse. Acta Otolaryngol 96:447–456

Park HJ, Shaukat S, Liu XZ et al (2003) Origins and frequencies of *SLC26A4 (PDS)* mutations in east and south Asians: global implications for the epidemiology of deafness. J Med Genet 40:242–248

Pelzl L, Fakhri H, Umbach AT et al (2013) Sgk1 sensitive pendrin expression in murine platelets. Cell Physiol Biochem 32:210–220

Pryor SP, Madeo AC, Reynolds JC et al (2005) *SLC26A4/PDS* genotype-phenotype correlation in hearing loss with enlargement of the vestibular aqueduct (EVA): evidence that Pendred syndrome and non-syndromic EVA are distinct clinical and genetic entities. J Med Genet 42:159–165

Rillema JA, Hill MA (2003) Prolactin regulation of the pendrin-iodide transporter in the mammary gland. Am J Physiol Endocrinol Metab 284:E25–E28

Royaux IE, Suzuki K, Mori A et al (2000) Pendrin, the protein encoded by the Pendred syndrome gene *(PDS)*, is an apical porter of iodide in the thyroid and is regulated by thyroglobulin in FRTL-5 cells. Endocrinology 141:839–845

Royaux IE, Wall SM, Karniski LP et al (2001) Pendrin, encoded by the Pendred syndrome gene, resides in the apical region of renal intercalated cells and mediates bicarbonate secretion. Proc Natl Acad Sci U S A 98:4221–4226

Royaux IE, Belyantseva IA, Wu T et al (2003) Localization and functional studies of pendrin in the mouse inner ear provide insight about the etiology of deafness in pendred syndrome. J Assoc Res Otolaryngol 4:394–404

Rozenfeld J, Tal O, Kladnitsky O et al (2012) The pendrin anion exchanger gene is transcriptionally regulated by uroguanylin: a novel enterorenal link. Am J Physiol Renal Physiol 302:F614–F624

Scott DA, Karniski LP (2000) Human pendrin expressed in *Xenopus laevis* oocytes mediates chloride/formate exchange. Am J Physiol Cell Physiol 278:C207–C211

Scott DA, Wang R, Kreman TM et al (1999) The Pendred syndrome gene encodes a chloride-iodide transport protein. Nat Genet 21:440–443

Singh R, Wangemann P (2008) Free radical stress mediated loss of Kcnj10 protein expression in *stria vascularis* contributes to deafness in Pendred syndrome mouse model. Am J Physiol Renal Physiol 294:F139–F148

Soleimani M, Greeley T, Petrovic S et al (2001) Pendrin: an apical $Cl^-/OH^-/HCO_3^-$ exchanger in the kidney cortex. Am J Physiol Renal Physiol 280:F356–F364

Steel KP, Barkway C (1989) Another role for melanocytes: their importance for normal *stria vascularis* development in the mammalian inner ear. Development 107:453–463

Sugiura M, Sato E, Nakashima T et al (2005) Long-term follow-up in patients with Pendred syndrome: vestibular, auditory and other phenotypes. Eur Arch Otorhinolaryngol 262:737–743

Suzuki K, Royaux IE, Everett LA et al (2002) Expression of *PDS/Pds*, the Pendred syndrome gene, in endometrium. J Clin Endocrinol Metab 87:938–941

Suzuki H, Oshima A, Tsukamoto K et al (2007) Clinical characteristics and genotype-phenotype correlation of hearing loss patients with *SLC26A4* mutations. Acta Otolaryngol 127: 1292–1297

Tsukada K, Nishio SY, Hattori M et al (2015) Ethnic-specific spectrum of *GJB2* and *SLC26A4* mutations: their origin and a literature review. Ann Otol Rhinol Laryngol 124(Suppl 1):61S–76S

Unsold B, Kerst G, Brousos H et al (2000) KCNE1 reverses the response of the human K^+ channel KCNQ1 to cytosolic pH changes and alters its pharmacology and sensitivity to temperature. Pflugers Arch 441:368–378

Vanoni S, Nofziger C, Dossena S et al (2013) The human pendrin promoter contains two N(4) GAS motifs with different functional relevance. Cell Physiol Biochem 32:238–248

Vidarsson H, Westergren R, Heglind M et al (2009) The forkhead transcription factor Foxi1 is a master regulator of vacuolar H-ATPase proton pump subunits in the inner ear, kidney and epididymis. PLoS One 4, e4471

Wangemann P (2006) Supporting sensory transduction: cochlear fluid homeostasis and the endocochlear potential. J Physiol 576:11–21

Wangemann P (2013) Mouse models for pendrin-associated loss of cochlear and vestibular function. Cell Physiol Biochem 32:157–165

Wangemann P, Liu J, Marcus DC (1995) Ion transport mechanisms responsible for K^+ secretion and the transepithelial voltage across marginal cells of *stria vascularis in vitro*. Hear Res 84:19–29

Wangemann P, Shen Z, Liu J (1996) K^+-induced stimulation of K^+ secretion involves activation of the I_{sK} channel in vestibular dark cells. Hear Res 100:201–210

Wangemann P, Itza EM, Albrecht B et al (2004) Loss of KCNJ10 protein expression abolishes endocochlear potential and causes deafness in Pendred syndrome mouse model. BMC Med 2:30

Wangemann P, Nakaya K, Wu T et al (2007) Loss of cochlear HCO_3^- secretion causes deafness via endolymphatic acidification and inhibition of Ca^{2+} reabsorption in a Pendred syndrome mouse model. Am J Physiol Renal Physiol 292:1345–1353

Wangemann P, Kim HM, Billings S et al (2009) Developmental delays consistent with cochlear hypothyroidism contribute to failure to develop hearing in mice lacking *Slc26a4*/pendrin expression. Am J Physiol Renal Physiol 297:F1435–F1447

Yamasaki M, Komune S, Shimozono M et al (2000) Development of monovalent ions in the endolymph in mouse cochlea. ORL J Otorhinolaryngol Relat Spec 62:241–246

Yamauchi D, Nakaya K, Raveendran NN et al (2010) Expression of epithelial calcium transport system in rat cochlea and vestibular labyrinth. BMC Physiol 10:1

Yang T, Vidarsson H, Rodrigo-Blomqvist S et al (2007) Transcriptional control of *SLC26A4* is involved in Pendred syndrome and nonsyndromic enlargement of vestibular aqueduct (DFNB4). Am J Hum Genet 80:1055–1063

Yuan Y, You Y, Huang D et al (2009) Comprehensive molecular etiology analysis of nonsyndromic hearing impairment from typical areas in China. J Transl Med 7:79

Zalewski CK, Chien WW, King KA et al (2015) Vestibular dysfunction in patients with enlarged vestibular aqueduct. Otolaryngol Head Neck Surg 153:257–262

Zhou G, Gopen Q (2011) Characteristics of vestibular evoked myogenic potentials in children with enlarged vestibular aqueduct. Laryngoscope 121:220–225

Zhu J, Cao Q, Zhang N et al (2015) A study of deafness-related genetic mutations as a basis for strategies to prevent hereditary hearing loss in Hebei, China. Intractable Rare Dis Res 4:131–138

The *Slc26a4*^{loop} Mouse Model for Pendred's Syndrome and Nonsyndromic Deafness

3

Amiel A. Dror and Karen B. Avraham

Abstract

Pendred's syndrome results from bi-allelic pathogenic variants in the *SLC26A4* gene and is characterized by sensorineural deafness and a partial thyroid iodine organification defect, with or without goiter. Vestibular dysfunction is occasionally present, although it is not part of the classic clinical definition of Pendred's syndrome. The pendrin protein, encoded by the *SLC26A4* gene, is responsible for transporting anions across cell membranes, including in the inner ear and thyroid. In the ear, pendrin functions as a HCO_3^-/Cl^- exchanger and has a key role in maintaining endolymph homeostasis. In the thyroid, pendrin function is still not clear, and efforts are being made to identify the molecular mechanisms that underlie the broad thyroid phenotypic variation of Pendred's patients. While environmental factors, such as an iodine-deficient diet, contribute to phenotypic heterogeneity, other strong genetics factors should be considered. Several mouse models bearing *Slc26a4* mutations have been generated. Unique calcium oxalate minerals have been found in the vestibule of *loop*, a mouse model with a hypofunctional allele of *Slc26a4*. The abnormal mineralized bodies were detected ectopically within the vestibular labyrinth. The thyroids of *Slc26a4*$^{loop/loop}$ mice are small with atrophic follicles. In this chapter, we summarize the unique findings of *Slc26a4*$^{loop/loop}$ mice in the vestibular system and thyroid gland. The existence of phenotypic variation observed among different *Slc26a4* mouse models is encouraging evidence for further investigating mutation-specific mechanisms for Pendred's syndrome. The ability to predict specific phenotypes for a given mutation will contribute to better clinical management and treatment, with an emphasis on preventive medicine.

A.A. Dror • K.B. Avraham (✉)
Department of Human Molecular Genetics and Biochemistry, Sackler Faculty of Medicine and Sagol School of Neuroscience, Tel Aviv University, Tel Aviv 6997801, Israel
e-mail: amieldro@post.tau.ac.il; karena@post.tau.ac.il

S. Dossena, M. Paulmichl (eds.), *The Role of Pendrin in Health and Disease*,
DOI 10.1007/978-3-319-43287-8_3

3.1 Introduction

Pendred's syndrome was first described in 1896 by Dr. Vaughan Pendred, who documented his observation of two sisters with familial deafness and goiter (Pendred 1896). Although goiter was originally described as part of the syndrome, only some Pendred's patients present with an enlarged gland (Reardon et al. 1997). A consistent thyroid defect of Pendred's syndrome is partial iodine organification, measured by abnormal excessive discharge of iodide from the thyroid gland, following potassium perchlorate administration (Fraser et al. 1960; Morgan and Trotter 1958). The serum level of thyroid hormone in Pendred's patients, however, may remain within normal range (euthyroid) or be abnormally low (hypothyroid) (Blons et al. 2004).

In contrast to the variable thyroid size seen in Pendred's syndrome, the anatomic malformation of the inner ear is consistent with common radiological hallmarks of enlarged vestibular aqueducts (EVA) (Valvassori and Clemis 1978). Pendred's syndrome is the most common form of deafness associated with EVA, and accounts for approximately 10 % of all cases of hereditary hearing loss (Fraser 1965; Park et al. 2003; Reardon et al. 1997). The hearing impairment is classically characterized with early onset, prior to acquisition of speech or during early childhood (Choi et al. 2009; Pryor et al. 2005). Cases of progressive hearing loss in adolescence have been reported as well, in association with head injury, infection, or delayed secondary hydrops (Luxon et al. 2003). The hearing loss of Pendred's patients is sensorineural in nature, and often progresses with a fluctuating pattern or stepwise hearing decrements, eventually leading to profound deafness (Jackler and De La Cruz 1989; Levenson et al. 1989). Despite the close proximity of auditory and vestibular systems within the inner ear, the penetrance of vestibular dysfunction in Pendred's syndrome is partial, and only some of the affected individuals are symptomatic. When symptoms are present, they are usually mild and include motor developmental delay and vertigo episodes that are rarely debilitatingg (Zalewski et al. 2015). A recent study demonstrated that vestibular dysfunction among patients with bi-allelic *SLC26A4* mutations is strongly linked to hearing loss at low frequencies. A potential correlation of vestibular phenotype with allelic predisposing factor has been excluded (Jung et al. 2016).

The cloning of the Pendred's syndrome gene, *SLC26A4*, has established the molecular foundation for further investigating the role of its encoded protein pendrin (Everett et al. 1997). As a member of a family of solute carrier transporters, pendrin functions as an electroneutral exchanger capable of transporting HCO_3^-, Cl^-, and I^- (Royaux et al. 2001; Scott et al. 1999; Soleimani et al. 2001). Pendrin is expressed in several tissues including the inner ear, thyroid, and kidney (Everett et al. 1997). In humans however, mutations in *SLC26A4* predominantly affect the inner ear and thyroid gland. In the inner ear, pendrin is expressed in the cochlea, vestibule, and endolymphatic sac, where it is localized to the apical surface of various nonsensory epithelial cells (Dou et al. 2004; Royaux et al. 2003; Wangemann et al. 2004). In the inner ear, pendrin functions as a HCO_3^-/Cl^- exchanger, generating an efflux of HCO_3^- ions from cells (Wangemann et al. 2007). The extracellular reservoir of HCO_3^- ions buffers the endolymphatic fluids of the inner ear and prevents its acidification. In the absence of pendrin, as seen in *Slc26a4*-null mice, a series of pathological developmental processes results in endolymphatic acidification, extracellular calcium accumulation, enlargement of the endolymphatic compartments, and loss of endocochlear potential (Kim and Wangemann 2011; Wangemann et al. 2004, 2007). A recent study of the conditional *Slc26a4* mouse model showed that delayed deletion of pendrin results in fluctuating hearing loss episodes, mimicking human EVA-related deafness (Ito et al. 2015). The study showed that the fluctuating pattern of hearing loss is attributed to cyclic pathological events that involve the failure and recovery of *stria vascularis* function. This research provides a robust mouse model to study an interventional approach for preventing EVA-related deafness during the critical time window prior to the irreversible loss of hearing. More details on the molecular pathogenetic mechanisms of pendrin mutations in the mouse inner ear are reviewed elsewhere ((Wangemann 2013) and Chap. 2).

Despite the vast knowledge gathered on pendrin function within the auditory system, in the thyroid gland, pendrin function has yet to be fully elucidated (see also Chap. 7). The localization of pendrin to the apical membrane of thyroid follicular cells, together with its iodide transport capacity, led to the hypothesis that pendrin participates in iodide efflux into the follicular lumen (Gillam et al. 2004; Royaux et al. 2000; Yoshida et al. 2002). However, the fact that some Pendred's patients do not develop a hypothyroid state argues against a major role of pendrin in maintaining follicular iodide level, which is essential for thyroid hormone production.

In this chapter, we discuss the unique pathological findings of the *Slc26a4*$^{loop/loop}$ mouse mutant that is informative for the variable thyroid and vestibular phenotype of Pendred's syndrome (Dror et al. 2010). In terms of auditory dysfunction, *loop* mice are phenocopies of other conventional *Slc26a4* mouse models. In the vestibular system and the thyroid gland, however, *loop* mice show distinctive findings. In the vestibule, giant pathological stones were found in the gravity sensory organs, the utricle and saccule, of *loop* mutants, instead of thousands of small otoconia that are normally detected. Furthermore, whereas the mineralized fraction of normal otoconia is composed of calcium carbonate, a pathological composition of calcium oxalate is found in the saccules of *Slc26a4*$^{loop/loop}$. Moreover, these giant otoconia are often dislocated from the *maculae* to ectopic positions within the vestibule, including the semicircular canal and the *cristae ampullares*. The thyroid gland pathology of *Slc26a4*$^{loop/loop}$ mice includes atrophic thyroid tissue with a majority of microfollicles (Dror et al. 2014). This phenotypic variability among *Slc26a4* mouse mutants emphasizes the importance of studying different mutations to fully elucidate the function of Pendred's syndrome gene. Understanding mutation-specific mechanisms will ultimately enable better prediction of a thyroid-related outcome, thereby providing clinicians with tools for establishing tailored approaches for better patient management.

3.2 The *Slc26a4*^{loop} Mouse Is a Model for Pendred's Syndrome

The *loop* mice were generated in an ENU mutagenesis screening program (Hrabe de Angelis et al. 2000). These mice failed to respond to the click box test, a screening tool for an initial hearing examination. A pure tone is produced for a short duration, and the response of the mice is observed. When hearing is normal, the click box test will trigger a prominent startle response, whereas deaf mice will continue their activity regardless of the sound stimulus. The vestibular perception of *loop* mutants was evaluated by several behavioral tests. In contrast to the consistent finding of profound hearing loss in all *loop* mutants, their vestibular impairment is highly variable, ranging from mild to severe dysfunction. The auditory and vestibular defects of *loop* mice segregate in a recessive mode of inheritance that mapped to mouse chromosome 7. Further sequencing of the candidate genes within this locus identified the causative mutation in the *Slc26a4* gene (Dror et al. 2010). The pathogenic variant c.1223C>T leads to a single amino acid change p.Ser408Phe, in a highly conserved transmembrane domain of the pendrin protein, according to bioinformatics analysis. The substitution of a serine (polar) to a phenylalanine (aromatic) amino acid significantly reduces pendrin transport activity as demonstrated by an in vitro functional assay of the *loop* mutation (Dror et al. 2010). The auditory brainstem recording shows that *loop* mice are profoundly deaf, with a threshold higher than 100 dB at all tested frequencies (8, 16, and 32 kHz). The anatomical malformation of *loop* inner ears includes enlargement of the vestibular aqueduct (EVA), expanded endolymphatic duct and sac, and increased endolymphatic compartment of the cochlea. Histological sections of the cochlea show a significant increase in the size of the scala media. On the other hand, the perilymphatic compartments, the scala vestibuli and the scala tympani, are significantly smaller. This gross malformation of the inner ear is similar to the previously described phenotype of the *Slc26a4*-null mice and resembles the EVA hallmark of Pendred's syndrome patients (Everett et al. 2001). *Loop* mice are phenocopies of *Slc26a4*-null mice in term of auditory manifestations; the unique findings of *Slc26a4*$^{loop/loop}$ vestibule and thyroid gland are discussed in the following sections.

3.3 *Slc26a4*^{loop} Is Informative for the Variable Vestibular Phenotype of Pendred's Syndrome

The most penetrant inner ear anomaly related to Pendred's syndrome is EVA, which associates with severe to profound hearing loss of affected individuals. Interestingly, despite the close anatomical proximity of the vestibular system to the auditory organ, vestibular symptoms among Pendred's patient are found in only some of the patients (Berrettini et al. 2005). When vestibular perception is affected, the symptoms range from mild to severe vertigo episodes with associated nystagmus (Sugiura et al. 2005). Similar to the variable vestibular dysfunction of humans, the *Slc26a4*$^{loop/loop}$ mouse model exhibits inconsistent vestibular deficits, including unsteady gait and circling behavior of various degrees of severity (Dror et al. 2010). The normal vestibular function depends on five sensory organs within the vestibular system, which contains specialized mechanosensory cells essential for transforming body movements into neurostimulation. Three *cristae ampullares* are responsible for perceiving rotational movements, and two gravity receptors, the utricle and saccule, are responsible for sensation of linear acceleration and deceleration. An extracellular matrix, associated with each sensory organ, is essential for hair cell activation upon mechanical stimulus. The saccule and utricle depend on a unique extracellular gelatinous matrix, covered with otoconia, which are calcite mineralized bodies that bear weight on the sensory epithelium. In-depth characterization of the vestibular system of *loop* mice at different ages revealed a gross pathological appearance of the saccule and utricle. Instead of thousands of otoconia, giant stones were apparent over the sensory *maculae* of postnatal day 0 (P0) *loop* mice (Fig. 3.1). Despite the significant increase in size, 50–100 times larger than a single normal otoconium, the giant stone morphology of newborn mutant mice was similar to that of wild-type otoconia and contained the same mineral composition of calcite. Interestingly, an age-related change of the mutant stone morphology and mineral composition is found exclusively in the saccule of 10-month-old *loop* mice. This progressive process results in the formation of unique calcium oxalate stones with a distinctive envelope shape morphology. Further characterization of the saccule stones at different time points between P0 and 10 months old reveals an intermediate form of mineralized bodies with a pitted surface, consistent with calcite mineral dissolution in an acidic environment (Dror et al. 2010).

The acidic environment has a major influence on the stability of crystals, depending on the composition of mineralized structures (Lund et al. 1975). Calcium oxalate will remain crystalized in a range of acidic solution that will lead to complete calcium carbonate dissolution. A comprehensive study on the function of pendrin as a Cl^-/HCO_3^- exchanger in the inner ear showed that pendrin dysfunction leads to endolymphatic fluid acidification due to the diminished supply of luminal HCO_3^- ions (Wangemann et al. 2007). It has been proposed that the originally formed calcium carbonate otoconia are gradually dissolved when the endolymphatic pH decreases as a result of pendrin dysfunction. Subsequently, the increase in free Ca^{+2} released from dissolved otoconia favors the crystallization with oxalate forming calcium oxalate stones that are resistant to endolymph acidification as compared to calcite otoconia.

Another pathological feature of the giant vestibular minerals in *Slc26a4^loop/loop^* mice is related to their anatomical position within the vestibular labyrinth. Whereas the native position of otoconia is confined to the saccule and utricle, in *Slc26a4^loop/loop^* mice giant ectopic otoconia were found throughout the vestibule. Prominently, giant otoconia at the *cristae ampullares* (cupulolithiasis) and within the semicircular canal (canalithiasis) of *Slc26a4^loop/loop^* mice were observed. It is reasonable to suggest that such spatial and temporal changes in the position and composition of the abnormal giant minerals can directly affect the characteristics of the observed vestibular behavior. The observation of recurrent benign paroxysmal positional vertigo (BPPV) in Pendred's patients has been attributed to the EVA labyrinthine-related disorder (Manzari 2008). Other studies have linked the paroxysmal attack of episodic vertigo and balance disorders to displacement of otoconia and occlusion of the hydrodynamics of semicircular canals (Epley 1995; Ko et al. 2014). This proposed BPPV mechanism led to the development of physical treatments of repositioning maneuvers (Dorigueto et al. 2005). The observed ectopic otoconia across the vestibular system of *Slc26a4* loop mice may represent the possible mechanism for BPPV in Pendred's syndrome.

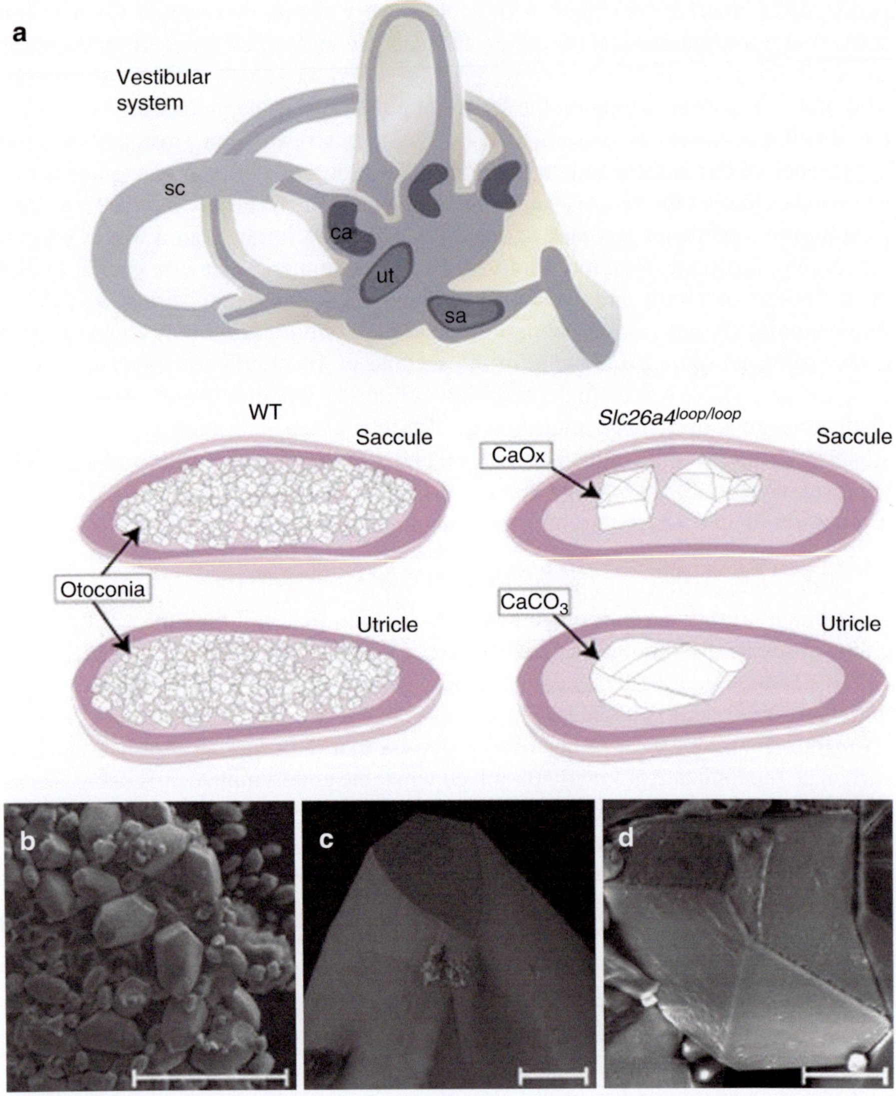

Fig. 3.1 Pathological giant otoconia of *Slc26a4*[loop] mutant mice. (**a**) Schematic drawing of the vestibular system is shown. The sensory receptors of the vestibule include three *cristae* and two gravity receptors (saccule and utricle). The saccule and utricle of wild-type mice are covered with thousands of small otoconia, whereas giant otoconia appear over the utricular and saccular *maculae* of mutant mice. (**b**) Scanning electron microscopy of wild-type otoconia composed of calcium carbonate crystals. (**c**) A giant calcium carbonate stone isolated from the utricle of mutant mice. (**d**) Of note is the pathological composition of calcium oxalate stones within the saccule of adult mutant mice. The scale bars represent 20 μm (**b**) and 100 μm (**c**, **d**). *sc* semicircular canal, *ca crista ampullaris*, *ut* utricle, *sa* saccule, *CaOx* calcium oxalate, *$CaCO_3$* calcium carbonate

3.4 *Slc26a4^loop^* Is Informative for the Variable Thyroid Phenotype of Pendred's Syndrome

The thyroid phenotype of Pendred's syndrome ranges from normal-sized thyroid glands with preserved hormone synthesis to goiter and congenital hypothyroidism in its most severe form (Reardon et al. 1997). Congenital hypothyroidism is the most common endocrine deficiency detected in the newborn period, with an incidence of 1 in 3,000–4,000 births (Haddow et al. 1999; Waller et al. 2000). Thus, neonatal screening programs are essential to ensure the early detection and replacement of the missing hormone (Buyukgebiz 2013). Intervention soon after birth is crucial to prevent later episodes of thyroid hormone deficiency that leads to cretinism, a condition of developmental delay and mental retardation (Syed 2015). Based on its pathophysiology, primary congenital hypothyroidism is divided into two main subclasses ((Agrawal et al. 2015); for a detailed description, see Chap. 6). Thyroid dysgenesis, which is usually a sporadic disorder, comprises 85 % of the cases and is referred to abnormal development of the gland. In this subclass, the thyroid gland is either missing, significantly small or in an ectopic position. In the remaining 15 % of cases, classified as thyroid dyshormonogenesis, hormone synthesis is impaired, but the thyroid gland is either normal in size or enlarged (goiter). This subclass is generally familial and results from inherited mutations in key genes for thyroid hormone biosynthesis (Grasberger and Refetoff 2011). One example for dyshormonogenesis is found in Pendred's syndrome that results from bi-allelic mutation in the *SLC26A4* gene encoding pendrin.

The initial approach to pendrin function in the thyroid is based on its known transport capacity of iodide (Scott et al. 1999). The positive thyroid perchlorate discharge test in Pendred's patients indicates that pendrin function is important for iodide organification, a key step of thyroid hormone synthesis (Fraser et al. 1960; Morgan and Trotter 1958). The synthesis of thyroid hormone is dependent on an iodine-rich environment of the colloid inside the follicular lumen. The route of iodide flows from blood capillaries into the colloid and passes through layers of thyrocytes, the thyroid follicular cells. At the baso-lateral side of the thyrocyte, the sodium-iodide symporter (NIS) facilitates iodide uptake from blood circulation into the cytoplasm ((Portulano et al. 2014); see also Chap. 6). The molecular identity of the apical iodide transport protein, by which iodide leaves the thyrocyte to the follicular lumen, is still under investigation (see Chap. 7). The localization of pendrin to thyrocyte apical membrane, together with its known iodide transport capacity, led to the hypothesis that pendrin functions as an apical iodide transporter in the thyroid (Royaux et al. 2000). On the other hand, the fact that only a portion of Pendred's syndrome patients are hypothyroid questions pendrin's putative function as the sole apical iodide transporter. It has been suggested that other proteins are redundant for pendrin function in the thyroid and compensate for it loss. These candidate proteins included the cystic fibrosis transmembrane conductance regulator (CFTR) and the chloride channel CLC-5 (Li et al. 2010; van den Hove et al. 2006). Another promising candidate is Anoctamin 1 (ANO1), a Ca^{2+}-activated chloride channel, which was shown to function as an apical iodide transporter of thyroid (Iosco et al. 2014;

Twyffels et al. 2014). A recent study shows that iodide excess increases pendrin abundance and insertion to thyrocyte plasma membrane and extends pendrin's half-life (Calil-Silveira et al. 2016). The iodide excess condition also led to an increase in iodide efflux from PCCl3 cells, suggesting that pendrin has a potential role in mediating iodide efflux in thyroid follicular cells. More details on the molecular mechanism of thyroid apical iodide transport are reviewed elsewhere ((Silveira and Kopp 2015) and Chap. 7).

In vitro studies of different pendrin mutations demonstrated the deleterious effect of various alleles on pendrin cellular localization, function, and expression (see Chap. 11). As a transmembrane protein, pendrin must reach the cell membrane in order to exert its function (Rotman-Pikielny et al. 2002). Mutations in the conserved transmembrane domains of pendrin lead to accumulation in the endoplasmic reticulum (Brownstein et al. 2008; Taylor et al. 2002). Other mutations have been shown to have a negative effect on pendrin transport capacity or ion affinity (Dossena et al. 2009; Pera et al. 2008). It has been proposed that residual transport capacity of a hypofunctional pendrin allele is consistent with euthyroid patients, whereas loss-of-function alleles lead to hypothyroidism (Scott et al. 2000). The consistency of this association has been refuted with the identification of pendrin-null alleles in patients with normal thyroid function, whereas significant goiter was documented in the presence of pendrin partial transport function (Ishinaga et al. 2002; Tsukamoto et al. 2003). Additional studies have been attempting to identify a genotype–phenotype correlation for the thyroid phenotypic variation among Pendred's syndrome patients, but no conclusive correlations have been found to date (Miyagawa et al. 2014; Suzuki et al. 2007). It is possible, however, that epigenetic modifications yet to be identified contribute to the wide spectrum of the SLC26A4-related thyroid phenotype (Xing et al. 2003). Moreover, mutations in transcriptional regulatory elements in the *SLC26A4* promoter may expand the molecular pathogenesis of Pendred syndrome in a dosage-dependent model (Yang et al. 2007). This model suggests that mutations in the *SLC26A4* promoter and the transcription factor *FOXI1*, along with mutations in *SLC26A4* or other genes, may lead to a reduction in gene expression, which would lead to a hearing or thyroid phenotype.

Several mouse models for Pendred's syndrome have been generated for pendrin functional investigation (see Chap. 2). The first mouse thyroid characterization came from a *Slc26a4*-null mouse mutant that shows a normal thyroid function with preserved gland histology (Everett et al. 2001). Furthermore, an iodine-deficient diet did not induce goiter in *Slc26a4*-null mice, which highlighted the redundant role of pendrin in thyroid iodine transport in the mouse (Iwata et al. 2011). This observation is consistent with the subgroup of Pendred's patients that are euthyroid. The *Slc26a4*$^{loop/loop}$ mice, however, show an atrophic smaller gland in comparison with littermate controls ((Dror et al. 2014) and Fig. 3.2). This unexpected structural defect came as a surprise and in contrast to the enlarged thyroid gland that is classically seen in Pendred's syndrome patients. Interestingly, a recent study aimed to identify new genetic mutations underlying thyroid dysgenesis discovered *SLC26A4* bi-allelic mutations in two patients with small and absent thyroid tissue (Kuhnen et al. 2014). This abnormal hypoplastic thyroid tissue stands in contrast to the

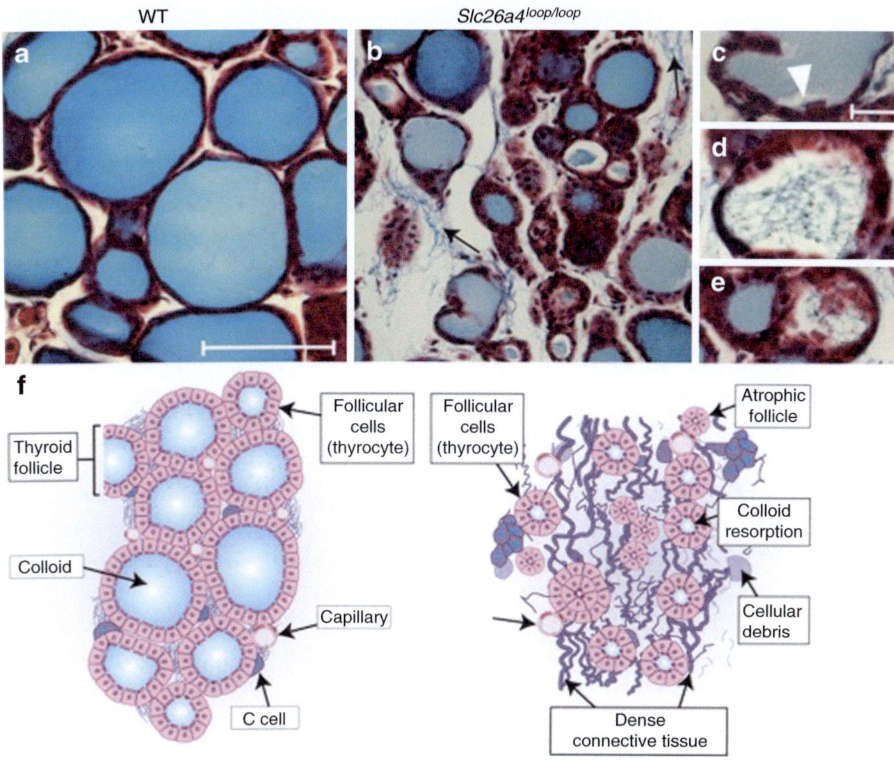

Fig. 3.2 Thyroid atrophy in *Slc26a4loop* mutant mice. Masson's trichrome staining of serial paraffin sections taken from equivalent anatomical positions within the thyroid gland of wild-type and mutant mice. (**a**) A cluster of thyroid follicles filled with colloid (*blue*) is shown. Each colloid is surrounded by a monolayer of thyroid follicular cells (*red*). The round nuclei of the follicular cells are shown (*black*). The thin space between neighboring follicles (*white*) is composed of extracellular lose lamellar connective tissue to support the follicles. (**b**) The majority of the *loop* mutant gland is composed of atrophic follicles, which are unevenly distributed. The follicles are separated by thick fibrotic bands of dense connective tissue (*arrows*). (**c**) A peripheral lacuna of colloid resorption surrounding the brush border of follicular cells is seen in a number of small follicles (*arrowhead*). (**d–e**) Reduced homogeneous colloid and progressive atrophy of thyroid follicles is seen in the *loop* mutant gland. (**f**) Schematic illustration of the pathological changes observed in mutant thyroid tissue (*right*) compared to normal appearance of wild-type thyroid tissue (*left*). The scale bars represent 100 μm (**a** applies to **b**) and 30 μm (**c** applies to **d**, **e**)

common goiter of Pendred's patients. One possible explanation for thyroid atrophy as a result of pendrin dysfunction is related to the increased intracellular free radicals due to an elevated TSH stimulatory effect (Resch et al. 2002). Another study showed that pendrin deficiency mediates a cellular compensatory response, which includes the up regulation of CLC-5 expression that eventually leads to apoptosis via activated caspase-6 (Senou et al. 2010). A different mechanistic explanation for thyroid atrophy is related to a possible role of pendrin as an apical Cl^-/HCO_3^- transporter in the thyroid. Measurement of thyroid follicular lumen acidification in

Slc26a4$^{-/-}$ supports a potential role for pendrin in maintaining luminal pH within physiological range (Wangemann et al. 2009). It is possible that decompensated acidification of thyroid colloid can induce tissue cell death and thyroid hypoplasia. Further investigation of pendrin thyroid function is therefore essential to understand the molecular basis that underlies thyroid phenotypic variability among Pendred's patients.

3.5 Summary

Pendred syndrome is one of the most common syndromic forms of deafness. While hearing impairment predominantly affects all patients, only some Pendred syndrome individuals report vestibular dysfunction and hypothyroidism. Several environmental factors have been attributed to this variability. In the thyroid, for example, an iodide-deficient diet may affect thyroid hormone production and accelerate development of hypothyroidism in affected individuals. While environmental factors at least in part explain the phenotypic variability, other genetic factors probably underlie the extreme cases of congenital hypothyroidism in some Pendred syndrome patients. In this chapter, we summarized the recent studies that investigate the pathological and molecular basis for Pendred syndrome phenotypic heterogeneity, with an emphasis on the vestibular system and thyroid gland. Additional knowledge gathered from the work on *Slc26a4* mouse models for Pendred syndrome is informative for this phenotypic heterogeneity. The unique observations of *Slc26a4*^{loop} thyroid and vestibular systems highlight the importance to study multiple alleles to understand possible mutation-specific mechanisms in Pendred syndrome.

Acknowledgments Research in the K.B. Avraham laboratory is supported by the National Institute of Health (NIH)/NIDCD R01DC011835, I-CORE Gene Regulation in Complex Human Disease Center No. 41/11, and United States-Israel Binational Science Foundation (BSF) 2013027.

References

Agrawal P, Philip R, Saran S, Gutch M, Razi MS, Agroiya P, Gupta K (2015) Congenital hypothyroidism. Indian J Endocrinol Metab 19(2):221–227

Berrettini S, Forli F, Bogazzi F, Neri E, Salvatori L, Casani AP, Franceschini SS (2005) Large vestibular aqueduct syndrome: audiological, radiological, clinical, and genetic features. Am J Otolaryngol 26(6):363–371

Blons H, Feldmann D, Duval V, Messaz O, Denoyelle F, Loundon N, Sergout-Allaoui A, Houang M, Duriez F, Lacombe D, Delobel B, Leman J, Catros H, Journel H, Drouin-Garraud V, Obstoy MF, Toutain A, Oden S, Toublanc JE, Couderc R, Petit C, Garabedian EN, Marlin S (2004) Screening of *SLC26A4 (PDS)* gene in Pendred's syndrome: a large spectrum of mutations in France and phenotypic heterogeneity. Clin Genet 66(4):333–340

Brownstein ZN, Dror AA, Gilony D, Migirov L, Hirschberg K, Avraham KB (2008) A novel *SLC26A4 (PDS)* deafness mutation retained in the endoplasmic reticulum. Arch Otolaryngol Head Neck Surg 134(4):403–407

Buyukgebiz A (2013) Newborn screening for congenital hypothyroidism. J Clin Res Pediatr Endocrinol 5(Suppl 1):8–12

Calil-Silveira J, Serrano-Nascimento C, Kopp PA, Nunes MT (2016) Iodide excess regulates its own efflux: a possible involvement of pendrin. Am J Physiol Cell Physiol 310(7):C576–C582

Choi BY, Stewart AK, Madeo AC, Pryor SP, Lenhard S, Kittles R, Eisenman D, Kim HJ, Niparko J, Thomsen J, Arnos KS, Nance WE, King KA, Zalewski CK, Brewer CC, Shawker T, Reynolds JC, Butman JA, Karniski LP, Alper SL, Griffith AJ (2009) Hypo-functional *SLC26A4* variants associated with nonsyndromic hearing loss and enlargement of the vestibular aqueduct: genotype-phenotype correlation or coincidental polymorphisms? Hum Mutat 30(4):599–608

Dorigueto RS, Gananca MM, Gananca FF (2005) The number of procedures required to eliminate positioning nystagmus in benign paroxysmal positional vertigo. Braz J Otorhinolaryngol 71(6):769–775

Dossena S, Rodighiero S, Vezzoli V, Nofziger C, Salvioni E, Boccazzi M, Grabmayer E, Botta G, Meyer G, Fugazzola L, Beck-Peccoz P, Paulmichl M (2009) Functional characterization of wild-type and mutated pendrin (SLC26A4), the anion transporter involved in Pendred syndrome. J Mol Endocrinol 43(3):93–103

Dou H, Xu J, Wang Z, Smith AN, Soleimani M, Karet FE, Greinwald JH Jr, Choo D (2004) Co-expression of pendrin, vacuolar H^+-ATPase alpha4-subunit and carbonic anhydrase II in epithelial cells of the murine endolymphatic sac. J Histochem Cytochem 52(10):1377–1384

Dror AA, Politi Y, Shahin H, Lenz DR, Dossena S, Nofziger C, Fuchs H, Hrabe de Angelis M, Paulmichl M, Weiner S, Avraham KB (2010) Calcium oxalate stone formation in the inner ear as a result of an *Slc26a4* mutation. J Biol Chem 285(28):21724–21735

Dror AA, Lenz DR, Shivatzki S, Cohen K, Ashur-Fabian O, Avraham KB (2014) Atrophic thyroid follicles and inner ear defects reminiscent of cochlear hypothyroidism in *Slc26a4*-related deafness. Mamm Genome 25(7-8):304–316

Epley JM (1995) Positional vertigo related to semicircular canalithiasis. Otolaryngol Head Neck Surg 112(1):154–161

Everett LA, Glaser B, Beck JC, Idol JR, Buchs A, Heyman M, Adawi F, Hazani E, Nassir E, Baxevanis AD, Sheffield VC, Green ED (1997) Pendred syndrome is caused by mutations in a putative sulphate transporter gene (*PDS*). Nat Genet 17(4):411–422

Everett LA, Belyantseva IA, Noben-Trauth K, Cantos R, Chen A, Thakkar SI, Hoogstraten-Miller SL, Kachar B, Wu DK, Green ED (2001) Targeted disruption of mouse *Pds* provides insight about the inner-ear defects encountered in Pendred syndrome. Hum Mol Genet 10(2):153–161

Fraser GR (1965) Association of Congenital Deafness with Goitre (Pendred's Syndrome): A study of 207 families. Ann Hum Genet 28:201–249

Fraser GR, Morgans ME, Trotter WR (1960) The syndrome of sporadic goitre and congenital deafness. Q J Med 29:279–295

Gillam MP, Sidhaye AR, Lee EJ, Rutishauser J, Stephan CW, Kopp P (2004) Functional characterization of pendrin in a polarized cell system. Evidence for pendrin-mediated apical iodide efflux. J Biol Chem 279(13):13004–13010

Grasberger H, Refetoff S (2011) Genetic causes of congenital hypothyroidism due to dyshormonogenesis. Curr Opin Pediatr 23(4):421–428

Haddow JE, Palomaki GE, Allan WC, Williams JR, Knight GJ, Gagnon J, O'Heir CE, Mitchell ML, Hermos RJ, Waisbren SE, Faix JD, Klein RZ (1999) Maternal thyroid deficiency during pregnancy and subsequent neuropsychological development of the child. N Engl J Med 341(8):549–555

Hrabe de Angelis MH, Flaswinkel H, Fuchs H, Rathkolb B, Soewarto D, Marschall S, Heffner S, Pargent W, Wuensch K, Jung M, Reis A, Richter T, Alessandrini F, Jakob T, Fuchs E, Kolb H, Kremmer E, Schaeble K, Rollinski B, Roscher A, Peters C, Meitinger T, Strom T, Steckler T, Holsboer F, Klopstock T, Gekeler F, Schindewolf C, Jung T, Avraham K, Behrendt H, Ring J, Zimmer A, Schughart K, Pfeffer K, Wolf E, Balling R (2000) Genome-wide, large-scale production of mutant mice by ENU mutagenesis. Nat Genet 25(4):444–447

Iosco C, Cosentino C, Sirna L, Romano R, Cursano S, Mongia A, Pompeo G, di Bernardo J, Ceccarelli C, Tallini G, Rhoden KJ (2014) Anoctamin 1 is apically expressed on thyroid follicular cells and contributes to ATP- and calcium-activated iodide efflux. Cell Physiol Biochem 34(3):966–980

Ishinaga H, Shimizu T, Yuta A, Tsukamoto K, Usami S, Majima Y (2002) Pendred's syndrome with goiter and enlarged vestibular aqueducts diagnosed by *PDS* gene mutation. Head Neck 24(7):710–713

Ito T, Nishio A, Wangemann P, Griffith AJ (2015) Progressive irreversible hearing loss is caused by *stria vascularis* degeneration in an *Slc26a4*-insufficient mouse model of large vestibular aqueduct syndrome. Neuroscience 310:188–197

Iwata T, Yoshida T, Teranishi M, Murata Y, Hayashi Y, Kanou Y, Griffith AJ, Nakashima T (2011) Influence of dietary iodine deficiency on the thyroid gland in *Slc26a4*-null mutant mice. Thyroid Res 4(1):10

Jackler RK, De La Cruz A (1989) The large vestibular aqueduct syndrome. Laryngoscope 99(12):1238–1242, discussion 1242–1233

Jung J, Seo YW, Choi JY, Kim SH (2016) Vestibular function is associated with residual low-frequency hearing loss in patients with bi-allelic mutations in the *SLC26A4* gene. Hearing Res 335:33–39

Kim HM, Wangemann P (2011) Epithelial cell stretching and luminal acidification lead to a retarded development of *stria vascularis* and deafness in mice lacking pendrin. PLoS One 6(3), e17949

Ko KM, Song MH, Kim JH, Shim DB (2014) Persistent spontaneous nystagmus following a canalith repositioning procedure in horizontal semicircular canal benign paroxysmal positional vertigo. JAMA Otolaryngol Head Neck Surg 140(3):250–252

Kuhnen P, Turan S, Frohler S, Guran T, Abali S, Biebermann H, Bereket A, Gruters A, Chen W, Krude H (2014) Identification of PENDRIN (*SLC26A4*) mutations in patients with congenital hypothyroidism and "apparent" thyroid dysgenesis. J Clin Endocrinol Metab 99(1):E169–E176

Levenson MJ, Parisier SC, Jacobs M, Edelstein DR (1989) The large vestibular aqueduct syndrome in children. A review of 12 cases and the description of a new clinical entity. Arch Otolaryngol Head Neck Surg 115(1):54–58

Li H, Ganta S, Fong P (2010) Altered ion transport by thyroid epithelia from *CFTR*$^{-/-}$ pigs suggests mechanisms for hypothyroidism in cystic fibrosis. Exp Physiol 95(12):1132–1144

Lund K, Fogler HS, McCune CC, Ault JW (1975) Acidization-II. The dissolution of calcite in hydrochloric acid. Chem Eng Sci 30(8):825–835

Luxon LM, Cohen M, Coffey RA, Phelps PD, Britton KE, Jan H, Trembath RC, Reardon W (2003) Neuro-otological findings in Pendred syndrome. Int J Audiol 42(2):82–88

Manzari L (2008) Enlarged vestibular aqueduct (EVA) related with recurrent benign paroxysmal positional vertigo (BPPV). Med Hypotheses 70(1):61–65

Miyagawa M, Nishio SY, Usami S, Deafness Gene Study C (2014) Mutation spectrum and genotype-phenotype correlation of hearing loss patients caused by *SLC26A4* mutations in the Japanese: a large cohort study. J Hum Genet 59(5):262–268

Morgan ME, Trotter WR (1958) Association of congenital deafness with goiter; The nature of the thyroid defect. Lancet 1:607–609

Park HJ, Shaukat S, Liu XZ, Hahn SH, Naz S, Ghosh M, Kim HN, Moon SK, Abe S, Tukamoto K, Riazuddin S, Kabra M, Erdenetungalag R, Radnaabazar J, Khan S, Pandya A, Usami SI, Nance WE, Wilcox ER, Riazuddin S, Griffith AJ (2003) Origins and frequencies of *SLC26A4* (*PDS*) mutations in east and south Asians: global implications for the epidemiology of deafness. J Med Genet 40(4):242–248

Pendred V (1896) Deaf-mutism and goitre. Lancet ii:532

Pera A, Dossena S, Rodighiero S, Gandia M, Botta G, Meyer G, Moreno F, Nofziger C, Hernandez-Chico C, Paulmichl M (2008) Functional assessment of allelic variants in the *SLC26A4* gene involved in Pendred syndrome and nonsyndromic EVA. Proc Natl Acad Sci USA 105(47):18608–18613

Portulano C, Paroder-Belenitsky M, Carrasco N (2014) The Na^+/I^- symporter (NIS): mechanism and medical impact. Endocr Rev 35(1):106–149

Pryor SP, Madeo AC, Reynolds JC, Sarlis NJ, Arnos KS, Nance WE, Yang Y, Zalewski CK, Brewer CC, Butman JA, Griffith AJ (2005) *SLC26A4/PDS* genotype-phenotype correlation in hearing loss with enlargement of the vestibular aqueduct (EVA): evidence that Pendred syndrome and non-syndromic EVA are distinct clinical and genetic entities. J Med Genet 42(2):159–165

Reardon W, Coffey R, Phelps PD, Luxon LM, Stephens D, Kendall-Taylor P, Britton KE, Grossman A, Trembath R (1997) Pendred syndrome--100 years of underascertainment? QJM 90(7):443–447

Resch U, Helsel G, Tatzber F, Sinzinger H (2002) Antioxidant status in thyroid dysfunction. Clin Chem Lab Med 40(11):1132–1134

Rotman-Pikielny P, Hirschberg K, Maruvada P, Suzuki K, Royaux IE, Green ED, Kohn LD, Lippincott-Schwartz J, Yen PM (2002) Retention of pendrin in the endoplasmic reticulum is a major mechanism for Pendred syndrome. Hum Mol Genet 11(21):2625–2633

Royaux IE, Suzuki K, Mori A, Katoh R, Everett LA, Kohn LD, Green ED (2000) Pendrin, the protein encoded by the Pendred syndrome gene (*PDS*), is an apical porter of iodide in the thyroid and is regulated by thyroglobulin in FRTL-5 cells. Endocrinology 141(2):839–845

Royaux IE, Wall SM, Karniski LP, Everett LA, Suzuki K, Knepper MA, Green ED (2001) Pendrin, encoded by the Pendred syndrome gene, resides in the apical region of renal intercalated cells and mediates bicarbonate secretion. Proc Natl Acad Sci USA 98(7):4221–4226

Royaux IE, Belyantseva IA, Wu T, Kachar B, Everett LA, Marcus DC, Green ED (2003) Localization and functional studies of pendrin in the mouse inner ear provide insight about the etiology of deafness in pendred syndrome. J Assoc Res Otolaryngol 4(3):394–404

Scott DA, Wang R, Kreman TM, Sheffield VC, Karniski LP (1999) The Pendred syndrome gene encodes a chloride-iodide transport protein. Nat Genet 21(4):440–443

Scott DA, Wang R, Kreman TM, Andrews M, McDonald JM, Bishop JR, Smith RJ, Karniski LP, Sheffield VC (2000) Functional differences of the *PDS* gene product are associated with phenotypic variation in patients with Pendred syndrome and non-syndromic hearing loss (DFNB4). Hum Mol Genet 9(11):1709–1715

Senou M, Khalifa C, Thimmesch M, Jouret F, Devuyst O, Col V, Audinot JN, Lipnik P, Moreno JC, Van Sande J, Dumont JE, Many MC, Colin IM, Gerard AC (2010) A coherent organization of differentiation proteins is required to maintain an appropriate thyroid function in the Pendred thyroid. J Clin Endocrinol Metab 95(8):4021–4030

Silveira JC, Kopp PA (2015) Pendrin and anoctamin as mediators of apical iodide efflux in thyroid cells. Curr Opin Endocrinol Diabetes Obes 22(5):374–380

Soleimani M, Greeley T, Petrovic S, Wang Z, Amlal H, Kopp P, Burnham CE (2001) Pendrin: an apical $Cl^-/OH^-/HCO_3^-$ exchanger in the kidney cortex. Am J Physiol Renal Physiol 280(2):F356–F364

Sugiura M, Sato E, Nakashima T, Sugiura J, Furuhashi A, Yoshino T, Nakayama A, Mori N, Murakami H, Naganawa S (2005) Long-term follow-up in patients with Pendred syndrome: vestibular, auditory and other phenotypes. Eur Arch Otorhinolaryngol 262(9):737–743

Suzuki H, Oshima A, Tsukamoto K, Abe S, Kumakawa K, Nagai K, Satoh H, Kanda Y, Iwasaki S, Usami S (2007) Clinical characteristics and genotype-phenotype correlation of hearing loss patients with *SLC26A4* mutations. Acta Otolaryngol 127(12):1292–1297

Syed S (2015) Iodine and the "near" eradication of cretinism. Pediatrics 135(4):594–596

Taylor JP, Metcalfe RA, Watson PF, Weetman AP, Trembath RC (2002) Mutations of the *PDS* gene, encoding pendrin, are associated with protein mislocalization and loss of iodide efflux: implications for thyroid dysfunction in Pendred syndrome. J Clin Endocrinol Metab 87(4):1778–1784

Tsukamoto K, Suzuki H, Harada D, Namba A, Abe S, Usami S (2003) Distribution and frequencies of *PDS (SLC26A4)* mutations in Pendred syndrome and nonsyndromic hearing loss associated with enlarged vestibular aqueduct: a unique spectrum of mutations in Japanese. Eur J Hum Genet 11(12):916–922

Twyffels L, Strickaert A, Virreira M, Massart C, Van Sande J, Wauquier C, Beauwens R, Dumont JE, Galietta LJ, Boom A, Kruys V (2014) Anoctamin-1/TMEM16A is the major apical iodide channel of the thyrocyte. Am J Physiol Cell Physiol 307(12):C1102–C1112

Valvassori GE, Clemis JD (1978) The large vestibular aqueduct syndrome. Laryngoscope 88:723–728

van den Hove MF, Croizet-Berger K, Jouret F, Guggino SE, Guggino WB, Devuyst O, Courtoy PJ (2006) The loss of the chloride channel, ClC-5, delays apical iodide efflux and induces a euthyroid goiter in the mouse thyroid gland. Endocrinology 147(3):1287–1296

Waller DK, Anderson JL, Lorey F, Cunningham GC (2000) Risk factors for congenital hypothyroidism: An investigation of infant's birth weight, ethnicity, and gender in California, 1990–1998. Teratology 62(1):36–41

Wangemann P (2013) Mouse models for pendrin-associated loss of cochlear and vestibular function. Cell Physiol Biochem 32(7):157–165

Wangemann P, Itza EM, Albrecht B, Wu T, Jabba SV, Maganti RJ, Lee JH, Everett LA, Wall SM, Royaux IE, Green ED, Marcus DC (2004) Loss of KCNJ10 protein expression abolishes endocochlear potential and causes deafness in Pendred syndrome mouse model. BMC Med 2:30

Wangemann P, Nakaya K, Wu T, Maganti RJ, Itza EM, Sanneman JD, Harbidge DG, Billings S, Marcus DC (2007) Loss of cochlear HCO_3^- secretion causes deafness via endolymphatic acidification and inhibition of Ca^{2+} reabsorption in a Pendred syndrome mouse model. Am J Physiol Renal Physiol 292(5):F1345–F1353

Wangemann P, Kim HM, Billings S, Nakaya K, Li X, Singh R, Sharlin DS, Forrest D, Marcus DC, Fong P (2009) Developmental delays consistent with cochlear hypothyroidism contribute to failure to develop hearing in mice lacking *Slc26a4*/pendrin expression. Am J Physiol Renal Physiol 297(5):F1435–F1447

Xing M, Tokumaru Y, Wu G, Westra WB, Ladenson PW, Sidransky D (2003) Hypermethylation of the Pendred syndrome gene *SLC26A4* is an early event in thyroid tumorigenesis. Cancer Res 63(9):2312–2315

Yang T, Vidarsson H, Rodrigo-Blomqvist S, Rosengren SS, Enerback S, Smith RJ (2007) Transcriptional control of *SLC26A4* is involved in Pendred syndrome and nonsyndromic enlargement of vestibular aqueduct (*DFNB4*). Am J Hum Genet 80(6):1055–1063

Yoshida A, Taniguchi S, Hisatome I, Royaux IE, Green ED, Kohn LD, Suzuki K (2002) Pendrin is an iodide-specific apical porter responsible for iodide efflux from thyroid cells. J Clin Endocrinol Metab 87(7):3356–3361

Zalewski CK, Chien WW, King KA, Muskett JA, Baron RE, Butman JA, Griffith AJ, Brewer CC (2015) Vestibular dysfunction in patients with enlarged vestibular aqueduct. Otolaryngol Head Neck Surg 153(2):257–262

Pendrin-Linked Deafness in Humans

4

Sebastian Roesch, Miklós Tóth, and Gerd Rasp

Abstract

Hearing loss is a well-known clinical sign in the context of pendrin malfunction. Pendred syndrome (PS) and nonsyndromic hearing loss (DFNB4) with enlarged vestibular aqueduct (EVA) are two distinct, genetically determined clinical entities of which hearing loss is the main feature. Knowledge on clinical and genetic properties of both entities is constantly rising. In order to improve understanding of inner ear malfunction in connection with pendrin malfunction, we provide an overview on inner ear structure and function, with a clear focus on structures possibly affected in the setting of PS/DFNB4.

4.1 Introduction

Of all human senses, hearing is the major tool for communication between human beings. In modern society, hearing and speaking give the opportunity to interact with each other face to face, as well as over great distances through the help of electronic devices. The process of learning how to hear and speak properly takes at least 4 years. Once disturbed, the window of opportunity for compensation closes rapidly. It is important to emphasize that skills in reading and writing usually go along with hearing and speaking. Depending on the specific language, the amount of words varies greatly between spoken language (75,000 up to 600,000 words) and sign language (~19,000 words in the case of German sign language). Moreover, aspects such as intonation may only be recognized through adequate hearing. Besides communication, hearing allows for the recognition of potential danger

S. Roesch, MD (✉) • M. Tóth, MD, PhD • G. Rasp, MD
Department of Otorhinolaryngology, Head and Neck Surgery, Paracelsus Medical University, Müllner Hauptstraße 48, Salzburg 5020, Austria
e-mail: sebastian.roesch@pmu.ac.at

S. Dossena, M. Paulmichl (eds.), *The Role of Pendrin in Health and Disease*,
DOI 10.1007/978-3-319-43287-8_4

during everyday life, as well as the joy of cultural aspects, such as music. Eyes may be shut, but ears always remain open.

The sense of hearing includes the ears, representing the primary sensory organ, and the central processing, located in the brain. Hearing loss is a common feature of aging and may be tolerated in aging individuals. However, hearing loss in newborns has major negative impacts throughout life and therefore necessitates great alertness.

Hearing loss is the most common sensory deficiency in newborns. Considering all possible causes, 0.1 % of newborns are affected by a functional deficiency of the inner ear of variable degree—50 % of which are due to genetics. Besides clinical history, clinical status, audiometry and imaging, genetic investigations have become a valuable instrument for diagnostic work-up in cases of sensorineural hearing loss (SNHL). With the detection of novel genetic variations, the challenge of interpreting their true impact on hearing loss rises.

This chapter explains the clinical aspects and investigations required for diagnosis of impaired hearing, with special emphasis on the inner ear, which is the organ affected the most by pendrin dysfunction.

4.2 Hearing in Humans

4.2.1 Embryology

Embryology of the human ear starts with the development of the inner ear at about embryonic day 22 post fertilization, Carnegie Stage 9/10. It originates from the neuroectoderm, specifically from the otic placode. The otic placode invaginates to become the otic pit, and after fusion of the surface epithelium, the otic pit detaches, forming the otocyst or otic vesicle (Fig. 4.1). In the following weeks, the dorsal part of the otic vesicle becomes the vestibular part of labyrinth, and the ventral part forms the cochlea. During further differentiation, three folds of the wall of the otic vesicle can be observed (Fig. 4.1e), which subdivide it into the main parts of the final otic labyrinth. The main parts are the endolymphatic duct and sac, the utricle with the semicircular canals and the saccule with the *ductus reuniens* and the cochlear duct (Bast and Anson 1949, Tóth and Csillag 2005, Bianchi and Fuchs 2010).

Fold I (Fig. 4.1e) gradually deepens and separates the endolymphatic duct from the rest of the otic vesicle. This is the largest and deepest of all three folds. During further development, its tip forms a free flap of tissue, which extends into the utricle and separates the mouth of the utricular duct from the utricular cavity (Bast 1928).

Fold II (Fig. 4.1e) is a rather small notch on the medial side of the vesicle between the saccule and the opening of the endolymphatic duct. It separates the endolymphatic duct from the saccule.

Fold III (Fig. 4.1e), between the utricle and saccule, forms the utricular duct and the saccular duct.

With ongoing differentiation of the otic vesicle into the membranous labyrinth, sensory structures such as the *cristae ampullares* and the *maculae*, are developed. This process is induced by nerve fibers, growing from the neural crest and getting into contact with the otic vesicle. A similar process can be observed in the cochlea,

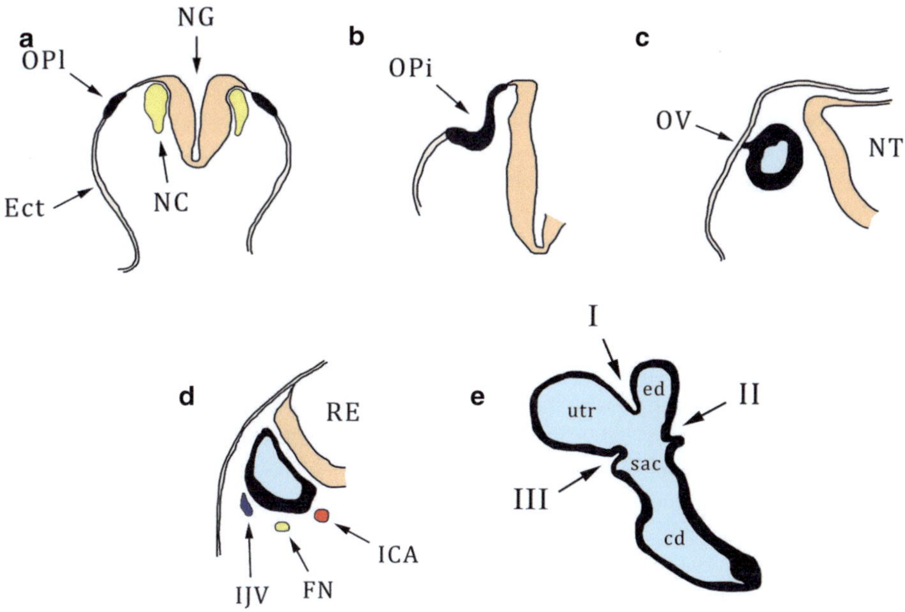

Fig. 4.1 Development of the membranous labyrinth; transverse sections of human embryos at 9 somites (**a**), 16 somites (**b**), approximately 4 mm length (**c**), 7 mm length (**d**), and 22.8 mm length (**e**). *OPl* otic placode, *OPi* otic pit, *OV* otic vesicle, *Ect* ectoderm, *NG* neural groove, *NT* neural tube, *NC* neural crest, *RE* rhombencephalon, *IJV* internal jugular vein, *FN* facial nerve, *ICA* internal carotid artery, *utr* utricule, *sac* saccule, *cd* cochlear duct, *ed* endolymphatic duct. Numbers in (**e**) denote the three folds of the wall of the otic vesicle (see Sect. 4.2.1)

during development of the organ of Corti. The cochlear nerve penetrates the cochlear duct epithelium, inducing inner hair cell differentiation first, followed by outer hair cell differentiation (Kelley et al. 2005).

The mature middle ear shows complex morphology. Embryological development is completely independent from the inner ear. The auditory ossicles derive from the first and second pharyngeal arches (Richany et al. 1954, Anthwal and Thompson 2016). The tympanic cavity and the mastoid air cells derive from the first pharyngeal pouch, which is connected directly to the nasopharynx through the auditory tube. The external auditory canal develops from the first pharyngeal cleft (Anthwal and Thompson 2016). Ectodermal cells proliferate and fill the entire lumen of the pharyngeal cleft to form the so-called meatal plug. In the center of the plug, the cells disappear continuously from lateral to medial. The *pinna* (auricle) is derived from six hillocks, three on each of pharyngeal arch one and two.

4.2.2 Anatomy

The human hearing organ is imbedded in the temporal bone and forms the central part of the lateral skull base. The organ of hearing, as well as the vestibular organ, is an evolutionary derivative of the lateral line canals of early aquatic vertebrates.

The vestibular organ is limited to the inner ear in contrast to the hearing organ, which spreads along the entire temporal bone up to the external surface of the head and consists of three main segments, the outer, middle and inner ear.

The outer ear consists of two parts, the *pinna* and the external auditory canal. The *pinna* collects and amplifies sound and guides it through the external auditory canal to the tympanic membrane. The tympanic membrane is a funnel-shaped disk and belongs to the middle ear. It is the only mobile part of the tympanic cavity, capable for transmitting sound toward the auditory ossicles. The three small bones (malleus, incus, stapes) as a whole conduct sound vibration from the tympanic membrane to the fluid-filled inner ear, where mechanical vibrations are converted into nerve impulses and sent to the brain.

The inner ear comprises a system of membranous ducts containing receptor structures for hearing and balance, the so-called membranous labyrinth. A thick bony shell, called the osseous labyrinth, encloses the membranous labyrinth. There are two orifices (oval window and round window) and two channels (*aquaeductus vestibuli* or vestibular aqueduct (VA) and *canaliculus cochleae* or cochlear *canaliculus*) within the osseous labyrinth, providing communication with surrounding structures (Bast and Anson 1949, Gulya and Schuknecht 1995, Tóth and Csillag 2005).

The transduction of sound waves into electrochemical impulses occurs in the cochlea. It is the most ventral part of the inner ear, and its bony covering resembles truly a snail shell, making about 2.5 turns around the modiolus. The membranous labyrinth of the cochlea consists of three *scalae*: the *scala vestibuli*, the *scala tympani*, and the *scala media*. *Scala vestibuli* and *scala tympani* are filled with perilymph, *scala media* with endolymph. The most important structure of the *scala media* is the organ of Corti, which is stimulated by fluid waves and sends electrical signs along the auditory nerve to the brain. The organ of Corti is a complex organ, built up by specific cells (inner and outer hair cells, pillar cells, Hensen's cells, Claudius cells, and the interdental cells), and the tectorial membrane.

The morphologically larger part of the inner ear is the vestibular organ. It is divided into the vestibule and the semicircular canals. The vestibule, which is part of the osseous labyrinth, encloses two small, irregularly shaped sacs, the saccule, and utricule. Both, the saccule and the utricule, show a well-defined thickening, the so-called *macula*. The saccular *macula* is positioned vertically and the utricular *macula* horizontally. The *macula* is a vestibular sensory epithelium containing the sensory and supporting cells embedded into the gelatinous otolithic membrane. Special particles of both *maculae* are the otoliths, lying in the superficial layer of the otolithic membrane. The otoliths are relatively heavy particles and allow the sense of gravity and motion.

There are three semicircular canals, lying perpendicular to each other. One can distinguish a horizontal semicircular canal and two further ones, positioned vertically – the superior (anterior) and posterior semicircular canals. The semicircular canals have an ampullated and a nonampullated end. The ampullated end is wider and contains sensory epithelium.

The endolymphatic duct extends from the junction of the saccular and utricular duct to an expanded portion, the endolymphatic sac (Fig. 4.2). The

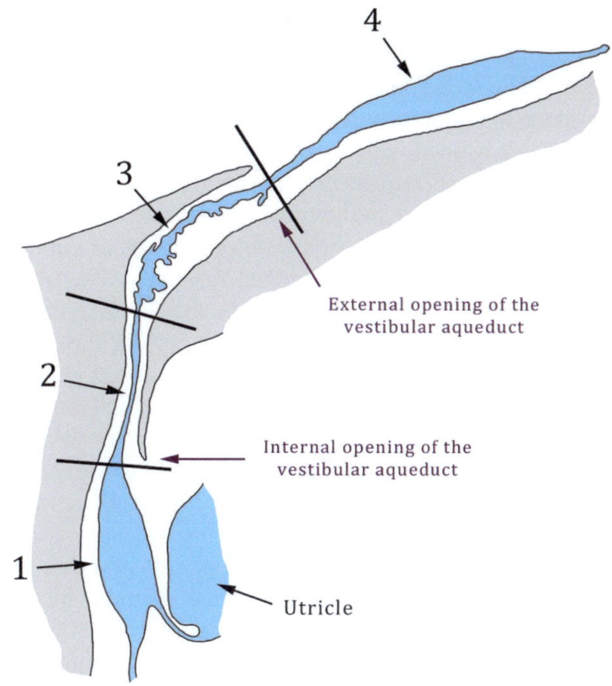

Fig. 4.2 Schematic drawing of the endolymphatic duct and sac (depicted in *blue*) lying within the vestibular aqueduct (after Anson & Bast) into the temporal bone (depicted in *gray*). The first part of the endolymphatic duct is the sinus (*1*) and may be found within the vestibule. Second part is the isthmus (*2*). The lateral wall of the bony covering of the isthmus is always a thin bony shell and forms the internal opening of the vestibular aqueduct. The rugose portion of the endolymphatic sac (*3*) is covered by bone as well. Only the distal, smooth portion of the endolymphatic sac (*4*) is exposed on the posterior surface of the temporal bone without any bony covering. *Black lines* illustrate limits between the different portions

endolymphatic duct runs through the temporal bone within the vestibular aqueduct. The endolymphatic sac projects posterior-laterally of the temporal bone between the two layers of the *dura mater* and can be found close to the posterior semicircular canal. The most proximal part of the endolymphatic duct shows a dilated area, the so-called *sinus* of the endolymphatic duct. In the course, the duct narrows and forms the *isthmus* of the endolymphatic duct. Next segment is the rugose portion of the endolymphatic sac, still within the vestibular aqueduct. As the external opening of the vestibular aqueduct is reached, the sac widens again and terminates as the proper endolymphatic sac (smooth portion of the endolymphatic sac) (Bast and Anson 1949).

The course of the endolymphatic duct and sac is fairly straight during early fetal life. In the second half of fetal life, the originally straight course starts to change. The *isthmus* of the endolymphatic duct and the proximal part of the endolymphatic sac develop a curved course toward a caudal direction. From birth, the sac is hanging downward from the posterior edge of the temporal bone.

The *sinus* of the endolymphatic duct lies within the periotic tissue of the otic capsule and demonstrates large variation. It may be constricted, moderately dilated, or widely distended. The loose periotic tissue allows volume changes of the sinus. The sinus begins at the union of the utricular and saccular duct and terminates where it leaves the vestibule. It is lined with a smooth surface of cuboidal cells. Structural irregularities may occur, most frequently in its distal part.

The *isthmus* is the narrowest part of the endolymphatic duct, which runs through the vestibular aqueduct. The lining of the *isthmus* also consists of cuboidal epithelium, forming a smooth surface with little longitudinal rugosities. The *isthmus* part is not surrounded by periotic tissue. During early fetal life, the lumen of the isthmus is quite wide but shrinks with ongoing ossification of the otic capsule during further development. After the 16th fetal week, a considerable amount of perichondrial bone arises on the dural surface of the otic capsule, additionally. Both structural reorganizations result in a lengthening of the vestibular aqueduct and its content as a whole.

The endolymphatic sac is the enlarged terminal segment of the endolymphatic duct system. The proximal and smaller part of the sac lies within the vestibular aqueduct. The distal and larger part can be found between the internal and external layers of the *dura mater*. The proximal part is also called rugose portion of the endolymphatic sac.

The rugose portion of the endolymphatic sac is the intermediate part between the isthmus of the endolymphatic duct and the proper endolymphatic sac. Its lumen is wide, and the lining epithelium is strongly folded. The epithelium is surrounded by vascular and connective tissues. The folding is a constant structural feature of this part, not only in adults but also in children and even fetuses.

The proper endolymphatic sac (smooth portion) represents about two thirds of the endolymphatic sac system, covered laterally by the periosteum of the temporal bone and medially by the *dura mater* of the brain. Histological sections of this part give a fairly smooth and collapsed impression (Fig. 4.3d). The lumen of this part is commonly larger than that of the rugose portion. The size apparently depends on the amount of endolymphatic fluid contained. The epithelial lining is low cuboidal type, which is typical for fluid-filled cavities.

In the middle of fetal life, the whole endolymphatic sac measures about 3 mm in width and 5 mm in length. From third postnatal year, it may reach a size of about 5 mm width and 10 mm length.

4.2.3 Physiology

Sound mostly spreads in waves through the air to reach the ear. In its simplest form, it is a sinus tone defined by amplitude and frequency. Considering the human range of hearing ability, usually from 20 Hz to 20 kHz, the wavelength (λ) of sound that can be perceived by humans is given by Eq (4.1),

$$\lambda = \frac{c}{f} \tag{4.1}$$

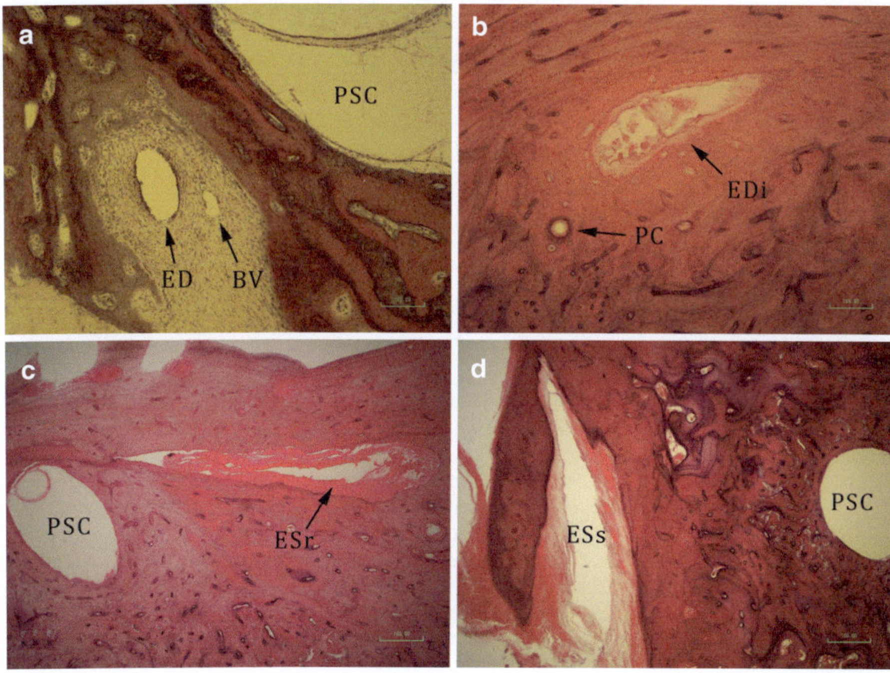

Fig. 4.3 Histological slides of the endolymphatic duct and endolymphatic sac (hematoxylin and eosin staining). (**a**) The developing endolymphatic duct (*ED*) and a blood vessel (*BV*) can be observed in a 26-week-old fetus near the posterior semicircular canal (*PSC*). The vestibular aqueduct is rather wide, and it is filled with loose fibrous tissue (stroma) around the ED. (**b**) In adults, the isthmus of the endolymphatic duct (*EDi*) runs in a narrow bony canal commonly accompanied by one or two canals, the so-called paravestibular canaliculi (*PC*). The proximal (rugose) portion of the endolymphatic sac (*ESr*) can be found within the vestibular aqueduct near the PSC (**c**), but the distal (smooth) portion (*ESs*) lies between the two layers of the *dura mater* on the posterior surface of the temporal bone (**d**). Scale bar 100 μM

λ = wavelength (m); c = speed of sound (340 m/s in air, 20 °C); f = frequency in Hz (1/s)

and therefore is between 1.7×10^{-2} m and 17 m.

The human ear may detect sound pressure levels (L_p) from 2×10^{-5} to 20 Pa (=1 N/m^2). As human hearing is a nonlinear feature, loudness is usually given in decibel sound pressure level (dB SPL), in audiograms usually corrected as decibel for hearing level (dB HL) for a linear display of hearing levels (Fig. 4.4).

L_p is given by Eq. (4.2).

$$L_p = 20 \log_{10}\left(\frac{p}{p_0}\right) \mathrm{dB} \tag{4.2}$$

p = actual sound pressure (Pa); p_0 = reference sound pressure (20 μPa)

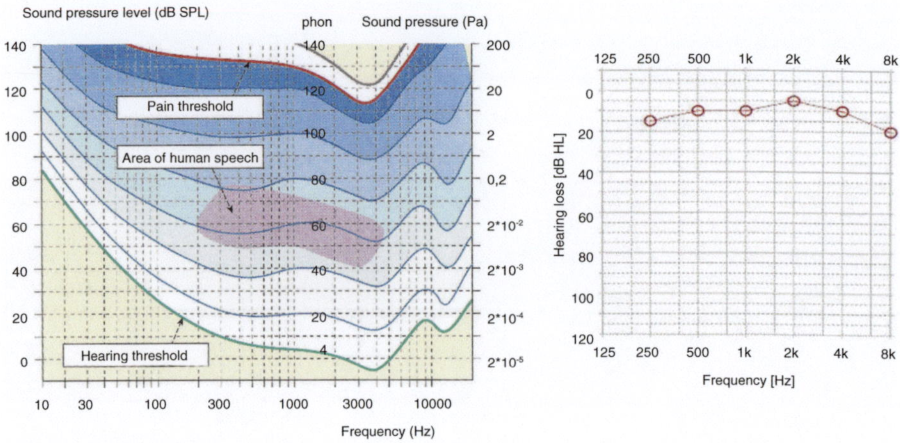

Fig. 4.4 *Left*: Hearing thresholds of the human ear. The sound pressure level (dB SPL) is displayed on the *y*-axis, and the frequency is shown on the *x*-axis. Each line represents hearing perception for a specific amount of subjective loudness threshold given in phon. The *green curve* shows the hearing threshold. The *red curve* represents the pain threshold. The *purple area* marks the sound spectrum of human speech (Adapted from Schmidt et al. 2011, with permission from Springer-Verlag Berlin Heidelberg). *Right*: Example of an audiogram, indicating hearing loss in a linear scale (*y*-axis) with sound pressure given in dB HL. The *red line* exemplifies a normal hearing threshold

Perception of the same loudness for different frequencies requires frequency-specific sound pressure levels (dB SPL), leading to a curved line for each amount of loudness perception described in phon (Fig. 4.4).

The human ear can detect sound pressure levels between 0 and 120 dB. 0 dB resembles absolute silence, and 120 dB is the so-called pain threshold. Humans have a level of comfort being somewhere between 30 and 80 dB. Missing sound, as well as loud noise, may have undesired effects on the human body and mind. This is true for normal hearing ears. It is crucial to understand that any damage in the human ear results in a raise of the threshold for sound detection as well as a lowering of the limit for noise tolerance. Therefore, the wide dynamic range may be decreased from 120 to 20 dB in a case of profound hearing loss (for a definition of the different degrees of hearing loss, see Chap. 4.3.1). The tuning curve of the human ear with its physiological range is shown in Fig. 4.4.

The longitudinal sound wave passes the outer ear canal and reaches the tympanic membrane. The wave is conducted from a gas medium (air) into solid ossicles and transferred into the fluid-filled inner ear. In the inner ear, the transverse, fluid-membrane waves (Voss et al. 2013) are converted into electrical signals that pass along the auditory nerve. The pathway of a sound signal through the human ear is shown in Fig. 4.5.

The transmission of mechanical forces through different aggregation states usually results in a certain loss of power. The specific structure of the middle ear components, especially the tympanic membrane, reduces reflection of sound power (Voss et al. 2013).

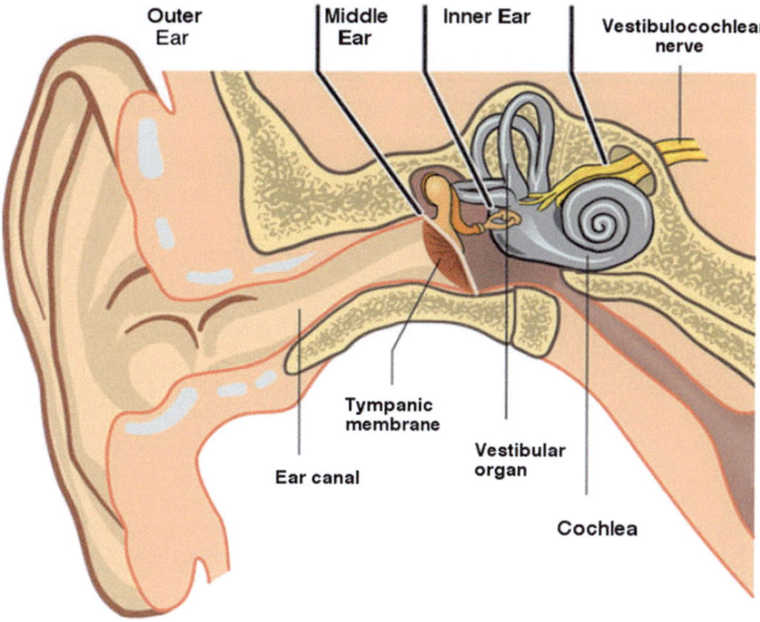

Fig. 4.5 Pathway of a sound signal through the human ear. In the ear canal, the airwave travels to the tympanic membrane. In the middle ear, the wave is conducted by the ossicles. In the cochlea, the wave is transmitted as a fluid wave and finally transformed into electrical impulses. These impulses travel via the vestibulocochlear nerve to the brain (Adapted from Schmidt et al. 2011, with permission from Springer-Verlag Berlin Heidelberg)

Within the cochlea, the fluid-membrane waves force a movement of the basilar membrane against the tectorial membrane. This movement leads to a passive mechanical deflection of the stereocilia of the inner hair cells. Additional amplification and fine-tuning is achieved by the three rows of outer hair cells, being in contact with the tectorial membrane. The dynamic range of the human ear reaches about 120 dB. Almost half of this range is achieved by the activity of the outer hair cells. Outer hair cells are able to contract actively and thus influence the movement of the tectorial membrane and the basilar membrane. They may even inhibit it by contracting in the opposite phase of the sound wave (Fig. 4.6). There is a clear tonotopy in the inner ear. The detection of different frequencies is spread out from the highest frequencies near the round window membrane to the lowest frequencies in the apex of the cochlea.

The deflection of stereocilia opens a potassium channel on the apical surface of the inner hair cells and allows the influx of K^+ ions from the endolymph into the cell. This leads to an intermittent depolarization of the cell and induces the secretion of glutamate into the synaptic space. Selective binding of the transmitter at the postsynaptic membrane leads to a nerve action potential. The signal is subsequently transmitted to the brain (Fig. 4.7).

Each hair cell holds synapses of different sensitivity and thus reacts specifically to different amplitudes of incoming signals. Moreover, there is a mechanism for

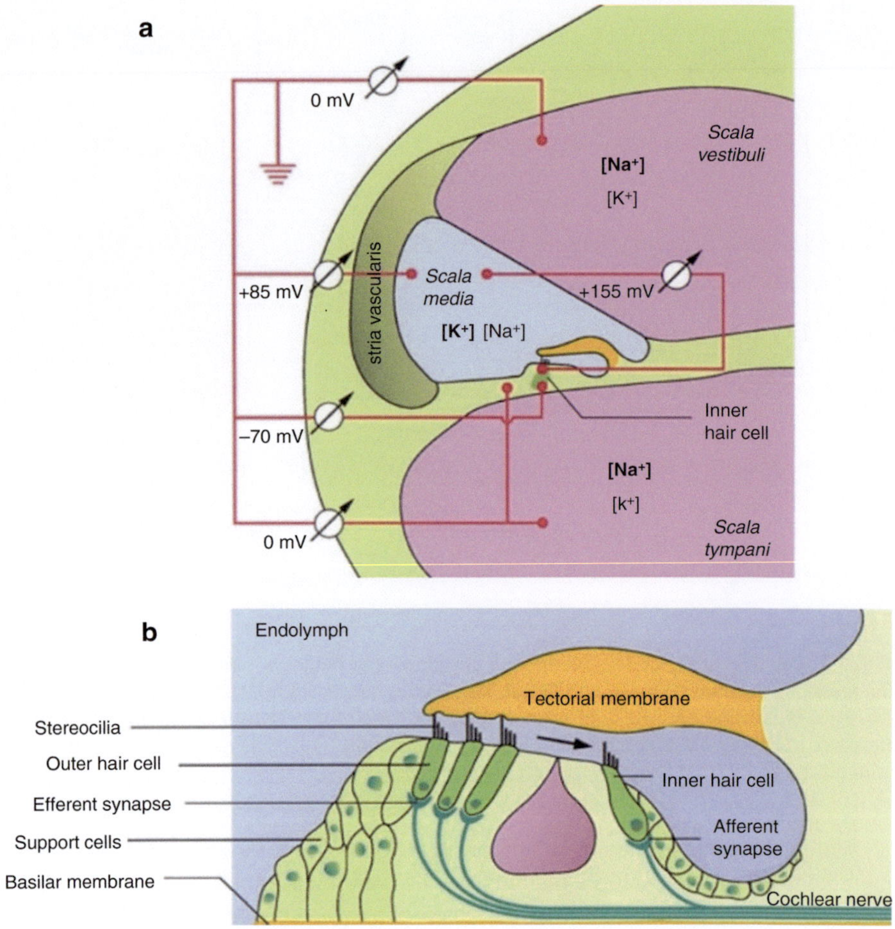

Fig. 4.6 Organ of Corti. Situated between the *scala vestibuli* and the *scala tympani*, both filled with perilymph (*pink color*), the organ of Corti is surrounded by endolymph (*gray color*). (**a**) Electrolyte concentration of the endolymph is maintained through active and passive secretion by the *stria vascularis*. (**b**) Movement of the basilar membrane is transmitted to the inner hair cells within the organ of Corti. The outer hair cells actively influence these movements and act as an amplifier for sound pressure levels below 60 dB. The inner hair cells send signals to the cochlear nerve (Adapted from Schmidt et al. 2011, with permission from Springer-Verlag Berlin Heidelberg)

further extension of amplitude discrimination through activation of neighboring hair cells. The rate of action potentials rises with increasing sound pressure levels up to a maximum rate for each cell. Further increase of sound amplitudes can be analyzed by a simultaneous activation of neighboring cells for higher and lower frequencies near the best frequency cell (Fig. 4.8). Specific potential patterns of neighboring hair cells allow an identification of best frequency cell through subsequent analysis in the central hearing regions of the central nervous system.

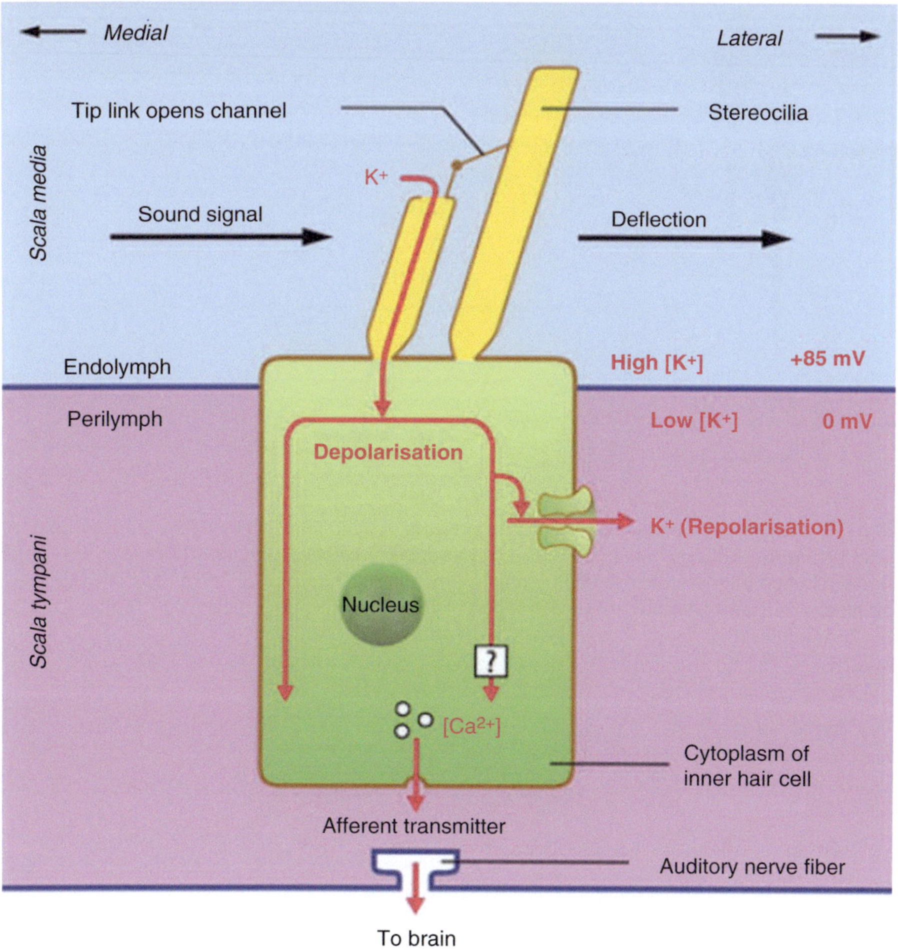

Fig. 4.7 Depolarization of inner hair cell. Deflection of stereocilia by the fluid wave in the inner ear leads to an influx of K^+ ions. This leads to a release of the specific transmitter in the synapse and finally to an action potential in the afferent nerve fiber (Adapted from Schmidt et al. 2011, with permission from Springer-Verlag Berlin Heidelberg)

On the way to these central hearing regions, the signals pass the cochlear nuclei. The following olive nuclei are important for the calculation of different signal latencies and amplitudes in order to spatialize the origin of sound. Passing the *lemnisci* and *colliculus*, the signals are finally transmitted to the cortical regions. Within these regions, the incoming information is compared with preexisting knowledge, and understanding of acoustic information takes place (Fig. 4.9) (Schmidt et al. 2011).

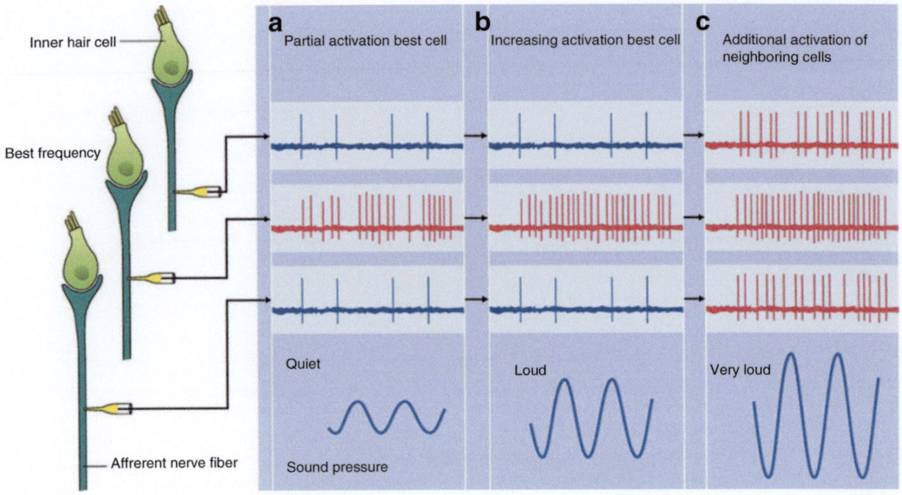

Fig. 4.8 Activation of frequency-specific hair cells in case of quiet (**a**), loud (**b**), and very loud (**c**) sound pressure. Afferent nerve fibers of neighboring hair cells are activated in case of very loud signals (**c**) and thus overcome a ceiling effect of action potentials of the afferent fiber of the best frequency hair cell (Adapted from Schmidt et al. 2011, with permission from Springer-Verlag Berlin Heidelberg)

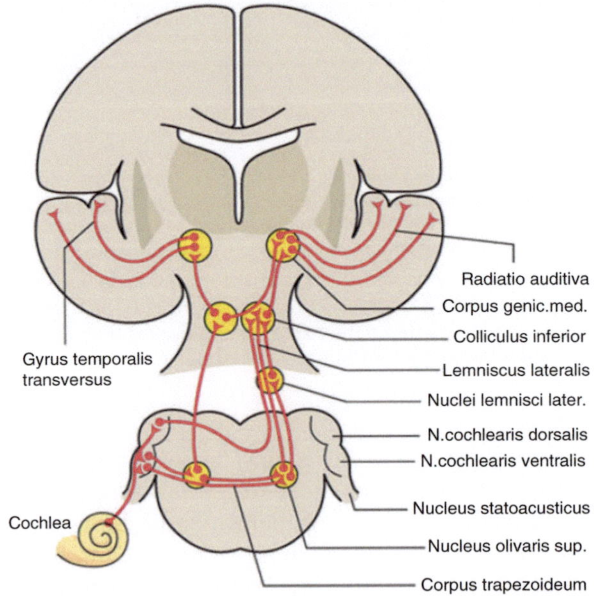

Fig. 4.9 Central hearing pathway. Signal is led through dorsal and ventral cochlear nuclei and passed to the superior olive nuclei; the signal then partly crosses to the contralateral *nuclei lemnisci* and reaches the auditory cortical regions via *colliculus inferior* and *corpus geniculatum* (Adapted from Schmidt et al. 2011, with permission from Springer-Verlag Berlin Heidelberg)

4.3 Impaired Hearing

4.3.1 Hearing Loss in General

Impaired hearing represents a clinical symptom realized by a patient or diagnosed with the help of different audiologic investigations.

As previously mentioned, the organ of hearing as a whole consists of three general compartments as well as specific regions of the central nervous system, capable for hearing. Following the natural pathway of sound, the first compartment is the external ear, followed by the middle ear and the inner ear. Sound reaches the inner ear through the oval window, a natural orifice of the cochlea. The inner ear includes the cochlea and the vestibular organ. Within the cochlea, sound waves are transformed into electrical signals that are transmitted to the central hearing regions in the brain with the help of the eighth cranial nerve.

Depending on the anatomical site of impaired function, hearing loss is defined as conductive or sensorineural hearing loss (SNHL). Conductive hearing loss is forced by any malfunction within the external and/or the middle ear. SNHL is the result of any malfunction in the regions from the inner ear to central nervous system.

Further classification of hearing loss is based on audiometric findings and the point of onset. Audiometric findings describe severity and allow a description of clinical development over time in case of repeated audiologic testing. Severity of hearing loss is categorized as mild, moderate, severe, and profound, based on pure-tone audiometric testing and resulting dB HL values. According to the World Health Organization, "hearing loss" is categorized as mild for dB HL values between 26 and 40, moderate between 41 and 60, severe between 61 and 80, and profound for values above 80 (www.who.int/pbd/deafness/hearing_impairment_grades/). "Deafness" means no measurable sound perception in any audiologic investigation.

The onset of hearing loss is described as pre- or perilingual, depending on speech development at time of onset.

The transmembrane protein pendrin is known to be located in different regions of the human inner ear, e.g., the cochlea, endolymphatic duct, and endolymphatic sac. Consequently, hearing loss in case of pendrin dysfunction is triggered through inner ear malfunction, and is therefore categorized as sensorineural in the majority of cases. The onset of hearing loss occurs at birth or within early childhood, depending on factors not to be known, yet.

The prevalence of congenital hearing loss in general is defined as 1.2–1.4 of 1,000 newborns, based on data from Germany and the United States (Gross et al. 2000, Centers for Disease Control and Prevention 2009). A genetic cause may be found in about 60 % of all cases (Burke et al. 2014), of which 70 % are nonsyndromic. The remaining cases belong to a group of manifold syndromic diseases, which include the clinical symptom of congenital hearing loss.

4.3.2 Hearing Loss in Case of Pendrin Dysfunction

4.3.2.1 Diagnosis

Hearing loss is a characteristic clinical symptom in the context of pendrin dysfunction. This applies to Pendred syndrome (PS) as well as nonsyndromic hearing (DFNB4) loss with enlarged vestibular aqueduct. Hearing loss in case of enlarged vestibular aqueduct is defined either as congenital sensorineural or congenital mixed (conductive + sensorineural), depending on auditory results (Seo et al. 2015, Rah et al. 2014, Merchant et al. 2007).

Diagnostic approach is based on specific, age-adapted audiometric testing for newborns and children. Early detection of possible hearing impairment provides best possible conditions for individual hearing support and possible rehabilitation. Based on this knowledge, newborn hearing screening was established in many countries of the Western world during the last decades. This objective investigation, based on measurement of transient otoacoustic emissions of the outer hair cells, may give early hint for possible hearing impairment within days after birth. Results depend on severity of hearing loss, which may be small at the moment of testing in case of PS or DFNB4. Therefore, normal screening results do not exclude existing hearing impairment or later development. In many cases, congenital SNHL is recognized for the first time in case of missing speech development. Conspicuous screening results or missing speech development requires further objective audiometric investigations after clinical exclusion of conductive hearing loss. Measurement of electrical potentials, triggered through acoustic stimulation, gives information on the function of the inner ear. Brainstem auditory evoked response (BAER) test is commonly used for definite diagnosis of SNHL in newborns and young children.

With ongoing adolescence, pure-tone audiometry for specific frequencies (0.25–8 kHz), an investigation depending on direct feedback from the patient, becomes the standard diagnostic tool for audiologic evaluation. Repeated pure-tone audiometry allows statements on development of hearing impairment, which may show progression, fluctuation, or steady state (see below for further explanation).

Identification of enlarged vestibular aqueduct (EVA), a radiologic finding, is provided through computed tomography (CT) or magnetic resonance imaging (MRI) of the temporal bone.

Application of imaging techniques is not part of primary diagnostic work-up for SNHL, especially in children with regard to radiation or necessity of anesthesia for MRI. Therefore, EVA in children is usually recognized during preoperative imaging prior to cochlear implant operation in case of severe hearing loss (see Sect. 4.4). In adult patients, imaging is used more frequently for further causative clarification of hearing loss, especially for exclusion of tumors or disease of the central nervous system.

As previously mentioned, the vestibular aqueduct (VA) is a bony canal within the temporal bone, containing a membranous tube, called the endolymphatic duct (ED) (Fig. 4.2). The VA runs almost horizontally from the labyrinth portion of the inner ear to the posterior cranial fossa in a mediolateral fashion (Fig. 4.10). Therefore, VA may be identified in axial planes of imaging techniques. A normal VA is hardly

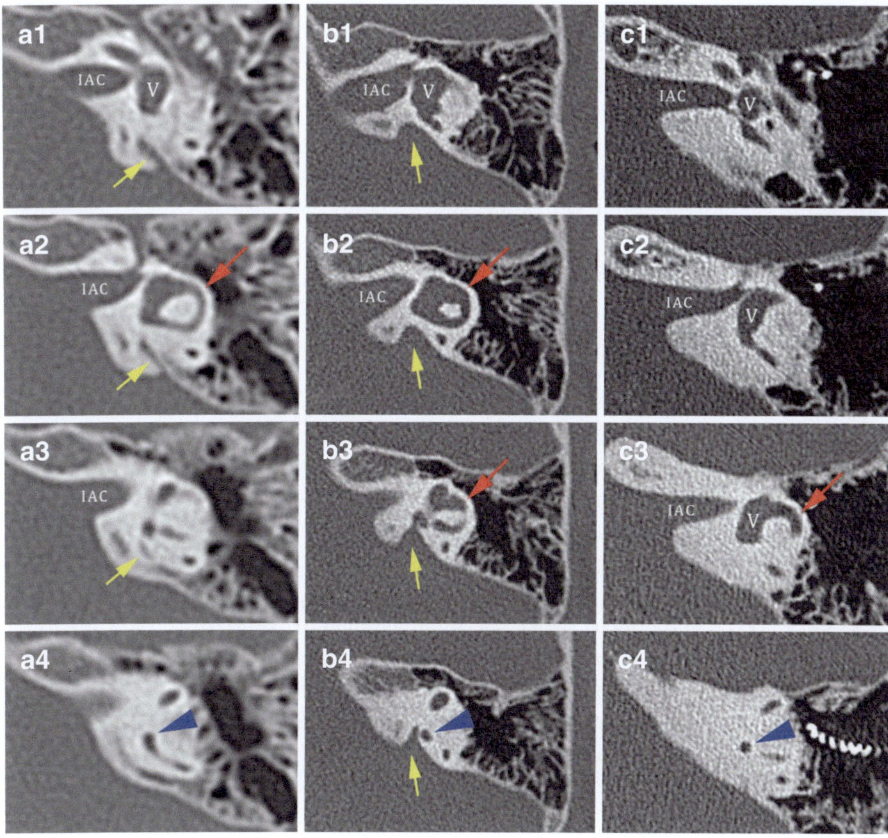

Fig. 4.10 Series of CT scans of left temporal bones at different levels (axial plane). The first level (**A–C1**) represents the internal auditory canal (*IAC*) and the vestibule (*V*) just beneath the lateral semicircular canal, seen in the following sections. The second level (**A–C2**) runs through the lateral semicircular canal (*red arrow*) and the third one (**A–C3**) right above. The fourth level (**A–C4**) shows the *common crus* of the anterior and posterior semicircular canal (*blue arrowhead*) in cross section. (**A1–4**) represent the normal anatomy of the vestibular aqueduct (*yellow arrow*), which can be identified between the medial wall of the vestibule and the posterior surface of the temporal bone. (**B1–4**) demonstrate a typical case of enlarged vestibular aqueduct (*yellow arrow*), and (**C1–4**) reveal an unusual finding of a missing vestibular aqueduct from a patient with profound sensorineural hearing loss and vestibular symptoms

visible due to its narrow appearance. An enlarged VA is defined differently by different authors. According to Valvassori and Clemis (Valvassori and Clemis 1978), the vestibular aqueduct can be defined as "enlarged" (EVA) when the midpoint diameter (measured halfway between the internal opening of the VA toward the labyrinth, also described as *fundus* in the literature, and the external opening toward the posterior cranial fossa, also described as *porus*) is $\geq$ 1.5 mm. Recently, Muskett and collaborators (Muskett et al. 2015) redefined the midpoint diameter criterion to more than 1.0 mm. Vijayasekaran (Vijayasekaran et al. 2007) considered a further parameter and defined the VA as "normal" when the midpoint diameter is $\leq$0.9 mm and the external opening is $\leq$1.9 mm. Further definitions, found in literature but not regularly used, vary in numeric reference values or include the overall appearance of VA.

EVA may be found unilaterally or bilaterally. A clinical exclusion of possible PS/DFNB4 based on unilateral appearance of EVA is not valid due to proven biallelic pathogenic mutations of the *SLC26A4* gene encoding pendrin in patients with unilateral EVA (Chattaraj et al. 2013). Moreover, size of EVA does not correlate with severity of hearing loss (Griffith et al. 1996, King et al. 2010). Agenesis of VA is a radiological finding (Fig. 4.10), not mentioned in the context of pendrin malfunction, yet. However, any malformation of inner ear structures related to pendrin expression, going along with hearing loss and vestibular symptoms needs to be taken into consideration, especially in situations of unclear genotype-phenotype correlations (King et al. 2010).

In summary, EVA represents a clear radiological hint for possible pendrin malfunction, potentially resulting from an altered morphogenesis of the inner ear during embryology. There are further malformations of the temporal bone, being discussed in the context of pendrin malfunction, such as Mondini malformation (Arcand et al. 1991, Valvassori and Clemis 1978), also known as incomplete partition II (Sennaroglu and Saatci 2002), and altered radiological appearance of compartments of the inner ear such as the vestibule (Goldfeld et al. 2005). A common feature of most variations is an increase in luminal size, leading to a reduction of the surrounding bony structure. To date, there exists no clear correlation between morphological variations of the temporal bone and actual pendrin malfunction (King et al. 2010).

The vestibular organ is embedded within the inner ear in direct connection with the hearing organ. Testing of vestibular function in case of suspected pendrin malfunction is not regarded as standard procedure, yet. Vestibular symptoms are commonly described in patients with EVA (Grimmer and Hedlund 2007, Noordman et al. 2015, Zalewski et al. 2015), irrespectively from actual pendrin malfunction and without definition of appropriate testing protocol. In a recent prospective study, including 106 patients with uni- or bilateral EVA, Zalewski and collaborators reported on 45 % of the patients with vestibular symptoms and only 44 % of those with actual pathological videonystagmography (VNG) results. In contrast to studies on hearing impairment, there was no correlation between number of pendrin mutant alleles and vestibular symptoms, neither abnormal VNG results. However, these authors found an association between history of head injury and vestibular signs as well as abnormal VNG results (Zalewski et al. 2015).

Vestibular symptoms in humans with pendrin dysfunction seem not to be as common as described in mice models for Pendred syndrome (Everett et al. 2001; see also Chap. 1). However, selective anamnesis for vestibular symptoms, as well as VNG, can provide more insight into actual inner ear function in case of suspected pendrin malfunction, particularly in case of EVA.

After audiologic, vestibular, and radiologic assessment, next diagnostic step is genetic testing for mutations of the *SLC26A4* gene, giving proof of possible genetic variations and subsequent pendrin malfunction (see Chap. 5).

Table 4.1 Clinical categorization of hearing impairment development based on repeated audiologic testing, according to King et al. (2010)

Category	Definition
Stable	No significant fluctuation, improvement, or progression as defined below
Improving	10 dB improvement at any three frequencies or 15 dB improvement at any two frequencies or ≥20 dB improvement at any one frequency
Progressive	10 dB decline at any three frequencies or 15 dB decline at any two frequencies or ≥20 dB decline at any one frequency
Fluctuating	Interim audiogram shows 10 dB change at any three frequencies or 15 dB change at any two frequencies or ≥20 dB change at any one frequency, and final audiogram does not show significant improvement or progression from baseline
Fluctuating/Improving	Interim audiogram shows 10 dB change at any three frequencies or 15 dB change at any two frequencies or ≥20 dB change at any one frequency, and final audiogram shows significant improvement from baseline
Fluctuating/Progressive	Interim audiogram shows 10 dB change at any three frequencies or 15 dB change at any two frequencies or ≥20 dB change at any one frequency, and final audiogram shows significant progression from baseline

4.3.2.2 Clinical Course

The clinical course of hearing impairment in case of pendrin malfunction is generally quite heterogeneous. Time of onset, as well as progression or severity, seem to depend on number of *SLC26A4* mutant alleles (King et al. 2010), type of mutations (Rah et al. 2014), possible external influences and further factors not to be known, yet. Looking at the great variety of hearing phenotypes, an influence of multiple factors seems reasonable.

Time of onset is expected to be during early childhood, due to congenital appearance. Individual definition of onset depends on time of diagnosis. Clinical differentiation may be based on the occasion of possible speech development, resulting in pre- or postlingual hearing impairment.

There is a great individual change in severity of hearing loss observed over time. Clinical categorizations depend on repeated pure-tone audiometry investigations in order to give evidence on development of impairment, stated in [dB] for each frequency tested (Table 4.1). A detailed categorization of hearing development has been described by King and collaborators in 2010. Besides stable or progressive hearing loss, fluctuating hearing loss is commonly described in conjunction with pendrin malfunction (King et al. 2010). The same authors also described improving hearing loss in a single patient with one *SLC26A4* mutant allele (King et al. 2010).

A further clinical feature is a sudden decrease of hearing ability, depicted as hearing drop. Hearing drops related to presence of EVA are well described and explained trough head injury, harming the inner ear with mechanical impact. A meta-analysis by Noordman and collaborators in 2015 concludes that hearing drops associated with head injury appear in only one third of patients with EVA. Moreover, all of the patients affected had fluctuating hearing, leading to the conclusion that only in patients with fluctuating hearing ability, higher awareness of head trauma is necessary. Therefore, head trauma may influence the clinical course of hearing impairment, too. The remaining cases of hearing drops lack a conclusive explanation.

Genetic factors seem to have greatest impact on hearing loss and its clinical course. Numerous observational studies provide data on correlation between genotypes and audiometric phenotypes (King et al. 2010, Chattaraj et al. 2013; Rah et al. 2014). A categorization depending on the number of mutant *SLC26A4* alleles (zero, one, or two, defined as M0, M1, or M2, respectively) allows for statements on clinical symptoms. M0 status is usually found in cohorts of EVA patients suffering from hearing loss, with a prevalence of more than 50 % (King et al. 2010). In case of M2 genotype, hearing thresholds appear to be poorer than in M1 or M0 individuals, at least in Caucasian populations.

Moreover, pathogenicity of each *SLC26A4* variant needs to be taken into consideration in order to predict possible clinical impact. This requires a categorization between pathogenic and nonpathogenic mutations (King et al. 2010).

Finally, ethnicity seems to have an influence too, since there are great differences between Caucasian and East Asian cohorts concerning the audiometric phenotype, independently from the number of mutant alleles (Rah et al. 2014). Rah and collaborators reported similar hearing impairment in M1 and M2 patients from East Asia, in contrast to the observations of King and collaborators (King et al. 2010).

The contribution of the different pathological derangements in causing hearing loss in case of pendrin mutations remains unclear, at least in humans. To date, most of the knowledge of pathogenesis of hearing loss due to pendrin malfunction derives from mouse models (see Chap. 2).

A loss in the endocochlear potential, as well as oxidative stress and a loss of expression of KCNJ10 channel within the *stria vascularis* seem to be concrete factors contributing to pathogenesis (Ito et al. 2014). Crucial time frame, when pendrin malfunction may profoundly affect hearing, seems to be during embryological development, at least in the mouse model (Choi et al. 2011).

From a clinical point of view, continuous audiologic testing and direct access to specialized clinics are mandatory in case of suspected or proven pendrin malfunction. Repeated evaluation of the clinical course of hearing ability allows for appropriate treatment in time, in order to achieve best possible communication skills and social development.

4.4 Treatment of Hearing Loss

4.4.1 Treatment of Hearing Loss in General

Treatment of hearing loss is based on the type of hearing impairment. As previously mentioned, hearing loss caused by deficits in sound perception is called sensorineural hearing loss. An impaired recognition of high frequency sounds only, like singing birds, usually does not require specific treatment due to minor restriction during everyday life; more influencing for humans is a deficit in understanding speech, resulting from deficits in multiple frequencies. Mild forms begin with problems in understanding conversations in a loud environment, and severe forms prevent development of speech skills in children and can lead to the avoidance of social contacts in adults. For humans, it is crucial to have two functional ears, as monaural hearing leads to deficits in localization of sound sources and understanding in noise.

The treatment of conductive hearing loss is usually based on the underlying causes. Obstructions of the outer ear canal can be treated surgically. A temporary or permanent improvement can also be achieved by the use of bone conduction hearing aids (Colquitt et al. 2011). If the middle ear is affected, manifold ear surgery procedures have been developed over the last 60 years (Wullstein 1960). Middle ear surgery includes closure of tympanic membranes and the restoration of a destroyed ossicular chain. Surgery is usually performed with a microscope; endoscopic techniques are under development (Kozin et al. 2014). Effusions in the middle ear may be drained by paracentesis or drainage tubes (Ciuman 2015). Effusions are a typical disease in preschool children usually recognized due to problems in speech development. Therefore, middle ear effusion is an important clinical entity to be excluded in case of suspected sensorineural hearing loss in children.

If surgical therapy with passive implants does not sufficiently improve hearing ability, active implants may offer additional possibilities (Braun et al. 2015). Many different systems were developed; the most frequently implanted system worldwide, called soundbridge, is shown in Fig. 4.11. Even a direct artificial sound wave transduction to the inner ear is possible, through the so-called acoustic cochlear implant (Lenarz et al. 2014).

If sensorineural hearing loss or combined hearing loss is evident, conventional hearing aids may be the first proper solution for the patient. Usually worn bilaterally, they can provide amplifications up to 30 dB in different frequency ranges. In case of mild to moderate hearing losses, this may be a sufficient solution (Fig. 4.12). Fitting and training are crucial to obtain patient compliance for regular use of the devices (Lavie et al. 2014).

For severe to profound sensorineural hearing loss, a lack of outer and inner hair cell function in the inner ear prevents any acoustic aids to achieve understanding words or sentences by amplification of sounds. For this condition, the implantation of an electrode into the cochlea leads to a restoration of hearing. The technique allows the production of electronic signals by an external speech processor, transmitted to the cochlea through an electrode (Fig. 4.13), independent from the function of the cochlea. The electrode is surgically placed into the *scala tympani* of the cochlea. The presentation of electrical potentials in striking distance to nerve fibers in the cochlea (as well as in the auditory nerve and the *nucleus cochlearis*) leads to open speech understanding in patients either implanted before the age of 2 years in congenital hearing loss as well as in patients with postlingual deafness. Prelingually deaf patients usually do not profit from implantation later than the fourth year of life.

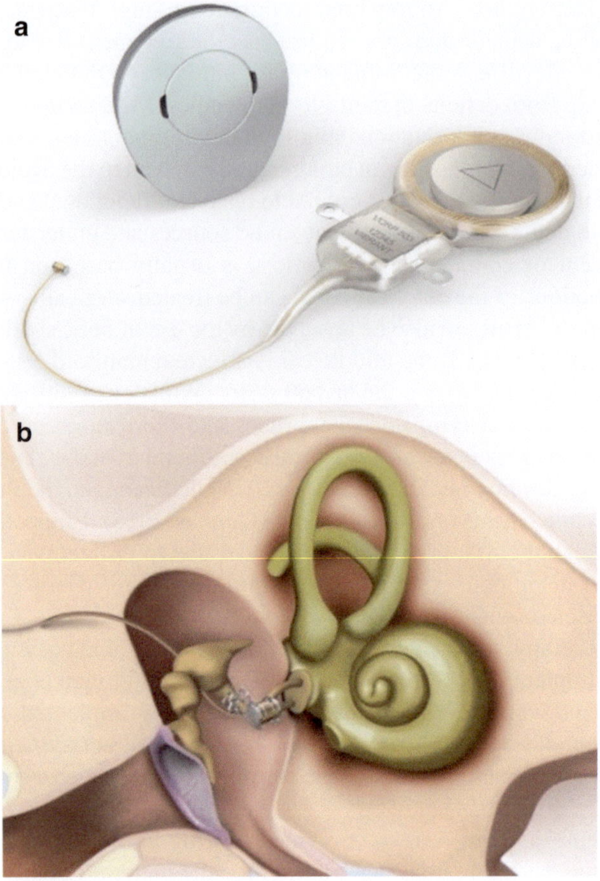

Fig. 4.11 The vibrant soundbridge system. (**a**) External sound processor (*left*) and an implantable receiver (*right*) with an electromagnetic actuator at the end of a cable. The actuator may be attached to the ossicular chain (**b**) as well as placed directly to the oval or round window (From MedEl corp. Innsbruck, Austria)

4.4.2 Treatment of Hearing Loss in Case of Pendrin Dysfunction

Cochlear implant supply in the context of pendrin dysfunction and accompanying inner ear malformations requires special alertness during surgery as well as surgical experience. A common phenomenon described during opening of the cochlea is the so-called Gusher, an effervescent discharge of perilymph. Though impressive at first sight, it is a circumstance surgically manageable through adequate sealing at the region of cochleostomy after insertion of the electrode. It is important to consider possible intraoperative difficulties related to altered anatomic situations linked to inner ear malformations, preoperatively. However, malformations of the inner ear in the context of pendrin malfunction, such as EVA, incomplete partition of the

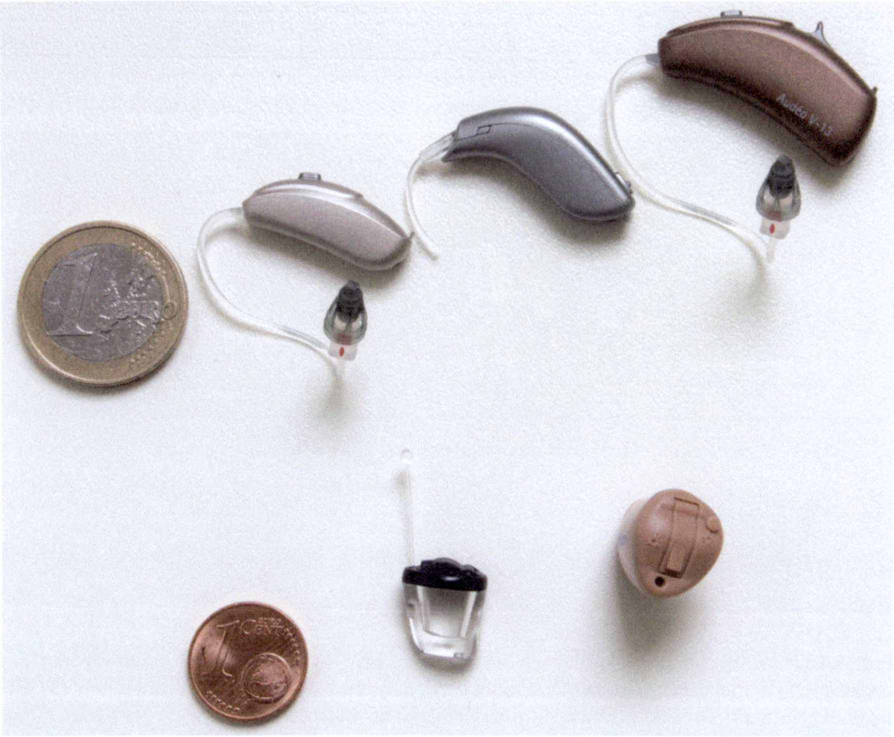

Fig. 4.12 Different types of hearing aids. Hearing aids may be placed behind the ear – *top*, three examples, or in the ear canal – *bottom*, two examples. Depending on individual amplification required, technical setup of sound transmission to the middle ear varies widely (From Neuroth AG, Austria)

cochlea type II or enlargement of the labyrinth (see Sect. 4.3.2.1), should not have any impact on the surgical steps, especially not on the insertion of the electrode of the cochlear implant.

Clinical outcome, concerning hearing ability and speech perception after cochlear implant supply in case of *SLC26A4* mutations, was evaluated recently by Busi and collaborators in 2015 (Busi et al. 2015). These authors concluded that *SLC26A4* sequence alterations have no negative impact on the outcome of cochlear implant supply. This clearly supports the great value of cochlear implant in case of deafness, linked to pendrin dysfunction.

Conclusions

In conclusion, to date, therapeutic options cover almost every situation of hearing loss. Concerning sensorineural hearing loss, first step therapy is a conventional hearing aid for sound amplification. In case of profound hearing loss or deafness, a cochlear implant represents the most valuable therapeutic approach. Clear indication and timely intervention are essential to obtain best possible hearing rehabilitation and satisfaction of the patient.

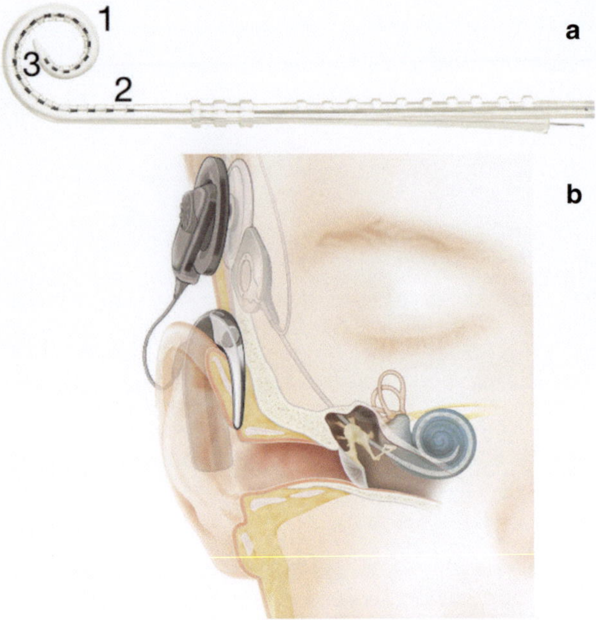

Fig. 4.13 Cochlear implant system (Nucleus® CI 512). (**a**) Electrode for insertion into the cochlea – (*1*) perimodiolar portion, (*2*) straight portion for the basal turn of the cochlea, and (*3*) soft tip of the electrode. (**b**) Complete cochlear implant system; external speech processor (*dark color*), transmitter coil lying under the skin (*light color*), and electrode running through the middle ear into the cochlea (*blue*) (From Cochlear corp. Lane cove, Australia)

References

Anthwal N, Thompson H (2016) The development of the mammalian outer and middle ear. J Anat 228:217–232. doi:10.1111/joa.12344

Arcand P, Desrosiers M, Dubé J, Abela A (1991) The large vestibular aqueduct syndrome and sensorineural hearing loss in the pediatric population. J Otolaryngol 20:247–250

Bast TH (1928) The utriculo-endolymphatic valve. Anat Rec 40:61–65. doi:10.1002/ar.1090400106

Bast TH, Anson BJ (1949) The temporal bone and the ear. Charles C. Thomas Co., Springfield

Bianchi L, Fuchs P (2010) The Oxford handbook of auditory science. Oxford University Press, Oxford

Braun K, Zenner H-P, Friese N, Tropitzsch A (2015) Differential indication of active middle ear implants. HNO 63:402–418. doi:10.1007/s00106-015-0018-5

Burke WF, Lenarz T, Maier H (2014) Hereditary hearing loss. HNO 62:759–770. doi:10.1007/s00106-014-2901-x

Busi M, Rosignoli M, Castiglione A et al (2015) Cochlear implant outcomes and genetic mutations in children with ear and brain anomalies. Biomed Res Int 2015:696281–19. doi:10.1155/2015/696281

Centers for Disease Control and Prevention (CDC) (2009) Early hearing detection and intervention (EHDI) hearing screening & follow-up survey (revised November 2011)

Chattaraj P, Reimold FR, Muskett JA et al (2013) Use of *SLC26A4* mutation testing for unilateral enlargement of the vestibular aqueduct. JAMA Otolaryngol Head Neck Surg 139:907–913. doi:10.1001/jamaoto.2013.4185

Choi BY, Kim H-M, Ito T et al (2011) Mouse model of enlarged vestibular aqueducts defines temporal requirement of Slc26a4 expression for hearing acquisition. J Clin Invest 121:4516–4525. doi:10.1172/JCI59353

Ciuman RR (2015) Middle ear drainage and ventilation: indications, complications and their treatment. Klin Padiatr 227:54–60. doi:10.1055/s-0034-1398537

Colquitt JL, Loveman E, Baguley DM et al (2011) Bone-anchored hearing aids for people with bilateral hearing impairment: a systematic review. Clin Otolaryngol 36:419–441. doi:10.1111/j.1749-4486.2011.02376.x

Everett LA, Belyantseva IA, Noben-Trauth K et al (2001) Targeted disruption of mouse *Pds* provides insight about the inner-ear defects encountered in Pendred syndrome. Hum Mol Genet 10:153–161

Goldfeld M, Glaser B, Nassir E et al (2005) CT of the ear in Pendred syndrome1. Radiology 235:537–540. doi:10.1148/radiol.2352031583

Griffith AJ, Arts HA, Downs C et al (1996) Familial large vestibular aqueduct syndrome. Laryngoscope 106:960–965. doi:10.1097/00005537-199608000-00009

Grimmer JF, Hedlund G (2007) Vestibular symptoms in children with enlarged vestibular aqueduct anomaly. Int J Pediatr Otorhinolaryngol 71:275–282. doi:10.1016/j.ijporl.2006.10.010

Gross M, Finchk-Krmer U, Spormann-Lagodzinski M (2000) Connatal hearing disorders in children. Part I: Connatally acquired hearing loss. HNO 12:879–886

Gulya AJ, Schuknecht HF (1995) Anatomy of the temporal bone with surgical implications. Parthenon Pub Group, New York

Ito T, Li X, Kurima K et al (2014) Slc26a4-insufficiency causes fluctuating hearing loss and *stria vascularis* dysfunction. Neurobiol Dis 66:53–65. doi:10.1016/j.nbd.2014.02.002

Kelley MW, Wu DK, Popper AN, Fay RR (2005) Development of the inner ear. doi:10.1007/0-387-30678-1

King KA, Choi BY, Zalewski C et al (2010) *SLC26A4* genotype, but not cochlear radiologic structure, is correlated with hearing loss in ears with an enlarged vestibular aqueduct. Laryngoscope 120:384–389. doi:10.1002/lary.20722

Kozin ED, Lehmann A, Carter M et al (2014) Thermal effects of endoscopy in a human temporal bone model: implications for endoscopic ear surgery. Laryngoscope 124:E332–E339. doi:10.1002/lary.24666

Lavie L, Banai K, Attias J, Karni A (2014) Better together: reduced compliance after sequential versus simultaneous bilateral hearing aids fitting. Am J Audiol 23:93–98. doi:10.1044/1059-0889(2013/13-0010)

Lenarz T, Schwab B, Maier H, Kludt E (2014) Direct acoustic cochlear stimulation for therapy of severe to profound mixed hearing loss: Codacs™ Direct Acoustic Cochlear Implant System. HNO 62:481–489. doi:10.1007/s00106-014-2893-6

Merchant SN, Nakajima HH, Halpin C et al (2007) Clinical investigation and mechanism of air-bone gaps in large vestibular aqueduct syndrome. Ann Otol Rhinol Laryngol 116:532–541. doi:10.1177/000348940711600709

Muskett JA, Chattaraj P, Heneghan JF et al (2015) Atypical patterns of segregation of familial enlargement of the vestibular aqueduct. Laryngoscope. doi:10.1002/lary.25737

Noordman BJ, van Beeck Calkoen E, Witte B et al (2015) Prognostic factors for sudden drops in hearing level after minor head injury in patients with an enlarged vestibular aqueduct: a meta-analysis. Otol Neurotol 36:4–11. doi:10.1097/MAO.0000000000000659

Rah YC, Kim AR, Koo J-W et al (2014) Audiologic presentation of enlargement of the vestibular aqueduct according to the *SLC26A4* genotypes. Laryngoscope 125:E216–E222. doi:10.1002/lary.25079

Richany SF, Bast TH, Anson BJ (1954) The development and adult structure of the malleus, incus and stapes. Ann Otol Rhinol Laryngol 63:394–434

Schmidt RF, Lang F, Heckmann M (2011) Physiologie des Menschen. doi:10.1007/978-3-642-01651-6

Sennaroglu L, Saatci I (2002) A new classification for cochleovestibular malformations. Laryngoscope 112:2230–2241. doi:10.1097/00005537-200212000-00019

Seo YJ, Kim J, Choi JY (2015) Correlation of vestibular aqueduct size with air-bone gap in enlarged vestibular aqueduct syndrome. Laryngoscope. doi:10.1002/lary.25664

Tóth M, Csillag A (2005) The organ of hearing and equilibrium. In: Csillag A (ed) Atlas of the sensory organs. Humana Press, Totowa, pp 1–83

Valvassori GE, Clemis JD (1978) The large vestibular aqueduct syndrome. Laryngoscope 88:723–728

Vijayasekaran S, Halsted MJ, Boston M et al (2007) When is the vestibular aqueduct enlarged? A statistical analysis of the normative distribution of vestibular aqueduct size. AJNR Am J Neuroradiol 28:1133–1138. doi:10.3174/ajnr.A0495

Voss SE, Nakajima HH, Huber AM, Shera CA (2013) Function and acoustics of the normal and diseased middle ear. In: Puria S, Fay RR, Popper AN (eds) The middle ear. Springer, New York, pp 67–91

Wullstein H (1960) Techniques of tympanoplasty I, II, and III. AMA Arch Otolaryngol 71:424–427. doi:10.1001/archotol.1960.03770030066014

Zalewski CK, Chien WW, King KA et al (2015) Vestibular dysfunction in patients with enlarged vestibular aqueduct. Otolaryngol Head Neck Surg 153:257–262. doi:10.1177/0194599815585098

Genetic Diagnosis of Deafness

5

Sueli Matilde da Silva Costa, Priscila Zonzini Ramos, Fábio Tadeu Arrojo Martins, and Edi Lúcia Sartorato

Abstract

Genetic testing can provide an accurate diagnosis, contributing to appropriate treatment, prognosis and precise genetic counseling for patients with hearing loss. It is estimated that approximately 300 genes could be involved in determining syndromic and nonsyndromic hearing loss, making the diagnosis of hearing loss highly challenging. Conventional methodologies have been used for the screening of the most frequent mutations and genes according to ethnic background, but they are impractical for the screening of all genes associated with the complex auditory system. The Human Genome Project has directly or indirectly stimulated the development of many technologies and platforms for molecular diagnostics of genetic diseases, including hereditary hearing loss. All these tools are not only applicable to molecular diagnostics, but also enable the identification of new genes, providing new insights into the mechanisms of hearing loss.

5.1 Causes of Hearing Loss

Hearing is a complex process; consequently, it should be no surprise that the causes of hearing loss are also complex. Hearing loss can be caused by a variety of environmental and genetic factors, or by a combination of both. The relative contribution of environmental causes is often determined by the economic and social conditions that affect health, such as education, physical circumstances and the

S.M. da Silva Costa • P.Z. Ramos • F.T.A. Martins • E.L. Sartorato (✉)
Human Molecular Genetics Laboratory, Molecular Biology and Genetic Engineering Center (CBMEG), University of Campinas (UNICAMP),
Edi Lúcia Sartorato 400, Candido Rondon Avenue, Campinas, São Paulo, Brazil
e-mail: silvacosta.sm@gmail.com; priscila.zonzini@yahoo.com.br; fabio.arrojo@cbmeg.unicamp.br; sartor@unicamp.br

S. Dossena, M. Paulmichl (eds.), *The Role of Pendrin in Health and Disease*,
DOI 10.1007/978-3-319-43287-8_5

provision of neonatal and postnatal medical care. For this reason, there is a greater prevalence of environmental factors in developing countries, whereas genetic factors are the main cause of hearing loss in countries with high standards of health care. Many of the environmental causes could be reduced or eliminated by means of immunization, better therapies for bacterial and viral infections, public health regulations and lifestyle changes.

5.1.1 Environmental Causes

Environmental factors include use of ototoxic drugs, frequent exposure to high-intensity sounds, trauma and infections. Nongenetic congenital hearing loss commonly results from intrauterine infections including toxoplasmosis, rubella, cytomegalovirus and herpes virus (TORCH organisms). Other conditions occurring at the time of birth or during the postnatal period that can cause hearing loss include anoxia, prematurity, complications associated with the Rh factor in the blood and bacterial meningitis. These conditions typically cause degrees of sensorineural hearing loss ranging from mild to profound. Acquired hearing loss in adults is most often attributed to environmental factors, but it is likely that a genetic component could influence susceptibility or the severity of acquired hearing impairment (genetic–environmental interactions) (Smith et al. 2014). Causes of hearing loss in adults include otosclerosis, Ménière's disease (a combination of sensorineural hearing loss, vertigo, tinnitus and sensitivity to loud sounds), autoimmune inner ear disease, ototoxic medications, exposure to high-intensity sounds, tumors, head injury and presbycusis (sensorineural hearing loss that occurs gradually later in life).

5.1.2 Genetic Causes

Elucidation of the genetic causes of deafness provides crucial information for diagnosis and for understanding the mechanisms of hearing impairment. It is estimated that at least 1 % of all human genes (about 300 genes) are involved in the hearing process (Nance 2003; Friedman and Griffith 2003). Mutations in any of these genes or in regulatory elements may cause hearing loss. In developed countries, around 60 % of prelingual hearing loss cases have genetic origins, and the remaining cases can be attributed to environmental and unidentified genetic factors (Morton and Nance 2006).

Although most forms of hereditary hearing loss are caused by mutations in a single gene, in some cases, phenotypic expression requires alterations in two or more genes. Mutations in some genes may not cause hearing loss directly; instead, they increase the risk of deafness due to environmental factors such as exposure to high-intensity sounds or antibiotics. Age-related and noise-induced hearing losses are the most frequent examples of complex genetic–environmental interactions resulting in hearing loss (Huyghe et al. 2008; Konings et al. 2008).

5.2 Hereditary Hearing Loss

Approximately 70 % of hereditary hearing loss cases are nonsyndromic, where hearing impairment is the only symptom. In the remaining 30 % of cases, deafness is associated with other clinical findings and is characterized as syndromic (Hilgert et al. 2009a).

5.2.1 Syndromic Hearing Loss

Over 400 syndromes that include hearing loss have been reported, and at least 40 genes have been linked to the main types of syndromic hearing loss (Hilgert et al. 2009a; Van Camp and Smith 2015). The commonest syndromic forms of hearing loss include the following syndromes: Usher (associated with *retinitis pigmentosa*), Pendred (related to goiter and inner ear malformations), Waardenburg (involving pigmentary abnormalities), Branchio-oto-renal (BOR; branchial, otologic and renal manifestations), Jervell and Lange-Nielsen (prolongation of the QT interval is a measure of the time between the start of the Q wave and the end of the T wave in the heart's electrical cycle.), Stickler (ocular, auditory, skeletal and orofacial abnormalities) and Alport (related to renal and ocular anomalies).

5.2.1.1 Autosomal Recessive Syndromic Hearing Loss

Usher syndrome Usher syndrome is the most frequent type of autosomal recessive syndromic hearing loss (Smith et al. 2014). This syndrome is characterized by sensorineural hearing loss associated with *retinitis pigmentosa* (progressive retinal degeneration that leads to visual loss). Usher syndrome has been classified into three main phenotypes. Type I (USH1) presents congenital severe-to-profound sensorineural hearing loss, vestibular dysfunction and onset of *retinitis pigmentosa* in the first decade of life. Type II (USH2) consists of congenital mild-to-severe sensorineural hearing loss, normal vestibular function and *retinitis pigmentosa* beginning in the first two decades of life. Finally, type III (USH3) is characterized by postlingual progressive hearing impairment, variable vestibular dysfunction and *retinitis pigmentosa* with a variable age of onset.

To date, ten genes have been reported to cause Usher syndrome. Type I has been associated with mutations in six different genes: *MYO7A* (Usher syndrome type 1B, USH1B), *USH1C*, also known as harmonin or *PDZ73* (USH1C), *CDH23* (USH1D), *PCDH15* (USH1F), *SANS* (USH1G) and *CIB2* (USH1J) (Smith et al. 1992; Weil et al. 1995; Bork et al. 2001; Ahmed et al. 2001; Alagramam et al. 2001; Weil et al. 2003; Riazuddin et al. 2012). For type II, three genes have been described: *USH2A* (USH2A), *VLGR1*, also known as *GPR98* (USH2C) and *WHRN*, also known as *DFNB31* (USH2D) (Eudy et al. 1998; Weston et al. 2004; Ebermann et al. 2007). Only one gene, *CLRN1*, has been identified as causing type III (Joensuu et al. 2001). Additionally, three other genes have been related to Usher syndrome. It was proposed that the *PDZD7* gene could contribute to digenic inheritance, together with *GPR98*, and may modify disease severity in

USH2A patients (Ebermann et al. 2010). A novel biallelic mutation in the *HARS* gene was identified in two patients and was associated with USH3B (Puffenberger et al. 2012). Recently, the *CEP250* gene was linked to an atypical Usher syndrome (Khateb et al. 2014).

Pendred syndrome Pendred syndrome is the second most common type of autosomal recessive syndromic hearing loss (Smith et al. 2014). Classical features include congenital severe-to-profound sensorineural hearing impairment, euthyroid goiter, inner ear malformations such as enlarged vestibular aqueduct (EVA) or Mondini dysplasia and occasionally vestibular dysfunction (Pendred 1896). Mutations in the *SLC26A4* gene can be responsible for Pendred syndrome and can cause autosomal recessive nonsyndromic hearing loss (ARNSHL) with EVA (DFNB4) (Everett et al. 1997; Li et al. 1998); see Sect. 5.2.2.1. It has been proposed that digenic inheritance involving *SLC26A4* and either *FOXI1* or *KCNJ10* genes may also cause Pendred syndrome (Yang et al. 2007, 2009), but studies seem to indicate that this is an extremely rare cause (Chen et al. 2012; Cirello et al. 2012; Landa et al. 2013).

Jervell and Lange-Nielsen syndrome Jervell and Lange-Nielsen syndrome is the third most common type of autosomal recessive syndromic hearing loss (Smith et al. 2014). This syndrome is characterized by congenital hearing impairment, prolongation of the QT interval, syncopal episodes due to arrhythmias and high risk of sudden death (Jervell and Lange-Nielsen 1957). Mutations in two genes, *KCNQ1* and *KCNE1*, have been reported in patients with Jervell and Lange-Nielsen syndrome (Neyroud et al. 1997; Tyson et al. 1997; Schulze-Bahr et al. 1997).

5.2.1.2 Autosomal Dominant Syndromic Hearing Loss

Waardenburg syndrome Waardenburg syndrome is the most frequent type of autosomal dominant syndromic hearing loss (Smith et al. 2014). The syndrome consists of congenital sensorineural hearing impairment and pigmentary abnormalities of the eyes, skin and hair (Waardenburg 1951). Four clinical types of Waardenburg syndrome are recognized. Type I (WS1) and type II (WS2) share some similarities, but WS1 is distinguished from WS2 by the presence of *dystopia canthorum* (lateral displacement of the inner *canthus* of the eye). Type III (WS3) is characterized by *dystopia canthorum* and upper limb malformations, and type IV (WS4) is associated with Hirschsprung disease, a digestive disorder related to decreased intestinal motility caused by a lack of nerve cells (Read and Newton 1997). Mutations in the *PAX3* gene are responsible for WS1 and WS3 (Tassabehji et al. 1992; Hoth et al. 1993). The *MITF* and *SNAI2* (also known as *SLUG*) genes are involved in some cases of WS2 (Tassabehji et al. 1994; Sánchez-Martín et al. 2002). Mutations in the *EDNRB*, *EDN3* and *SOX10* genes are associated with WS4 (Attié et al. 1995; Edery et al. 1996; Pingault et al. 1998).

BOR syndrome BOR syndrome is the second most common type of autosomal dominant syndromic hearing loss (Smith et al. 2014). It is characterized by branchial cysts or fistulae; conductive, sensorineural or mixed hearing impairment due to malformations of the external, middle and inner ear; and renal anomalies. Mutations in the *EYA1* gene are found in about 40 % of patients with the syndrome. Pathogenic variants in the *SIX1* and *SIX5* genes are associated with BOR syndrome and have been found in 4 and 5 % of cases, respectively (Abdelhak et al. 1997; Ruf et al. 2004; Hoskins et al. 2007).

Stickler syndrome Stickler syndrome is a clinically and genetically heterogeneous disorder characterized by ocular manifestations (myopia, cataract or retinal detachment), conductive or sensorineural hearing loss, cleft palate, spondyloepiphyseal dysplasia (disorder of bone growth that results in short stature and skeletal abnormalities) and osteoarthritis (Stickler et al. 1965). Stickler syndrome types I and II (STL1 and STL2) are associated with severe myopia and predisposition to retinal detachment, while type III (STL3) is a nonocular form. STL1, STL2 and STL3 are caused by mutations in the *COL2A1*, *COL11A1* and *COL11A2* genes, respectively, and are inherited as an autosomal dominant form (Ahmad et al. 1991; Vikkula et al. 1995; Richards et al. 1996). On the other hand, mutations in the *COL9A1*, *COL9A2* and *COL9A3* genes are inherited as autosomal recessive forms and are associated with Stickler syndrome types IV, V and VI (STL4, STL5 and STL6, respectively) (Van Camp et al. 2006; Baker et al. 2011; Faletra et al. 2014). STL4 is associated with moderate-to-severe sensorineural hearing loss, myopia with vitreoretinopathy, cataracts and epiphyseal dysplasia; STL5 features mild-to-moderate hearing loss, strong myopia and vitreoretinopathy; and STL6 manifests as moderate-to-severe sensorineural hearing loss, moderate-to-high myopia and midface retrusion.

5.2.1.3 X-Linked Syndromic Hearing Loss

Alport syndrome Alport syndrome is characterized by progressive sensorineural hearing loss of varying severity, progressive glomerulonephritis resulting in end-stage renal disease and variable ophthalmologic abnormalities (Alport 1927). This syndrome is a hereditary disorder of the basement membranes caused by mutations in type IV collagen genes. It consists of a genetically heterogeneous disorder that can be inherited in X-linked, autosomal recessive or autosomal dominant modes. X-linked inheritance is the most common form, accounting for about 85 % of cases, and is related to mutations in the *COL4A5* gene. Approximately 15 % of cases are autosomal recessive, while autosomal dominant inheritance is rare. Pathogenic variants in *COL4A3* and *COL4A4* genes account for both autosomal recessive and dominant patterns (Barker et al. 1990; Mochizuki et al. 1994; Jefferson et al. 1997).

5.2.1.4 Mitochondrial Syndromic Hearing Impairment

Mutations in the mitochondrial DNA have been implicated in a variety of diseases, and hearing impairment is an important feature in mitochondrial disorders. Forms of mitochondrial syndromic hearing impairment include Kearns–Sayre syndrome (KSS – ophthalmoplegia, pigmentary degeneration of the retina and cardiomyopathy), MELAS (mitochondrial myopathy, encephalopathy, lactic acidosis and stroke-like episodes), MIDD (maternally inherited diabetes and deafness) and MEERF (myoclonic epilepsy with ragged red muscle fibers), among others. A specific pathogenic variant, m.3243A > G, in the *MT-TL1* gene has been identified in 2–6 % of individuals with family history of diabetes mellitus in Japan. Additionally, this mutation has been detected in about 60 % of cases with diabetes mellitus and sensorineural hearing loss (Kadowaki et al. 1994). The same mutation is also the most frequent pathogenic variant causing MELAS syndrome (Majamaa et al. 1998).

5.2.2 Nonsyndromic Hearing Impairment

Most hereditary hearing loss cases are nonsyndromic. It is estimated that autosomal recessive hearing loss accounts for about 80 % of nonsyndromic cases, while dominant hearing loss occurs in 20 % of cases, and X-linked or mitochondrial inheritance cases account for 1–2 % of the total (Hilgert et al. 2009a).

The nomenclature used for the different loci associated with nonsyndromic deafness is based on the prefix DFN and the inheritance mode, with DFNA, DFNB and DFNX indicating autosomal dominant, autosomal recessive and X-linked, respectively, followed by a number indicating the order in which they were identified.

Nonsyndromic hearing loss is genetically extremely heterogeneous, with more than 90 genes identified and 150 loci mapped to date (Van Camp and Smith 2015). Studies have shown that a single gene may be related to either syndromic or nonsyndromic hearing loss, or even be associated with different patterns of inheritance, whether dominant or recessive.

5.2.2.1 Autosomal Recessive Nonsyndromic Hearing Loss

ARNSHL typically presents as prelingual, severe-to-profound, sensorineural and nonprogressive, and usually affects all frequencies (Smith et al. 2014). Currently, about 85 loci and 60 genes have been related to this mode of inheritance (Van Camp and Smith 2015).

Most cases of ARNSHL are caused by mutations in a single gene, *GJB2*, which encodes the gap junction protein connexin 26. Currently, more than 350 mutations have been identified. Interestingly, one specific mutation, c.35delG, is most frequent in Caucasian populations, accounting for up to 70 % of all *GJB2* mutations (Kelsell et al. 1997; Gasparini et al. 2000; Snoeckx et al. 2005). Two large deletions in the *GJB6* gene encoding connexin 30, del(*GJB6*-D13S1830) and del(*GJB6*-D13S1854), are usually found in compound heterozygosity with a *GJB2* coding mutation, and cause hearing loss (del Castillo et al. 2002, 2003, 2005). Other deletions involving the DFNB1 locus (*GJB2* and *GJB6* genes) have been described, although they appear to be relatively rare (Feldmann et al. 2009; Wilch et al. 2010). Mutations in *GJB2* and *GJB6* genes account for at least 50 % of ARNSHL cases (del Castillo et al. 2003; Hilgert et al. 2009b).

Other genes frequently associated with this mode of inheritance include *SLC26A4*, *MYO15A*, *OTOF*, *CDH23* and *TMC1* (Hilgert et al. 2009b). However, the genetic epidemiology is extremely variable among populations, and additional knowledge is needed to identify the genes for which mutation screening should be indicated.

SLC26A4 is the second most common gene associated with ARNSHL. Mutations in *SLC26A4* are responsible for either ARNSHL with EVA (DFNB4) or Pendred syndrome (Everett et al. 1997; Li et al. 1998); see Sect. 5.2.1.1. To date, almost 500 mutations in the *SLC26A4* gene have been reported for both nonsyndromic deafness DFNB4 and Pendred syndrome (Stenson et al. 2003. Human Gene Mutation Database. http://www.hgmd.cf.ac.uk) (see Chap. 11).

MYO15A is a large gene with 66 exons, which encodes an unconventional myosin XVa. Mutations in this gene cause severe-to-profound congenital hearing loss (DFNB3) (Wang et al. 1998). Almost 100 mutations have been identified.

Mutations in the *OTOF* gene encoding otoferlin lead to prelingual ARNSHL (DFNB9) and auditory neuropathy (Yasunaga et al. 1999). Patients with auditory neuropathy exhibit abnormal pure-tone audiometry and auditory results in the brainstem response test, which measures the overall auditory pathway, but appear normal in the otoacoustic emissions (OAE) test, which detects responses of the outer hair cells to environmental sounds (Varga et al. 2003); see Chap. 4 for hearing tests. At least 110 mutations have been reported to date. The p.Q829X mutation, which is the most frequent *OTOF* mutation identified in the Spanish population, is considered the third most common cause of ARNSHL in this ethnic group (Migliosi et al. 2002).

Mutations in the *CDH23* gene encoding cadherin 23 are associated with moderate-to-profound prelingual ARNSHL (DFNB12) or Usher syndrome type 1D (USH1D); see Sect. 5.2.1.1. Genotype–phenotype correlations suggest that missense or in-frame mutations cause ARNSHL, while nonsense or truncating mutations cause USH1D, although this relationship is not absolute (Bork et al. 2001). Approximately 210 mutations in the *CDH23* gene have been described for both nonsyndromic deafness DFNB12 and USH1D.

Mutations in the *TMC1* gene encoding transmembrane channel-like protein 1 are a common cause of ARNSHL (DFNB7/11) in North Africa, the Middle East and some parts of South Asia. Most of reported cases show a similar phenotype characterized by severe-to-profound prelingual hearing loss. A single founder mutation, c.100C>T (p.R34X), is the most frequent recessive mutation for hearing loss in the *TMC1* gene (Kurima et al. 2002; Ben Saïd et al. 2010).

5.2.2.2 Autosomal Dominant Nonsyndromic Hearing Loss

Autosomal dominant nonsyndromic hearing loss (ADNSHL) is usually postlingual, progressive and sensorineural, although a more variable phenotype is observed (Smith et al. 2014). Presently, approximately 60 loci and 30 genes have been associated with this mode of inheritance (Van Camp and Smith 2015).

The commonest genes implicated in ADNSHL are *TECTA*, *WFS1*, *KCNQ4*, *COCH* and *GJB2*, but none of the genes associated with this mode of inheritance accounts for a significant number of cases (Hilgert et al. 2009b; Hildebrand et al. 2011). However, few studies report the contribution of other genes involved in ADNSHL, and considerable variability is observed among different populations.

Unusual audiometric phenotypes have been related to some ADNSHL genes. The audio profile associated with *TECTA* mutations depends on the alpha-tectorin protein domain and the nature of the residues affected. Genotype–phenotype correlations indicate that missense mutations in the *zona pellucida* domain and the N-terminal region result in mid-frequency hearing loss, while missense mutations in the *zona adherens* lead to high-frequency hearing loss. If cysteine residues are affected, the loss is progressive, while if other amino acids are altered, the loss is stable (Hildebrand et al. 2011). *WFS1*-related hearing loss is associated with a low-frequency audio profile (Bespalova et al. 2001; Young et al. 2001). High-frequency hearing loss can be caused by mutations in the *KCNQ4*, *DFNA5*, *COCH* or *POU4F3* genes, for example (Van Laer et al. 1998; Robertson et al. 1998; Vahava et al. 1998; Kubisch et al. 1999). Dominant mutations in the *GJB2* gene have mainly been reported in Caucasians and are responsible for ADNSHL and syndromic hearing loss associated with skin disorders (Denoyelle et al. 1998; Richard et al. 1998).

5.2.2.3 X-Linked Nonsyndromic Hearing Loss

To date, six loci associated with X-linked nonsyndromic hearing loss have been mapped and only four causative genes have been identified: DFNX1 (*PRPS1* gene encoding phosphoribosylpyrophosphate synthetase I), DFNX2 (*POU3F4* gene encoding POU domain, class 3, transcription factor 4), DFNX3, DFNX4 (*SMPX* gene encoding small muscle protein, X-linked), DFNX5 and DFNX6 (*COL4A6* gene encoding collagen, type IV, alpha-6) (de Kok et al. 1995; Liu et al. 2010; Huebner et al. 2011; Schraders et al. 2011; Rost et al. 2014; Van Camp and Smith 2015). X-linked nonsyndromic hearing loss is usually progressive and can be either pre- or postlingual. Most forms are sensorineural, with the exception of DFNX2, which is characterized by sensorineural hearing loss with or without a conductive component resulting from stapes fixation. In contrast to other types of conductive hearing impairment, stapedotomy should not be performed in patients with *POU3F4* mutations because of the unavoidable complication of perilymphatic fluid gusher (de Kok et al. 1995) (see Chap. 4).

5.2.2.4 Mitochondrial Nonsyndromic Hearing Loss

Mitochondrial DNA mutations can lead to maternally inherited nonsyndromic hearing loss, mainly caused by mutations in either the *MT-RNR1* or *MT-TS1* genes, which encode for the 12S ribosomal RNA and the transfer $RNA^{Ser(UCN)}$, respectively (Prezant et al. 1993; Reid et al. 1994). Wide ranges of severity, age of onset of hearing loss and penetrance are observed in individuals with pathogenic mitochondrial

DNA variants, suggesting that modifier factors can modulate the phenotypic manifestation (Kokotas et al. 2007).

Of particular note, the m.1555A>G mutation in the *MT-RNR1* gene is a frequent cause of maternally inherited nonsyndromic hearing loss, and can be associated with susceptibility to aminoglycoside ototoxicity or late-onset sensorineural hearing loss (Prezant et al. 1993). Mutations in the *MT-TS1* gene are responsible for either nonsyndromic or syndromic hearing loss (Reid et al. 1994; Nakamura et al. 1995).

5.3 Molecular Genetic Testing

The etiologic diagnosis of hereditary hearing loss requires a variety of tests performed by qualified professionals. These include clinical assessment (family history evaluation, physical examination, otologic and audiologic evaluation) and molecular genetic testing. When environmental factors are not identified or if the cause of hearing loss is not discovered after completion of the clinical investigation, genetic evaluation of the individual should be requested in order to discover the region, gene or variant responsible for the phenotype.

With the advances made in the field of molecular diagnostics, many technologies and platforms are available for the screening of hearing loss for purposes including the identification of major changes in the population or the discovery of new variants and genes associated with hearing loss.

5.3.1 Newborn Hearing Screening

During the last decades, universal newborn hearing screening programs have been implemented in several countries around the world, aiming to reduce the age of detection and treatment of infants with hearing loss (see also Chap. 4). The Joint Committee on Infant Hearing (2007) recommends that all infants should have access to hearing screening using a physiological test at no later than 1 month of age. All infants who do not pass the initial hearing screening and a subsequent rescreening should have appropriate audiological and medical evaluations to confirm the presence of hearing loss, at no later than 3 months of age. Some programs incorporate screening protocols that may include genetic tests. Genetic screening plays an important role in identifying newborns with hearing loss that is too mild to be detected in current screening programs. It can help to identify children who were born with normal hearing but might develop progressive hearing loss, as well as contribute to the diagnosis of syndromic hearing loss before the onset of other clinical symptoms. Early identification and intervention can maximize opportunities for better language and speech development, improving the acquisition of motor, cognitive and social skills (Joint Committee on Infant Hearing 2007; Hilgert et al. 2009b; Nivoloni Kde et al. 2010).

5.3.2 Conventional and New Molecular Tools

The identification of genetic variants associated with deafness has made a crucial contribution to understanding the molecular basis, physiology and pathophysiology of the auditory system. The development of Sanger sequencing (Sanger et al. 1977), followed by implementation of the polymerase chain reaction (PCR), marked a great revolution in molecular genetics and formed the basis for the Human Genome Project, completed in 2001 (International Human Genome Sequencing Consortium et al. 2001).

Over the years, linkage analysis or homozygosity mapping followed by candidate gene sequencing has been used as the most effective methods for the identification of genetic loci and new genes. However, these traditional methods have been successfully applied for autosomal recessive disorders and particularly in the case of consanguineous pedigrees with multiple affected individuals (Lander and Botstein 1987). The limitations of linkage methods, the long time required and the high costs of gene identification left many cases unsolved, and such methods have been replaced by new genomic technologies.

Conventional technologies for genetic deafness testing are highly accurate and sensitive, but are extremely slow in terms of throughput, and despite great genetic heterogeneity, only a minority of deafness genes is currently investigated in routine genetic diagnostics. The genes commonly included corresponded to those most frequently implicated in hearing loss, associated with other recognizable features or related to recognizable audio profiles.

Many studies have suggested different cost-effective protocols, especially for congenital hearing loss, aiming to optimize the diagnosis of hearing loss (Mafong et al. 2002; Preciado et al. 2005; Morzaria et al. 2005; Ramos et al. 2013). As observed in numerous other heterogeneous disorders, the prevalence and frequencies of mutations are associated with ethnic background, and this should be taken into account in deciding which type of assay is most appropriate for a patient.

The genes most frequently implicated in ARNSHL are *GJB2*, *SLC26A4, MYO15A, OTOF, CDH23* and *TMC1* (Hilgert et al. 2009b), with the *GJB2* mutations being the most frequent cause of ARNSHL in most populations worldwide (see Sect. 5.2.2.1). Therefore, *GJB2* gene analysis should be the first step in the search for mutations associated with nonsyndromic sensorineural hearing loss. The *GJB2* c.35delG, c.167delT and c.235delC mutations are the most frequent in Caucasian, Ashkenazi Jew and Asian populations, respectively. The c.35delG mutation accounts for 28–63 % of the mutant alleles in Caucasians (Gasparini et al. 2000). The mutations in this gene can be identified using a variety of single-gene screening techniques, including PCR followed by restriction digestion, allele-specific PCR and real-time PCR. These methods are rapid, inexpensive, highly sensitive and specific, but are limited by the fact that only a small number of point mutations are investigated (Schrijver 2004). Direct sequencing, which is the most comprehensive and definitive method, should be used to confirm the presence of the most frequent point mutations, as well as to screen for other less frequent mutations.

In addition, two large deletions in the *GJB6* gene, del(*GJB6*-D13S1830) and del(*GJB6*-D13S1854), have been found associated with *GJB2* gene point mutations, reflecting a digenic heterozygosity condition (del Castillo et al. 2002, 2003, 2005). Therefore, *GJB6* gene deletions should be investigated in individuals showing no mutation or only a single *GJB2* gene mutation (Schrijver 2004). Other deletions involving the *GJB6* gene have been described in the DFNB1 locus, although they appear to be relatively rare (Feldmann et al. 2009; Wilch et al. 2010). These large deletions can be investigated using multiplex PCR approaches, since direct DNA sequencing is not suitable for detecting deletions of entire genes or exons.

Mutations in the *SLC26A4* gene are the second most common cause of ARNSHL. Molecular genetic screening of the *SLC26A4* gene should therefore be performed, especially when the hearing loss is accompanied by an EVA, Mondini dysplasia or goiter.

Since *OTOF* gene mutations have been suggested as the major cause of auditory neuropathy, the investigation of mutations in this gene should be considered when OAE are present in the absence of auditory brainstem responses. Mutations in the *PJVK* gene encoding pejvakin (DFNB59) have also been reported to cause auditory neuropathy in a few families (Delmaghani et al. 2006), and the screening of this gene should be considered in patients negative for *OTOF* mutations.

None of the genes related to ADNSHL (Sect. 5.2.2.2) accounts for a significant number of cases, which means that genetic analysis is not cost-effective. Nevertheless, some ADNSHL genes are associated with unusual audiometric phenotypes, and these recognizable audio profiles can be useful in developing strategies for molecular genetic testing. For example, the *WFS1* and *DIAPH1* (Lynch et al. 1997) genes are associated with low-frequency hearing loss, while the *CCDC50* gene (Modamio-Hoybjor et al. 2007) is responsible for low-to-mid-frequency hearing loss, and the *TECTA* gene is related to hearing impairment at mid or high frequencies. In the case of autosomal dominant auditory neuropathy, the *DIAPH3* gene (Kim et al. 2004) should be tested for mutations, and in the presence of progressive late-onset hearing loss combined with vestibular abnormalities, mutations in the *COCH* gene may be involved. Dominant mutations in the *GJB2* gene should be considered if hearing loss is associated with skin disorders.

When an X-linked pattern of inheritance is observed (Sect. 5.2.2.3), with a phenotype of mixed deafness associated with bony labyrinth defects (stapes fixation), mutations in the *POU3F4* gene may be suspected, especially when perilymphatic gusher occurs during stapes surgery (de Kok et al. 1995).

In the case of maternal inheritance and in the presence of aminoglycoside-induced hearing loss, mitochondrial mutations should be investigated. In particular, screening should include the m.1555A>G mutation in the *MT-RNR1* gene (Sect. 5.2.2.4), which is a common cause of maternally inherited nonsyndromic hearing loss and is also associated with aminoglycoside ototoxicity and hearing loss susceptibility. Early identification of patients with hearing loss due to mitochondrial DNA mutations can not only influence genetic counseling regarding maternal inheritance, but also enable avoidance of known risk factors. Mutation screening is especially important in countries where aminoglycosides are widely used. Specific point mutations can be investigated by restriction fragment length polymorphism PCR (RFLP-PCR) method, or a more comprehensive analysis can be performed by direct sequencing of mitochondrial genes.

It is estimated that approximately 300 genes could be involved in both syndromic and nonsyndromic hearing losses (Nance 2003; Friedman and Griffith 2003), making the molecular diagnosis of deafness highly challenging. The limitations of conventional methodologies, mainly due to the large size of several genes and the high cost of direct sequencing, make it impracticable to screen for mutations in all the deafness genes.

To overcome this problem, several research groups and companies have developed high-throughput technologies for large-scale screening that are affordable enough to be used in routine molecular screening for hearing loss.

These approaches have been successfully used to identify single-nucleotide polymorphisms (SNPs) in candidate genes and/or a larger number of known mutations. Single-nucleotide extension microarrays detect a mutation by hybridizing specifically designed primers to DNA sample, followed by single-nucleotide extension at the site of the mutation. The Hereditary Hearing Loss Arrayed Primer Extension (HHL APEX) diagnostic test was the first to employ this method for hearing loss and included 198 mutations across eight genes (*GJB2*, *GJB6*, *GJB3*, *GJA1*, *SLC26A4*, *SLC26A5*, *MT-RNR1* and *MT-TS1*) in a single test (Rodriguez-Paris et al. 2010).

The TaqMan® OpenArray™ Genotyping System (Applied Biosystems, USA) is a high-throughput technology, based on real-time PCR, which enables the simultaneous analysis of up to 3,072 SNVs, small deletions and insertions using a single genotyping plate. This platform also allows 2,688 gene expression assays to be performed per plate. Both genotyping and gene expression platforms must follow a default layout test offered by the company (TaqMan® OpenArray™ Genotyping System User Guide 2011, Applied Biosystems). A limitation of the technology is that, once the layout is customized, it is not possible to change the variants included in the layout, because the tests are preloaded in the chip (Martins et al. 2013).

Another genotyping method available is the Sequenom MassARRAY® iPLEX Platform (Sequenom Inc, San Diego, USA). The assay consists of an initial locus-specific PCR reaction (called iPLEX-Gold), followed by single base extension using mass-modified dideoxynucleotide terminators of an oligonucleotide primer that anneals immediately upstream of the polymorphic site of interest. Using matrix-assisted laser desorption/ionization time-of-flight (MALDI-TOF) mass spectrometry, the distinct mass of the extended primer identifies normal and mutated alleles. This platform offers an inexpensive means of performing extensive screening. It is simple to use and allows flexibility in experimental design, because it is possible to replace the primers and generate a new customized panel (Svidnicki et al. 2015).

However, use of these technologies requires previous standardization, and it is necessary to validate the results by direct sequencing (a gold standard technique). Moreover, these techniques are only applied to identify known mutations and are unable to assess duplications or deletions of entire genes.

Chromosomal imbalances such as aneuploidies, microdeletion/microduplication syndromes, unbalanced chromosomal rearrangements and copy-number variants (CNVs) can be identified by array-based comparative genomic hybridization (array-CGH). This molecular cytogenetic method is able to evaluate areas of the human genome for gains or losses of chromosome segments at a higher resolution compared to traditional karyotyping (Walsh et al. 2010). However, a few unsolved deafness cases have been investigated using array-CGH; consequently, the proportion of deafness caused by large duplications or deletions remains unknown (Brownstein et al. 2012).

Sequencing is undoubtedly the most comprehensive and definitive method for molecular investigation of genetic disorders. Next-Generation Sequencing (NGS) is a powerful technology available for use in molecular diagnostics. For example, it is possible to screen a large number of genes simultaneously, the whole exome or even the whole genome, using different NGS methodologies (Ng et al. 2010; Lupski et al. 2010).

However, these technologies generate massive quantities of data that require intensive bioinformatics analysis as well as validation of mutations by Sanger sequencing of specific regions (Brownstein et al. 2014). Data management and interpretation is especially complex for Whole-Exome Sequencing (WES) or Whole-Genome Sequencing (WGS). The use of targeted NGS technologies for the genetic screening of a subset of genes associated with hearing loss can be more efficient in terms of time and cost, and can therefore offer an alternative to WES or WGS. This strategy enables the analysis of a greater number of samples and reduces the complexity of bioinformatics analysis and data interpretation, because only the genes selected for screening are sequenced.

Several research groups have developed multi-gene screening panels using NGS technologies. However, these panels differ in terms of the techniques used and the numbers of mutations and genes investigated. In some cases, the genes targeted are already known to be associated with hearing loss, while in other studies, sequencing includes genes implicated in syndromic and nonsyndromic deafness, as well as candidate regions and orthologous genes related to hearing function (in mice) (Shearer et al. 2010; Brownstein et al. 2011, 2014; Yang et al. 2013; Shearer et al. 2014; Vozzi et al. 2014; Shearer and Smith 2015).

Currently, these approaches remain expensive and are mainly limited to research use, but their clinical availability is gradually increasing. Commercial targeted NGS-based gene panels, such as OtoSCOPE, Otogenetics and OtoGenome, enable the efficient and inexpensive detection of variants in human genes known to be associated with hearing loss.

OtoSCOPE (Otologic Sequence Capture Of Pathogenic Exons), first described by Shearer et al. (2010), was a pioneer in panels based on NGS, providing a simultaneous test for 54 known deafness genes, covering 0.02 % of the whole genome or 1 % of the whole exome, at a relatively low cost. This platform, currently being offered as a molecular screening service, is based on Targeted Genome Capture (TGC) associated with Massive Parallel Sequencing (MPS). The sixth version of the platform is now commercially available, providing 99 % specificity and including 116 genes known to cause NSHL, Pendred syndrome and Usher syndrome, as well as genes associated with hearing loss-related phenotypes (Shearer et al. 2010; Shearer et al. 2014).

The Otogenetics Deafness Gene Panel offered by Otogenetics Corporation (Norcross, USA) allows the screening of 129 genes known to cause both nonsyndromic and syndromic hearing losses. The analysis is performed by NGS using custom oligonucleotide-based target capture followed by Illumina HiSeq sequencing of the coding regions of the panel genes, with more than 100-fold coverage at every target base, providing a comprehensive hearing loss evaluation.

Another clinical test, including 87 genes and taking about 8–12 weeks, is the OtoGenome Test for Hearing Loss and Related Syndromes, developed and offered by Partners Healthcare (Boston, USA). The OtoGenome Test uses the SureSelect platform (Agilent, Santa Clara, USA) with Illumina sequencing to screen genes related to hearing loss and syndromes that present deafness as one of the clinical manifestations, such as the Usher, Pendred, Jervell and Lange-Nielsen, BOR, Deafness and Male Infertility (DIS), Perrault and Waardenburg syndromes. The company also offers specific subpanels for individuals with confirmed diagnosis or clinical suspicion of these syndromes, hence reducing the time required for screening and analysis.

It is important to highlight that NGS technologies are not only applicable to molecular diagnostics, but also enable the identification of new genes and better characterization of the known disease-associated genes.

The diagram presented in Fig. 5.1 illustrates a suggested strategy for choosing appropriate tests, aiming to maximize the probability of finding the genetic etiology of hearing loss in a time and cost-effective manner.

5.4 Conclusions and Perspectives

Over the last decades, genetic studies of hearing loss have provided important information regarding the molecular basis, development and function of the auditory system. Recent technological advances have helped clarify the molecular diagnosis of hearing loss in many cases. However, even with all this progress, new limitations have appeared, since genomic interactions are still not completely known. The elucidation of the complex network of genes involved in the auditory system and characterization of their respective proteins is crucial and can provide new insights into the biological mechanisms implicated in the pathophysiology of hearing loss. This information will be fundamental for the development of a translational medicine which converts molecular knowledge into clinical applications (bench to bedside) involving effective therapeutic strategies, rehabilitation, and consequently contributing for the well-being of patients.

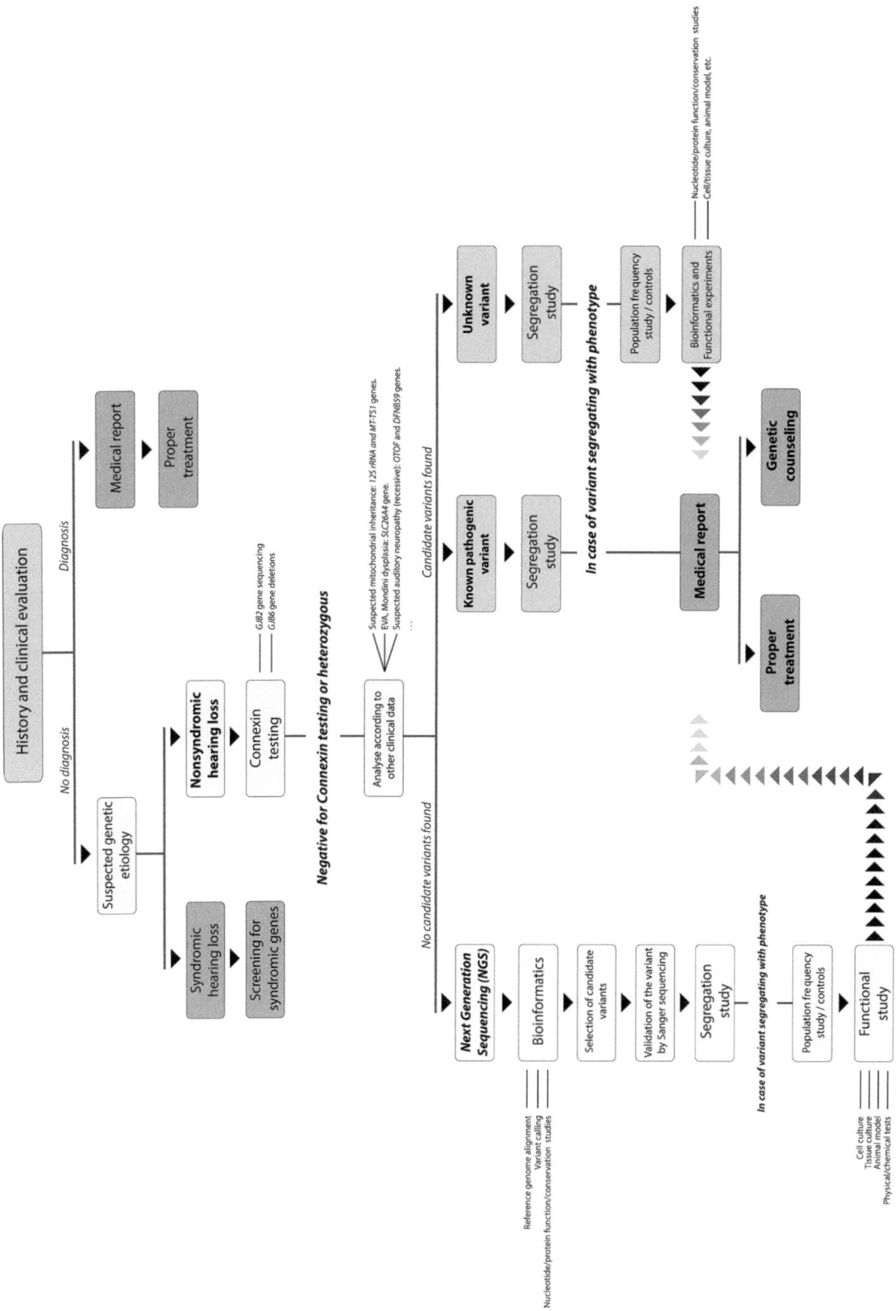

Fig. 5.1 Flow diagram with a general view of the molecular genetic analysis for patients with hearing loss. *EVA* Enlarged Vestibular Aqueduct

References

Abdelhak S, Kalatzis V, Heilig R et al (1997) A human homologue of the *Drosophila* eyes absent gene underlies branchio-oto-renal (BOR) syndrome and identifies a novel gene family. Nat Genet 15(2):157–164

Ahmad NN, Ala-Kokko L, Knowlton RG et al (1991) Stop codon in the procollagen II gene (*COL2A1*) in a family with Stickler syndrome (arthro-ophthalmopathy). Proc Natl Acad Sci U S A 88(15):6624–6627

Ahmed ZM, Riazuddin S, Bernstein SL et al (2001) Mutations of the protocadherin gene *PCDH15* cause Usher syndrome type 1F. Am J Hum Genet 69(1):25–34

Alagramam KN, Yuan H, Kuehn MH et al (2001) Mutations in the novel protocadherin *PCDH15* cause Usher syndrome type 1F. Hum Mol Genet 10(16):1709–1718

Alport AC (1927) Hereditary familial congenital haemorrhagic nephritis. Br Med J 1(3454): 504–506

Attié T, Till M, Pelet A et al (1995) Mutation of the endothelin-receptor B gene in Waardenburg-Hirschsprung disease. Hum Mol Genet 4(12):2407–2409

Baker S, Booth C, Fillman C et al (2011) A loss of function mutation in the *COL9A2* gene causes autosomal recessive Stickler syndrome. Am J Hum Genet 155A(7):1668–1672

Barker DF, Hostikka SL, Zhou J et al (1990) Identification of mutations in the *COL4A5* collagen gene in Alport syndrome. Science 248(4960):1224–1227

Ben Saïd M, Hmani-Aifa M, Amar I et al (2010) High frequency of the p.R34X mutation in the *TMC1* gene associated with nonsyndromic hearing loss is due to founder effects. Genet Test Mol Biomarkers 14(3):307–311

Bespalova IN, Van Camp G, Bom SJ et al (2001) Mutations in the Wolfram syndrome 1 gene (*WFS1*) are a common cause of low frequency sensorineural hearing loss. Hum Mol Genet 10(22):2501–2508

Bork JM, Peters LM, Riazuddin S et al (2001) Usher syndrome 1D and nonsyndromic autosomal recessive deafness DFNB12 are caused by allelic mutations of the novel cadherin-like gene *CDH23*. Am J Hum Genet 68(1):26–37

Brownstein Z, Friedman LM, Shahin H et al (2011) Targeted genomic capture and massively parallel sequencing to identify genes for hereditary hearing loss in Middle Eastern families. Genome Biol 12(9):R89

Brownstein Z, Bhonker Y, Avraham KB (2012) High-throughput sequencing to decipher the genetic heterogeneity of deafness. Genome Biol 13(5):245

Brownstein Z, Abu-Rayyan A, Karfunkel-Doron D et al (2014) Novel myosin mutations for hereditary hearing loss revealed by targeted genomic capture and massively parallel sequencing. Eur J Hum Genet 22(6):768–775

Chen K, Wang X, Sun L et al (2012) Screening of *SLC26A4*, *FOXI1*, *KCNJ10*, and *GJB2* in bilateral deafness patients with inner ear malformation. Otolaryngol Head Neck Surg 146(6):972–978

Cirello V, Bazzini C, Vezzoli V et al (2012) Molecular and functional studies of 4 candidate loci in Pendred syndrome and nonsyndromic hearing loss. Mol Cell Endocrinol 351(2):342–350

de Kok YJ, van der Maarel SM, Bitner-Glindzicz M et al (1995) Association between X-linked mixed deafness and mutations in the POU domain gene *POU3F4*. Science 267(5198): 685–688

del Castillo I, Villamar M, Moreno-Pelayo MA et al (2002) A deletion involving the connexin 30 gene in nonsyndromic hearing impairment. N Engl J Med 346:243–249

del Castillo I, Moreno-Pelayo MA, Del Castillo FJ et al (2003) Prevalence and evolutionary origins of the del(*GJB6*-D13S1830) mutation in the DFNB1 locus in hearing-impaired subjects: a multicenter study. Am J Hum Genet 73:1452–1458

del Castillo FJ, Rodriguez-Ballesteros M, Alvarez A et al (2005) A novel deletion involving the connexin-30 gene, del(GJB6-d13s1854), found *in trans* with mutations in the *GJB2* gene (connexin-26) in subjects with DFNB1 non-syndromic hearing impairment. J Med Genet 42:588–594

Delmaghani S, del Castillo FJ, Michel V et al (2006) Mutations in the gene encoding pejvakin, a newly identified protein of the afferent auditory pathway, cause DFNB59 auditory neuropathy. Nat Genet 38(7):770–778

Denoyelle F, Lina-Granade G, Plauchu H et al (1998) Connexin 26 gene linked to a dominant deafness. Nature 393(6683):319–320

Ebermann I, Scholl HP, Charbel Issa P et al (2007) A novel gene for Usher syndrome type 2: mutations in the long isoform of whirlin are associated with *retinitis pigmentosa* and sensorineural hearing loss. Hum Genet 121(2):203–211

Ebermann I, Phillips JB, Liebau MC et al (2010) *PDZD7* is a modifier of retinal disease and a contributor to digenic Usher syndrome. J Clin Invest 120(6):1812–1823

Edery P, Attié T, Amiel J et al (1996) Mutation of the endothelin-3 gene in Waardenburg-Hirschsprung disease (Shah-Waardenburg syndrome). Nat Genet 12(4):442–444

Eudy JD, Weston MD, Yao S et al (1998) Mutation of a gene encoding a protein with extracellular matrix motifs in Usher syndrome type IIa. Science 280(5370):1753–1757

Everett LA, Glaser B, Beck JC et al (1997) Pendred syndrome is caused by mutations in a putative sulphate transporter gene (*PDS*). Nat Genet 17(4):411–422

Faletra F, D'Adamo AP, Bruno I et al (2014) Autosomal recessive Stickler syndrome due to a loss of function mutation in the *COL9A3* gene. Am J Hum Genet 164A(1):42–47

Feldmann D, Le Maréchal C, Jonard L et al (2009) A new large deletion in the DFNB1 locus causes nonsyndromic hearing loss. Eur J Med Genet 52(4):195–200

Friedman TB, Griffith AJ (2003) Human nonsyndromic sensorineural deafness. Annu Rev Genomics Hum Genet 4:341–402

Gasparini P, Rabionet R, Barbujani G et al (2000) High carrier frequency of the 35delG deafness mutation in European populations. Genetic Analysis Consortium of *GJB2* 35delG. Eur J Hum Genet 8:19–23

Hildebrand MS, Morín M, Meyer NC et al (2011) DFNA8/12 caused by mutations in the most identified subtype of nonsyndromic autosomal dominant hearing loss. Hum Mutat 32(7):825–834

Hilgert N, Smith RJH, Van Camp G (2009a) Function and expression pattern of nonsyndromic deafness genes. Curr Mol Med 9(5):546–564

Hilgert N, Smith RJH, Van Camp G (2009b) Forty-six genes causing nonsyndromic hearing impairment: which ones should be analyzed in DNA diagnostics? Mutat Res 681(2–3):189–196

Hoskins BE, Cramer CH, Silvius D et al (2007) Transcription factor *SIX5* is mutated in patients with branchio-oto-renal syndrome. Am J Hum Genet 80(4):800–804

Hoth CF, Milunsky A, Lipsky N et al (1993) Mutations in the paired domain of the human *PAX3* gene cause Klein-Waardenburg syndrome (WS-III) as well as Waardenburg syndrome type I (WS-I). Am J Hum Genet 52(3):455–462

Huebner AK, Gandia M, Frommolt P et al (2011) Nonsense mutations in *SMPX*, encoding a protein responsive to physical force, result in X-chromosomal hearing loss. Am J Hum Genet 88(5):621–627

Huyghe JR, Van Laer L, Hendrickx JJ et al (2008) Genome-wide SNP-based linkage scan identifies a locus on 8q24 for an age-related hearing impairment trait. Am J Hum Genet 83:401–407

International Human Genome Sequencing Consortium et al (2001) Initial sequencing and analysis of the human genome. Nature 409(6822):860–921

Jefferson JA, Lemmink HH, Hughes AE et al (1997) Autosomal dominant Alport syndrome linked to the type IV collage alpha 3 and alpha 4 genes (*COL4A3* and *COL4A4*). Nephrol Dial Transplant 12(8):1595–1599

Jervell A, Lange-Nielsen F (1957) Congenital deaf-mutism, functional heart disease with prolongation of the Q-T interval and sudden death. Am Heart J 54(1):59–68

Joensuu T, Hämäläiner R, Yuan B et al (2001) Mutations in a novel gene with transmembrane domains underlie Usher syndrome type 3. Am J Hum Genet 69(4):673–684

Joint Committee on Infant Hearing (2007) Year 2007 position statement: principles and guidelines for early hearing detection and intervention programs. Pediatrics 120(4):898–921

Kadowaki T, Kadowaki H, Mori Y et al (1994) A subtype of diabetes mellitus associated with a mutation of mitochondrial DNA. N Engl J Med 330(14):962–968

Kelsell DP, Dunlop J, Stevens HP et al (1997) Connexin 26 mutations in hereditary non-syndromic sensorineural deafness. Nature 387:80–83

Khateb S, Zelinger L, Mizrahi-Meissonnier L et al (2014) A homozygous nonsense *CEP250* mutation combined with a heterozygous nonsense *C2orf71* mutation is associated with atypical Usher syndrome. J Med Genet 51(7):460–469

Kim TB, Isaacson B, Sivakumaran TA et al (2004) A gene responsible for autosomal dominant auditory neuropathy (AUNA1) maps to 13q14-21. J Med Genet 41(11):872–876

Kokotas H, Petersen MB, Willems PJ (2007) Mitochondrial deafness. Clin Genet 71(5):379–391

Konings A, Van Laer L, Michel S et al (2008) Variations in *HSP70* genes associated with noise-induced hearing loss in two independent populations. Eur J Hum Genet 17:329–335

Kubisch C, Schoroeder BC, Friedrich T et al (1999) *KCNQ4*, a novel potassium channel expressed in sensory outer hair cells, is mutated in dominant deafness. Cell 96(3):437–446

Kurima K, Peters LM, Yang Y et al (2002) Dominant and recessive deafness caused by mutations of a novel gene, *TMC1*, required for cochlear hair-cell function. Nat Genet 30(3):277–284

Landa P, Differ AM, Rajput K et al (2013) Lack of significant association between mutations of *KCNJ10* or *FOXI1* and *SLC26A4* mutations in Pendred syndrome/enlarged vestibular aqueducts. BMC Med Genet 14:85

Lander ES, Botstein D (1987) Homozygosity mapping: a way to map human recessive traits with the DNA of inbred children. Science 236:1567–1570

Li XC, Everett LA, Lalwani AK et al (1998) A mutation in *PDS* causes non-syndromic recessive deafness. Nat Genet 18(3):215–217

Liu X, Han D, Li J et al (2010) Loss-of-function mutations in the *PRPS1* gene cause a type of nonsyndromic X-linked sensorineural deafness, DFN2. Am J Hum Genet 86(1):65–71

Lupski JR, Reid JG, Gonzaga-Jauregui C et al (2010) Whole-genome sequencing in a patient with Charcot-Marie-Tooth neuropathy. N Engl J Med 362:1181–1191

Lynch ED, Lee MK, Morrow JE et al (1997) Nonsyndromic deafness DFNA1 associated with mutation of a human homolog of the *Drosophila* gene diaphanous. Science 278(5341): 1315–1318

Mafong DD, Shin EJ, Lalwani AK (2002) Use of laboratory evaluation and radiologic imaging in the diagnostic evaluation of children with sensorineural hearing loss. Laryngoscope 112:1–7

Majamaa K, Moilanen JS, Uimonen S et al (1998) Epidemiology of A3243G, the mutation for mitochondrial encephalomyopathy, lactic acidosis, and strokelike episodes: prevalence of the mutation in adult population. Am J Hum Genet 63(2):447–454

Martins FTA, Ramos PZ, Svidnicki MCCM et al (2013) Optimization of simultaneous screening of the main mutations involved in non-syndromic deafness using the TaqMan® OpenArray™ Genotyping Platform. BMC Med Genet 14:112

Migliosi V, Modamio-Hoybjor S, Moreno-Pelayo MA et al (2002) Q829X, a novel mutation in the gene encoding otoferlin (*OTOF*), is frequently found in Spanish patients with prelingual non-syndromic hearing loss. J Med Genet 39(7):502–506

Mochizuki T, Lemmink HH, Mariyama M et al (1994) Identification of mutations in the alpha 3 (III) and alpha 4 (IV) collagen genes in autosomal recessive Alport syndrome. Nat Genet 8(1):77–81

Modamio-Hoybjor S, Mencia A, Goodyear R et al (2007) A mutation in *CCDC50*, a gene encoding an effector of epidermal growth factor-mediated cell signaling, causes progressive hearing loss. Am J Hum Genet 80(6):1076–1089

Morton CC, Nance WE (2006) Newborn hearing screening – a silent revolution. N Engl J Med 354(20):2151–2164

Morzaria S, Westerberg BD, Kozak FK (2005) Evidence-based algorithm for the evaluation of a child with bilateral sensorineural hearing loss. J Otolaryngol 34:297–303

Nakamura M, Nakano S, Goto Y et al (1995) A novel point mutation in the mitochondrial tRNA(Ser(UCN)) gene detected in a family with MERRF/MELAS overlap syndrome. Biochem Biophys Res Commun 214(1):86–93

Nance WE (2003) The genetics of deafness. Ment Retard Dev Disabil Res Rev 9(2):109–119
Neyroud N, Tesson F, Denjoy I et al (1997) A novel mutation in the potassium channel gene *KVLQT1* causes the Jervell and Lange-Nielsen cardioauditory syndrome. Nat Genet 15(2):186–189
Ng SB, Buckingham KJ, Lee C et al (2010) Exome sequencing identifies the cause of a mendelian disorder. Nat Genet 42:30–35
Nivoloni Kde A, da Silva-Costa SM, Pomílio MC et al (2010) Newborn hearing screening and genetic testing in 8974 Brazilian neonates. Int J Pediatr Otorhinolaryngol 74(8):926–929
Pendred V (1896) Deaf mutism and goitre. Lancet II:532
Pingault V, Bondurand N, Kuhlbrodt K et al (1998) *SOX10* mutations in patients with Waardenburg-Hirschsprung disease. Nat Genet 18(2):171–173
Preciado DA, Lawson L, Madden C et al (2005) Improved diagnostic effectiveness with a sequential diagnostic paradigm in idiopathic pediatric sensorineural hearing loss. Otol Neurotol 26:610–615
Prezant TR, Agapian JV, Bohlman MC et al (1993) Mitochondrial ribosomal RNA mutation associated with both antibiotic-induced and non-syndromic deafness. Nat Genet 4(3):289–294
Puffenberger EG, Jinks RN, Sougnez C et al (2012) Genetic mapping and exome sequencing identify variants associated with five novel diseases. PLoS One 7(1), e28936
Ramos PZ, de Moraes VC, Svidnicki MC et al (2013) Etiologic and diagnostic evaluation: algorithm for severe to profound sensorineural hearing loss in Brazil. Int J Audiol 52(11):746–752
Read AP, Newton VE (1997) Waardenburg syndrome. J Med Genet 34(8):656–665
Reid FM, Vernham GA, Jacobs HT (1994) A novel mitochondrial point mutation in a maternal pedigree with sensorineural deafness. Hum Mutat 3(3):243–247
Riazuddin S, Belyantseva IA, Giese AP et al (2012) Alterations of the *CIB2* calcium- and integrin-binding protein cause Usher syndrome type 1J and nonsyndromic deafness DFNB48. Nat Genet 44(11):1265–1271
Richard G, White TW, Smith LE et al (1998) Functional defects of Cx26 resulting from a heterozygous missense mutation in a family with dominant deaf-mutism and palmoplantar keratoderma. Hum Genet 103(4):393–399
Richards AJ, Yates JR, Williams R et al (1996) A family with Stickler syndrome type 2 has a mutation in the *COL11A1* gene resulting in the substitution of glycine 97 by valine in alpha 1 (XI) collagen. Hum Mol Genet 5(9):1339–1343
Robertson NG, Lu L, Heller S et al (1998) Mutations in a novel cochlear gene cause DFNA9, a human nonsyndromic deafness with vestibular dysfunction. Nat Genet 20(3):299–303
Rodriguez-Paris J, Pique L, Colen T et al (2010) Genotyping with a 198 mutation arrayed primer extension array for hereditary hearing loss: assessment of its diagnostic value for medical practice. PLoS One 5(7):e11804
Rost S, Bach E, Neuner C et al (2014) Novel form of X-linked nonsyndromic hearing loss with cochlear malformations caused by a mutation in the type IV collagen gene *COL4A6*. Eur J Hum Genet 22(2):208–215
Ruf RG, Xu PX, Silvius D et al (2004) *SIX1* mutations cause branchio-oto-renal syndrome by disruption of *EYA1-SIX1*-DNA complexes. Proc Natl Acad Sci U S A 101(21):8090–8095
Sánchez-Martín M, Rodríguez-García A, Pérez-Losada J et al (2002) *SLUG* (*SNAI2*) deletions in patients with Waardenburg disease. Hum Mol Genet 11(25):3231–3236
Sanger F, Nicklen S, Coulson AR (1977) DNA sequencing with chain-terminating inhibitors. Proc Natl Acad Sci U S A 74:5463–5467
Schraders M, Haas SA, Weegerink NJ et al (2011) Next-generation sequencing identifies mutations of *SMPX*, which encodes the small muscle protein, X-linked, as a cause of progressive hearing impairment. Am J Hum Genet 88(5):628–634
Schrijver I (2004) Hereditary Non-syndromic sensorineural hearing loss. Transforming silence to sound. J Mol Diagn 6(4):275–284
Schulze-Bahr E, Wang Q, Wedekind H et al (1997) *KCNE1* mutations cause Jervell and Lange-Nielsen syndrome. Nat Genet 17(3):267–268

Shearer AE, Smith RJ (2015) Massively parallel sequencing for genetic diagnosis of hearing loss: the new standard of care. Otolaryngol Head Neck Surg 153(2):175–182
Shearer AE, DeLuca AP, Hildebrand MS et al (2010) Comprehensive genetic testing for hereditary hearing loss using massively parallel sequencing. Proc Natl Acad Sci U S A 107(49): 21104–21109
Shearer AE, Kolbe DL, Azaiez H et al (2014) Copy number variants are a common cause of non-syndromic hearing loss. Genome Med 6(5):37
Smith RJH, Lee EC, Kimberling WJ et al (1992) Localization of two genes for Usher syndrome type I to chromosome 11. Genomics 14(4):995–1002
Smith RJH, Shearer AE, Hildebrand MS et al (2014) Deafness and hereditary hearing loss overview. In: Pagon RA, Adam MP, Ardinger HH et al (eds) GeneReviews® [http://www.ncbi.nlm.nih.gov/books/NBK1434/]. University of Washington, Seattle. [Last revision January 9, 2014. Accessed 20 July 2015]
Snoeckx RL, Huygen PL, Feldmann D et al (2005) *GJB2* mutations and degree of hearing loss: a multicenter study. Am J Hum Genet 77:945–957
Stenson PD, Ball EV, Mort M et al (2003) The human gene mutation database (HGMD): 2003 Update. Hum Mutat 21(6):577–581. Available from: http://www.hgmd.cf.ac.uk. Accessed 22 Jan 2016
Stickler GB, Belau PG, Farrell FJ et al (1965) Hereditary progressive arthro-ophthalmopathy. Mayo Clin Proc 40:433–455
Svidnicki MCCM, Silva-Costa SM, Ramos PZ (2015) Screening of genetic alterations related to non-syndromic hearing loss using MassARRAY iPLEX® technology. BMC Med Genet 16:85
Tassabehji M, Read AP, Newton VE et al (1992) Waardenburg's syndrome patients have mutations in the human homologue of the *Pax-3* paired box gene. Nature 355(6361):635–636
Tassabehji M, Newton VE, Read AP (1994) Waardenburg syndrome type 2 caused by mutations in the human microphthalmia (*MITF*) gene. Nat Genet 8(3):251–255
Tyson J, Tranebjaerg L, Bellman S et al (1997) IsK and KvLQT1: mutation in either of the two subunits of the slow component of the delayed rectifier potassium channel can cause Jervell and Lange-Nielsen syndrome. Hum Mol Genet 6(12):2179–2185
Vahava O, Morell R, Lynch ED et al (1998) Mutation in transcription factor *POU4F3* associated with inherited progressive hearing loss in humans. Science 279(5358):1950–1954
Van Camp G, Smith RJH (2015) Hereditary Hearing Loss Homepage. http://hereditaryhearingloss.org. Accessed 21 Sept 2015
Van Camp G, Snoeckx RL, Hilgert N et al (2006) A new autosomal recessive form of Stickler syndrome is caused by a mutation in the *COL9A1* gene. Am J Hum Genet 79(3):449–457
Van Laer L, Huizing EH, Verstreken M et al (1998) Nonsyndromic hearing impairment is associated with a mutation in *DFNA5*. Nat Genet 20(2):194–197
Varga R, Kelley PM, Keats BJ et al (2003) Non-syndromic recessive auditory neuropathy is the result of mutations in the otoferlin (*OTOF*) gene. J Med Genet 40(1):45–50
Vikkula M, Mariman EC, Lui VC et al (1995) Autosomal dominant and recessive osteochondrodysplasias associated with the *COL11A2* locus. Cell 80(3):431–437
Vozzi D, Morgan A, Vuckovic D et al (2014) Hereditary hearing loss: a 96 gene targeted sequencing protocol reveals novel alleles in a series of Italian and Qatari patients. Gene 542(2):209–216
Waardenburg PJ (1951) A new syndrome combining developmental anomalies of the eyelids, eyebrows and nose root with pigmentary defects of the iris and head hair and with congenital deafness. Am J Hum Genet 3(3):195–253
Walsh T, Pierce SB, Lenz DR et al (2010) Genomic duplication and overexpression of *TJP2/ZO-2* leads to altered expression of apoptosis genes in progressive nonsyndromic hearing loss DFNA51. Am J Hum Genet 87:101–109
Wang A, Liang Y, Fridell RA et al (1998) Association of unconventional myosin *MYO15* mutations with nonsyndromic deafness DFNB3. Science 280(5368):1447–1451
Weil D, Blanchard S, Kaplan J et al (1995) Defective myosin VIIA gene responsible for Usher syndrome type 1B. Nature 374(6517):60–61

Weil D, El-Amraoui A, Masmoudi S et al (2003) Usher syndrome type I G (*USH1G*) is caused by mutations in the gene encoding SANS, a protein that associates with the USH1C protein, harmonin. Hum Mol Genet 12(5):463–471

Weston MD, Luijendijk MW, Humphrey KD et al (2004) Mutations in the *VLGR1* gene implicate G-protein signaling in the pathogenesis of Usher syndrome type II. Am J Hum Genet 74(2):357–366

Wilch E, Azaiez H, Fisher RA et al (2010) A novel DFNB1 deletion allele supports the existence of a distant *cis*-regulatory region that controls *GJB2* and *GJB6* expression. Clin Genet 78:267–274

Yang T, Vidarsson H, Rodrigo-Blomqvist S et al (2007) Transcriptional control of *SLC26A4* is involved in Pendred syndrome and nonsyndromic enlargement of vestibular aqueduct (DFNB4). Am J Hum Genet 80(6):1055–1063

Yang T, Gurrola JG II, Wu H et al (2009) Mutations of *KCNJ10* together with mutations of *SLC26A4* cause digenic nonsyndromic hearing loss associated with enlarged vestibular aqueduct syndrome. Am J Hum Genet 84(5):651–657

Yang T, Wei X, Chai Y et al (2013) Genetic etiology study of the non-syndromic deafness in Chinese Hans by targeted next-generation sequencing. Orphanet J Rare Dis 8:85

Yasunaga S, Grati M, Cohen-Salmon M et al (1999) A mutation in *OTOF*, encoding otoferlin, a FER-1-like protein, causes DFNB9, a nonsyndromic form of deafness. Nat Genet 21(4): 363–369

Young TL, Ives E, Lynch E et al (2001) Non-syndromic progressive hearing loss DFNA38 is caused by heterozygous missense mutation in the Wolfram syndrome gene *WFS1*. Hum Mol Genet 10(22):2509–2514

Part II

The Role of Pendrin in the Thyroid

Hypothyroidism, Subclinical Hypothyroidism and Related Diagnostic Tools

6

Gregor Schweighofer Zwink and Christian Pirich

Abstract

Thyroid hormones have a widespread effect on human metabolism. Their biochemical peculiarity is the need of iodine for their synthesis. Any lack of thyroid hormones is called hypothyroidism, the diagnosis of which can easily be established by testing a patient's serum. However, many different forms of hypothyroidism can be distinguished. Some of them occur already at birth, are usually caused by genetic defects like those encountered in Pendred syndrome and are called congenital forms of hypothyroidism. Others are acquired during lifetime and are usually caused by either iodine deficiency or several different types of thyroiditis. Also, some very distinct drugs can cause hypothyroidism. Another differentiation is to distinguish primary forms and central forms of hypothyroidism. The term primary refers to a lack of thyroid hormones due to an impaired production of hormones at the thyroid gland itself, whereas in central forms hypothalamic or pituitary disturbances lead to an absence of thyroid stimulation.

6.1 The Thyroid: Anatomy, Embryology and Histology

In Greek, the word thyroid means shield and describes the shape of the organ. The thyroid is an endocrine gland that consists of two lobes that are connected by an isthmus. It is located in the neck anterior to the trachea and between the jugular notch and cricoid cartilage of the larynx.

G.S. Zwink (✉) • C. Pirich
Universitätsklinik für Nuklearmedizin und Endokrinologie,
Müllner Hauptstraße 48, Salzburg 5020, Austria
e-mail: g.schweighofer-zwink@salk.at

S. Dossena, M. Paulmichl (eds.), *The Role of Pendrin in Health and Disease*,
DOI 10.1007/978-3-319-43287-8_6

Located dorsally to each lobe, the recurrent laryngeal nerves are in close proximity to the thyroid. Further, the parathyroid glands can be found dorsal to the thyroid and are of special interest during surgical removal of the thyroid (Miller 2003).

The gland is highly vascularized by four main arteries derived from the external carotid artery and subclavian artery. In 10 % of persons, an additional artery can be found that is derived from the aortic arch.

Lymphatic drainage from the upper and mid parts of the gland flows to the lateral cervical lymph nodes, and drainage from the caudal parts flows usually to the pre- and paratracheal lymph nodes.

The size of the gland varies significantly in adults, and it is common to describe it in milliliters. A thyroid that is greater than 18 mL in women or greater than 25 mL in men is considered enlarged. However, there is no generally defined lower size limit.

In up to 55 % of persons, ectopic thyroid tissue can be found located median at the neck. These additional thyroid lobules may or may not be connected to either the isthmus or the cranial parts of the main lobes. This accessory part of the thyroid is usually called the pyramidal lobe. The appearance of ectopic thyroid tissue can be explained by the development of the organ. Progenitor cells derive from the floor of the primitive pharynx just behind the tongue and migrate along the thyroglossal duct in a caudal direction in the neck (Braun et al. 2007).

Histologically, the highly irregular-shaped lobules of the thyroid gland consist of spherical follicles. The lumen of a follicle is filled with an amorphic substance called colloid. It mainly consists of a protein called thyroglobulin (Tg). Colloid serves as a reservoir for substances needed to produce thyroid hormones and for the thyroid hormones themselves being attached to Tg. Each follicle is bordered by a single layer of thyroid epithelial cells which produce colloid, as well as thyroid hormones.

Parafollicular cells can be found scattered between both the follicular cells and follicles. They are derived from the neural crest, and their progenitor cells join thyroid progenitors. The parafollicular or C cells produce the hormone calcitonin. This hormone plays a minor role in calcium-homeostasis and will not be further discussed in this chapter.

Each follicle is surrounded by connective tissue fibers and a vast network of fenestrated capillaries.

6.2 Thyroid Hormone Synthesis

The thyroid hormones, triiodothyronine (T3) and thyroxine (T4), are produced through an ether linkage between two iodinated tyrosine molecule residues. The thyroid follicular cells produce Tg, which contains numerous tyrosine molecules, and transport Tg into the follicular lumen. Atomic iodine (I^0) reacts with the tyrosine residues within Tg forming mono- and di-iodinated tyrosine. T4 involves the coupling of two di-iodinated tyrosine remnants within the Tg molecule. This leads to a molecule bearing four iodine atoms, for which the name T4 is derived. By linking a mono-iodinated remnant with a di-iodinated remnant, the result is either active T3 hormone or inactive reverse (r)T3. The difference is in the position of the missing iodine atom, as compared with T4.

The hormones are stored in the follicular lumen and are bound to processed Tg.

To liberate them, Tg is absorbed into the follicular cells via pinocytosis, and then, the large protein is partially disintegrated via proteolysis in the lysosomes. As lipophilic molecules, thyroid hormones are thought to diffuse passively through the basolateral cell membrane of thyrocytes. However, in recent years, it has been demonstrated that several transport proteins exist in different target tissues for thyroid hormones, as well as in the thyroid gland. These transporters seem to control cellular influx and efflux of thyroid hormones in the body. The major candidate transporters for follicular cells are monocarboxylate transporter (MCT) 8 and 10 (Brix et al. 2011).

As iodine is a crucial substance in thyroid hormone production (Fig. 6.1), thyroid follicle cells are highly capable of extracting iodide ions (I^-) out of blood via an active transport protein, the Na^+/I^- symporter (NIS/SLC5A5). I^- molecules diffuse through the intracellular space to the apical membrane, where they are transported *via* pendrin into the follicular lumen. Pendrin or SLC26A4 is a sodium-independent anion exchanger. However, doubts have arisen regarding iodide transport in recent studies. Although SLC26A4 definitely is capable of transporting iodide, it is unclear whether this exchanger transports I^- in physiological conditions or acts only as an assistant in the transport mechanism, in combination with other channel proteins (Bizhanova and Kopp 2009; Twyffels et al. 2011). For a detailed discussion of the role of pendrin in the thyroid, see Chap. 7.

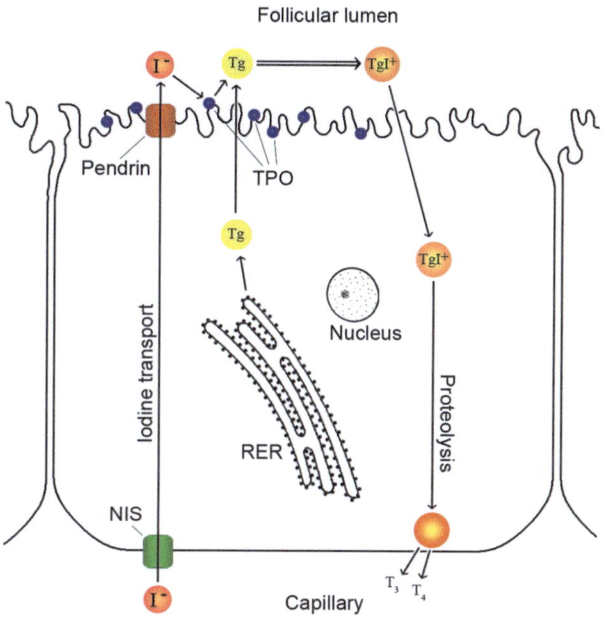

Fig. 6.1 A graphic of a follicular cell and thyroid hormone production

In the follicular lumen, I^- is oxidized with hydrogen peroxide via thyroid peroxidase (TPO) to atomic iodine or iodinium (I^+) that reacts with the tyrosine molecules of Tg.

6.3 Thyroid Hormone Metabolism

Thyroid hormone production and secretion is mostly regulated through thyroid-stimulating hormone (TSH). To a lesser degree, other growth factors such as IGF-1, EGF, TGF-beta, endothelins and various other cytokines also influence and modulate hormone production and release.

The thyroid hormone T4 is secreted in amounts approximately 20-fold greater than those of T3. Both hormones are lipophilic substances and are bound to plasma proteins in blood, such as albumin, thyroxine-binding globulin (TBG) and transthyretin. Those transport proteins increase the total amount of circulating hormones through a pooling effect. The transport proteins also delay thyroid hormone degradation and are thought to modulate the hormone transport to certain tissues.

Small amounts of unbound or so-called free hormones are thought to be biologically active (approximately 0.02 % of total T4 and 0.3 % of total T3). Those free proportions of the thyroid hormones (fT4 and fT3) are measured along with the TSH level to assess thyroid function (Dietrich et al. 2008).

T3 is a more potent hormone than T4, as the affinity of T3 to the thyroid hormone receptor exceeds that of T4. However, T4 has a much longer half-life than T3 and can therefore be seen as a precursor of T3 and as a type of storage form. The half-life of T4 is approximately 7 days, while the half-life of T3 is approximately 18 h.

T3 is produced either directly from Tg in the thyroid or, more commonly, by deiodination of T4 by deiodinases. These enzymes are capable of removing one iodine molecule from T4. There are several types of deiodinases. Type I is found in the thyroid, liver and kidneys. It has a slight affinity for T4 and is thought to act as a basal producer of active T3 for the whole body.

Type II deiodinase is found in the thyroid, pituitary gland, brain, heart, skeletal muscles and brown fat tissue. Type II deiodinase converts T4 into T3 in local tissues that are most dependent on thyroid hormones. The expression of type II allows for regulation of the local amount of active T3 in those dependent tissues.

Finally, type III deiodinase is the main actor in the inactivation of T3 and T4 and helps decarboxylases in the degradation of these hormones. After further deiodination, the products are called T2, T1 and T0, reflecting the number of remaining iodine atoms (Schweizer and Steegborn 2015).

After decarboxylation of either active T4 or T3 hormones or an already further deiodinated metabolite, the resulting molecules are called thyronamines. Thyronamines can activate a class of G protein–coupled receptors called trace amine–associated receptors or TAARs. They seem to partially antagonize some thyroid hormone effects (Brix et al. 2011).

Thyroid hormones enter target cells through active transport, mostly by monocarboxylate 8 transporters (MCT8/SLC16A2). The hormones mainly modulate gene transcription by binding to nuclear thyroid hormone receptors. By activating these receptors, they form either homodimers with each other or, more commonly, they form heterodimers with retinoic acid X receptors and bind to thyroid response elements at the DNA level. As a result, gene transcription in the target cells is modified. To a minor degree, thyroid hormones are able to stimulate responses directly acting on the plasma membrane, and they seem to modify the action of mitochondrial enzymes in target cells (Brix et al. 2011).

Thyroid hormones play a role in cell differentiation during the growth and development of a variety of tissues during embryonal and fetal development, as well as throughout life. In adults, they serve to maintain homeostasis of the metabolic rate of target tissues and affect protein, carbohydrate and fat metabolism in most parts of the body. They boost the sensitivity of the body to catecholamines and modulate the cardiovascular system. They also stimulate the body to produce heat (Klein and Danzi 2016).

6.4 Definition of Abnormal Thyroid Function

If free thyroid hormone levels decrease below a certain set point at the hypothalamus, the production of thyrotropin-releasing hormone (TRH) stimulates TSH formation and secretion from the pituitary gland. Then, TSH stimulates the formation and liberation of T4 and T3 in and out of the thyroid. TSH production and secretion is not only controlled by TRH exclusively, but also influenced by thyroid hormones as a form of feedback loop from the thyroid. TRH secretion itself is not only stimulated or suppressed by thyroid hormone levels, but also modified by several other stimuli, e.g., low body temperature. Recent studies suggest that there are several feedforward and feedback loops between the thyroid and the hypothalamus and pituitary gland. These regulatory loops are further tuned by other hormones as well as neurotransmitters. This complex regulatory system enables the body to respond dynamically to a great variability of environmental influences (Hoermann et al. 2015).

The hypothalamic set point, which defines the normal range of thyroid hormones within the blood stream, slightly differs from the need of the body for thyroid hormones. A little drop of serum hormones below the hypothalamic set point stimulates TRH and TSH production. The effect is an elevated TSH that can already be measured, although the free proportions of serum thyroid hormones are still within the normal range, according to laboratory references. The patient, at this state, normally does not demonstrate any sign or symptom of a lack of thyroid hormones. This state is then called subclinical hypothyroidism (elevated TSH, normal fT4/fT3; Table 6.1). If the free hormones further decrease below normal amounts, then the hypothyroidism becomes "overt" (elevated TSH, reduced fT4/fT3; Table 6.1). The

Table 6.1 Overview of primary thyroid dysfunction

	fT3 and fT4 level	TSH level
Euthyreosis (normal thyroid function)	Normal	Normal
Subclinical primary hypothyroidism	Normal	Elevated
Overt primary hypothyroidism	Decreased	Elevated
Subclinical primary hyperthyroidism	Normal	Decreased
Overt primary hyperthyroidism	Elevated	Decreased

higher TSH levels and the lower fT4 and fT3 levels are, the more likely the patient will demonstrate signs and symptoms of hypothyroidism.

However, if thyroid hormones increase above certain levels, TSH drops below normal, and a subclinically hyperthyroid state is defined (reduced TSH, normal fT4/fT3; Table 6.1). Finally, if fT4 and fT3 concentrations exceed the upper limits of the reference range, the hyperthyroidism can also become "overt" (suppressed TSH, elevated fT4/fT3; Table 6.1).

As the signs and symptoms of hypothyroidism and hyperthyroidism lack sensitivity and specificity, serum testing for TSH is strongly recommended whenever the slightest suspicion of thyroid dysfunction exists (Gaitonde et al. 2012; Indra et al. 2004).

6.5 Primary and Secondary Causes of Hypothyroidism

If the absence of thyroid hormones is caused by a disease of the thyroid gland itself or a thyroid dysfunction, it is considered primary hypothyroidism (PH). Further, if a malfunction in hormone transport or the hormone receptor system leads to the absence of thyroid hormone effects, this clinical state is referred to as PH as well but can also be called peripheral hypothyroidism.

If the thyroid has no structural damage but does not produce adequate hormone amounts because of decreased TSH stimulation, it is considered secondary hypothyroidism. Furthermore, tertiary hypothyroidism would exist if TRH secretion does not properly stimulate TSH production. As secondary and tertiary functional changes cannot easily and clearly be differentiated in every patient, they are usually referred to as central hypothyroidism (CH).

CH is about 1,000 times rarer than PH and is often combined with other hypothalamic and pituitary hormone disturbances. In Europe, the general prevalence of hypothyroidism is estimated to be 3.05 %, with an incidence rate of 226.2 per 100,000 per year. However, the prevalence of undiagnosed (and commonly subclinical) hypothyroidism is 4.94 % (Garmendia Madariaga et al. 2014). The different causes of hypothyroidism are presented in Table 6.2 (Khandelwal and Tandon 2012; Biondi and Wartofsky 2014).

6.6 Congenital Primary Hypothyroidism

If a thyroid hormone deficiency is present at birth, it is referred to as congenital hypothyroidism (Table 6.3). One can distinguish whether a disruption in the thyroid axis is caused by a "central" pathology in the hypothalamus or pituitary gland or whether the problem is with the thyroid.

From a pathophysiologic viewpoint, primary forms of congenital hypothyroidism can be related to a lack of thyroid tissue or dysfunctional hormone control and biosynthesis; misguided hormone biodistribution secondary to transport protein pathologies; and a loss of thyroid hormone receptor function. The latter two are also called peripheral hypothyroidism to indicate that the pathogenic mechanism does not reside within the thyroid gland itself.

Table 6.2 Overview of different causes of hypothyroidism

Primary hypothyroidism		Central hypothyroidism	
Congenital	Absent or malfunctioning gland; defects in thyroid hormone biosynthesis; TSH-receptor dysfunctions	**Congenital**	TRH and TSH defects; developmental disruption at the hypothalamus or pituitary gland level or combination of both
	Functional defects in thyroid hormone biodistribution, receptor defects or increased degradation	**Loss of pituitary tissue**	Invasive or compressive lesions, e.g., macroadenomas, craniopharyngiomas, Rathke's cleft cyst, metastasis, empty sella, carotid aneurysms, meningiomas, gliomas, etc.
Lack of iodine	Iodine deficiency and iodine excess		Infiltrative lesions, e.g., hemochromatosis, sarcoidosis and histiocytosis X
Loss of thyroid tissue	Thyroiditis, e.g., autoimmune thyroiditis (Hashimoto's thyroiditis), silent and postpartum thyroiditis, cytokine-induced thyroiditis, invasive fibrous thyroiditis (Riedel's thyroiditis), subacute thyroiditis (De Quervain thyroiditis) and acute infectious thyroiditis		Iatrogenic destruction of the pituitary gland secondary to surgery or radiation
	Surgical removal of thyroid parts or the whole thyroid		Injuries, e.g., head trauma
	Radioiodine ablation; inadvertent exposure to I-131		Vascular incidents, e.g., pituitary apoplexy, subarachnoid hemorrhage, postpartum necrosis and severe shock
	External neck irradiation therapy or radiation release from nuclear accidents		Autoimmune disease, e.g., lymphocytic hypophysitis and polyglandular autoimmune disease
	Infiltration or destruction of the thyroid by other (systemic) diseases, e.g., amyloidosis, sarcoidosis, lymphoma, hemochromatosis, systemic sclerosis, cystinosis and AIDS		Infections, e.g., tuberculosis, mycosis, syphilis, abscesses and toxoplasmosis

(continued)

Table 6.2 (continued)

Primary hypothyroidism		Central hypothyroidism	
Drugs	Antithyroid drugs inhibiting the thyroid hormone biosynthesis	**Drugs**	Bexarotene-, ipilimumab- and tremelimumab-induced hypophysitis
	Drugs interacting with iodine accumulation (sodium/potassium perchlorate, amiodarone and iodine-containing contrast media)	**Set point disease**	Infants born to mothers with poorly controlled hyperthyroidism
	Inadequate thyroid hormone replacement therapy	**Idiopathic**	No definite cause can be detected
	Others with variable effects such as lithium, cytokines, tyrosine kinase inhibitors, sulfonamides, thalidomide, ethionamide, 4-amino-salicyclic acid, amino-glutethimide, nivolumab, etc.		

Table 6.3 Overview of causes of congenital hypothyroidism

Primary congenital hypothyroidism	Thyroid dysgenesis: developmental anomaly of the organ like thyroid ectopia (wrong location of the organ), athyreosis (complete absence of thyroid tissue) or hypoplasia (thyroid too small to meet the needs of the body). In 98 % of cases, the cause is unknown or undetectable. In 2 %, mutations of genes such as TTF2 or PAX8 can be detected
	Thyroid dyshormonogenesis: the thyroid is present, but hormone production is impaired; associated with NIS gene mutations, TPO defects, pendrin defects, Tg defects and deiodinase defects
	TSH signaling problems: TSH-receptor defect; mutations in G proteins (the TSH signal cannot be transduced within the follicular cells)
Peripheral congenital hypothyroidism	Abnormalities in thyroid hormone transport mechanisms because of transthyretin, TBG or monocarboxylate transporter 8 (MCT8/SLC16A2) mutations
	Mutations leading to a loss of function or loss of number of the thyroid hormone receptors
	Increased thyroid hormone degradation by deiodinases
Central congenital hypothyroidism	TRH: defects in production and release; receptor mutations
	TSH: defects in production and release
	Impaired pituitary development; commonly a combination of several endocrine dysfunctions, e.g., septo-optic dysplasia
Potentially transient congenital hypothyroidism	Maternal and neonatal iodine deficiency or excess
	Maternal intake of antithyroid drugs that are passed through the placenta or breast milk
	Transplacental passage of antibodies blocking the TSH receptor
	Overexpression of deiodinase in a hemangioma or hemangioendothelioma

Thyroid dysfunction can be either isolated or combined with other pathologies in different syndromes.

From a clinical viewpoint, one can distinguish between non-goitrous forms of congenital hypothyroidism and goitrous forms. The non-goitrous forms account for approximately 85 %, while the goitrous forms occur in approximately 15 % of congenital hypothyroidism (Abduljabbar and Afifi 2012).

If the hypothyroidism is related to a lack of thyroid tissue or if no proper TSH stimulation exists, a non-goitrous form develops. If a problem with the biosynthesis, biodistribution or thyroid hormone receptor exists, the body tries to stimulate the thyroid using TSH. The elevated TSH causes growth of present thyroid cells, and this results in a goitrous form of congenital hypothyroidism.

Some infants can recover from congenital hypothyroidism, and in these circumstances, the condition is considered temporary. Depending on the hypothyroidism cause, several types are permanent disorders, necessitating life-long substitution therapy.

In studies, the incidence of congenital hypothyroidism varies from 1:800 to 1:10,000. The frequency rose significantly after commencing screening programs for all newborns. The incidence rates vary by geographic location and over time in the same location; there are significant differences between newborn race and sex, e.g., female newborns are diagnosed more often than male newborns (Abduljabbar and Afifi 2012). Differences in testing methodologies for congenital hypothyroidism can lead to variations in its incidence. It may be concluded that transient congenital hypothyroidism exceeds permanent hypothyroidism because prior to the instatement of neonatal screening programs for congenital hypothyroidism, clinical diagnosis and incidence had been reported to be between 1:7,000 and 1:10,000 (Abduljabbar and Afifi 2012).

The classic features of congenital hypothyroidism include a large tongue, a hoarse cry, facial puffiness, hypotonia, cold hands and feet, mottling skin, an umbilical hernia and general lethargy. Additional signs such as prolonged unconjugated hyperbilirubinemia, feeding difficulties, constipation, hypothermia and large fontanelles should raise the suspicion of a thyroid problem. However, these stated clinical signs and symptoms of congenital hypothyroidism lack in their sensitivity and specificity. Moreover, congenital hypothyroidism is further disguised by transport of maternal thyroid hormones through the placenta. Therefore, hypothyroidism develops slowly and gradually in most cases during a newborn's first few days of life.

If a complete blockage of thyroid hormone synthesis occurs or a newborn has absolutely no thyroid tissue (athyreosis), the symptoms are more severe than those of infants with an ectopic gland, leading to decreased hormones, but not their complete absence (Abduljabbar and Afifi 2012; Krude et al. 2015).

As thyroid hormones are essential for the maturation and growth of several organs, congenital hypothyroidism that is not detected as soon as possible can lead to severe pathologies. Therefore, newborn screening tests for hypothyroidism are performed in industrialized nations. The screening test is usually performed by collecting capillary blood on a filter paper with a heel prick. The dried blood can be tested for total T4, TSH and TBG. Total T4 and TBG testing enables an estimation of the fT4 value, which cannot be tested directly from dried blood. Umbilical cord blood samples can also be collected to perform newborn screening.

There are three different approaches for testing blood in congenital hypothyroidism. The first one involves testing primarily for T4 levels; TSH is only tested if the T4 levels are below a certain cutoff level. The second one is to perform the tests in the opposite order (testing primarily for TSH; if above a certain threshold, T4 is tested). The third option is to test simultaneously for T4 and TSH. The higher costs associated with testing for both TSH and T4 are a disadvantage of the latter strategy.

An advantage of testing for T4 before TSH is the potential to detect central forms of hypothyroidism, in addition to overt PH. A disadvantage is that subclinical forms of hypothyroidism can be missed. An advantage of testing for TSH before T4 is the potential to detect overt and subclinical forms of PH, but there is a possibility of missing central defects and transport protein defects for thyroid hormones. This strategy is also the method which is performed in Austria since 1976. The newborns' blood is screened centrally at the Allgemeines Krankenhaus der Stadt Wien for all Austrian newborns (Moslinger et al. 1997; Pollak and Kasper 2014).

Rarely, infants slowly develop a decreased T4 level and an increased TSH level after birth and appear to be normal upon newborn screening. This form is called atypical congenital hypothyroidism and commonly occurs in premature infants or babies with lesser severe forms of permanent hypothyroidism. In response to this possibility, a second screening should be performed a few weeks after birth if the slightest suspicion of hypothyroidism arises.

Whenever abnormalities in the newborn screening test occur, it is then recommended to perform a confirmatory test of TSH and free T4 levels out of a plasma sample collected by drawing venous blood. If a hypothyroid state is confirmed, treatment should and can be started immediately to prevent cretinism or growth failure.

An evaluation of bone age using radiographs can be performed to assess the duration and severity of hypothyroidism for a fetus in utero. If the hypothyroidism has been significant, the bone age would be low for gestational age, e.g., the ossification cores of the epiphysis would be missing. However, this test is not essential for hypothyroidism in clinical practice and lacks both sensitivity and specificity. Indeed, evaluating the thyroid gland is helpful in differentiating permanent forms from transient forms of hypothyroidism.

In neonates with abnormal blood levels of T4 or TSH, radionuclide scintigraphy with either I-123 or Tc-99 m pertechnetate is performed to locate the thyroid gland, to estimate its size and to check for its capability of trapping iodine. Four patterns of thyroid scintigrams can be recognized in neonates with congenital hypothyroidism: (a) normal, in patients with false-positive heel-stick screening but normal venous blood hormone levels; (b) hypoplasia-ectopia; (c) non-visualization, in patients with thyroid agenesis or fetal thyroid suppression by maternal antibodies or by antithyroid agents or in patients with defects of the sodium-iodide-symporter; and (d) dyshormonogenesis (markedly increased iodine or pertechnetate uptake) (Sfakianakis et al. 1999).

An additional ultrasound helps in confirming scintigraphic results on the location and size of thyroid tissue. By performing an ultrasound, goitrous forms of congenital hypothyroidism can be differentiated from non-goitrous forms.

To determine whether a malfunction in hormone production exists, an additional perchlorate discharge test can be performed. This situation is to be suspected if a goitrous thyroid is found in newborns with hypothyroidism and normal or increased iodine or pertechnetate uptake. The uptake can be easily calculated after the scintigraphy, comparing the amount of radiation emitted from the thyroid to the total dose of tracer which has been administered. An increase in tracer uptake suggests that the thyroid is still capable of upregulating the sodium-iodide-symporter (NIS/SLC5A5) in response to TSH stimulation, and the symporter function is normal, but the transport and organification of iodine or the biosynthesis of thyroid hormones show a loss of function.

A perchlorate discharge test is performed by administering an I-123 test dose to a patient, and the accumulation of I-123 and NIS function within the thyroid is confirmed with scintigraphy. About 2 h later, a sodium or potassium perchlorate dose is administered. The perchlorate works as an antagonist of iodide trapping within the thyroid cells and follicles. Iodine, which has not been organified by the thyroid (meaning oxidized and bound to Tg), leaves the thyroid in the presence of perchlorate. After administering perchlorate, a drop in iodine concentration within the thyroid is a positive discharge test. The drop can be measured using scintigraphy, resulting in a reduction in gamma radiation emitted from the thyroid. A positive test indicates a problem processing iodine molecules somewhere between the NIS and binding of oxidized iodine to Tg, as would be the case in Pendred syndrome (Meller et al. 1998). If no significant change in iodine concentration occurs in the thyroid after the intake of perchlorate, the cause of the hypothyroidism is located somewhere downstream after the iodine organification process.

During a clinical work-up, blood can be further tested for levels of Tg and for antibodies against thyroid antigens. In the case of a decreased total T4 but normal TSH and normal fT4 levels, TBG should be assessed. If a CH form is present, the other pituitary hormone axes should be tested to rule out combined pituitary dysfunctions.

The therapy of choice for all forms that result in a decreased thyroid hormone level is a high-dose T4 replacement therapy. In those rare cases without functioning deiodinases, an additional T3 replacement must be considered. In cases with defective transport plasma proteins also, T4 and T3 replacement can be of help. However, patients suffering from disturbances of the thyroid hormone cell membrane transporter (e.g., the MCT MTC8/SLC16A2), which leads to a peripheral lack of thyroid hormones, do not benefit from T4 replacement. Some studies tried to treat the patients with derivatives and metabolites of thyroid hormones but could not achieve a breakthrough. However, some patients with certain loss-of-function mutations of the thyroid hormone receptors sometimes benefit from additional T4 or metabolites of thyroid hormones that can still act as agonists at the receptor (Persani 2012; Abduljabbar and Afifi 2012; Krude et al. 2015).

6.7 Iodine Deficiency

Iodine is most essential in the biosynthesis of thyroid hormones. Nutritional absence of iodine leads to a decrease in thyroid hormones; as a result, TSH levels rise. This enables the thyroid to extract iodine more efficiently out of the blood and to recycle iodine after thyroid hormone degradation. This is mainly achieved by upregulating the NIS gene through TSH signaling.

Another effect of a deficient iodine supply is an increase in T3 production and a decrease in T4. T3 requires one less iodine atom; moreover, T3 has a more potent effect on thyroid hormone receptors.

An additional consequence of elevated TSH levels is the stimulation of existing thyroid follicular cell growth. Therefore, a goiter is the most common effect and clinical sign of iodine deficiency. The more important impacts of iodine deficiency, especially during fetal and neonatal development, are cognitive impairments because of hypothyroidism. However, as the thyroid has the capability of maximizing iodine uptake and recycling, an iodine deficiency has to be severe before the signs and symptoms of overt hypothyroidism occur.

To assess iodine supply and patients' status, urinary iodine concentrations are measured and combined with thyroid function tests (fT4, TSH, and Tg level), along with examination of the thyroid gland with an ultrasound and pertechnetate scintigraphy.

Measurement of the urinary iodine concentrations reflects the nutritional supply of iodine because the kidneys excrete approximately 90 % of the iodine that is ingested and absorbed, even if the NIS is working at its maximum. If the urinary iodine concentration is below 100 μg/L, the iodine intake and body supply are considered insufficient (Zimmermann and Boelaert 2015). A mild deficiency (urinary iodine concentration of 50–99 μg/L) commonly leads to the development of a diffuse goiter, as well as an increase in nodularity. Tg levels and thyroid size increase, but the TSH and fT4 levels usually remain within a normal range. In some populations with a mild iodine deficiency, TSH levels may even decrease. This can be explained by an increase in incidence and prevalence of autonomous nodules or of a toxic multinodular goiter. With a moderate iodine deficiency (urinary iodine concentration of 20–49 μg/L), slightly elevated TSH concentrations are more likely found. However, the thyroid employs compensational mechanisms, so fT4 levels and especially fT3 levels often remain within a normal range, and affected individuals develop only a subclinical form of hypothyroidism. With a decrease in iodine supply, the occurrence of a goiter and nodules gradually increases. Urinary iodine concentrations below 20 μg/L are considered a severe deficiency. Many affected patients develop a goiter and are most likely to experience overt hypothyroidism.

To achieve a sufficient iodine supply and normal urinary iodine concentrations, the WHO suggests a daily intake of 150 μg for teenagers and adults. The daily recommendation for pregnant and lactating women is 250 μg per day (WHO Guidelines Approved by the Guidelines Review Committee 2014). In the year 2014, the iodine intake is thought to be adequate in 112 countries of the world but is still deficient in 29 countries and excessive in 11 countries (excessive = a urinary iodine concentration >300 μg/L) (Zimmermann and Boelaert 2015; WHO Guidelines Approved by the Guidelines Review Committee 2014). A decade ago, the iodine intake was adequate in 67 countries only. An increase in adequate iodine intake has mainly been achieved by selective iodine supplementation.

Although more developing countries, usually with large populations, can be found on the list of iodine-deficient countries, several highly developed and industrialized countries such as Denmark, Italy and the UK can be found among them. Further, surveys suggest that in several high-income countries, iodine intake has decreased over the past decades. This is mainly thought to be caused by a decrease in iodine concentrations in dairy products (Zimmermann and Boelaert 2015).

To prevent and treat chronic iodine deficiencies in large populations, it is common to add potassium iodide to salt. This method is proven to be efficient, especially if the food industry uses iodinated salt. Moreover, it is inexpensive and medically safe (WHO Guidelines Approved by the Guidelines Review Committee 2014). Iodized nutrition is also administered to milk cows and helps to enhance nutritional iodine intake of a population through dairy products. In Austria, the law states that salt, which is iodized, should contain between 15 and 20 mg of iodide per kg of salt. Only iodized salt is allowed to be called full salt or table salt. To treat and prevent iodine deficiency in individuals, several dietary supplements are available, especially for pregnant and lactating women. In Austria, additional intake of 150–200 μg of iodine per day for pregnant and lactating women is commonly suggested (Zimmermann 2012).

6.8 Thyroiditis

The most common cause of hypothyroidism in countries with sufficient iodine supply is follicular cell loss because of inflammatory destruction of the thyroid gland. Several different types and causes of thyroid inflammation can be distinguished. Autoimmune and non-autoimmune entities are differentiated pathophysiologically.

Affected patients demonstrate great variability in clinical presentation. Some experience severe local pain with or without a general feeling of malaise. However, more commonly, thyroiditis occurs without any local symptoms, despite slow and gradual organ destruction. This causes subclinical and overt hypothyroidism at a later time. Depending on how quickly destruction occurs, each form of thyroiditis can lead to a transient hyperthyroidism state. The cause is the rapid release of preformed hormones from destroyed thyroid follicles. The more severe and quick the destruction is, the more frequently and severely the patient develops the signs and symptoms of hyperthyroidism. Consequently, the affected patient is more likely to develop the signs and symptoms of hypothyroidism thereafter because destroyed and lost thyroid follicles cannot be replaced in an appropriate amount. The slower the destruction, the less likely it is for an affected patient to present with the typical signs and symptoms of hypothyroidism because the body performs adaptive mechanisms. The inflammatory process can lead to either an atrophic thyroid gland or a goiter. Rarely, fibrosis can occur within an inflamed thyroid.

Considering the wide range of clinical presentations, a proper thyroid examination is necessitated to determine whether an inflammatory process is ongoing and which form of thyroiditis is present. Further, a physical examination, neck palpation, an ultrasound, and a Tc-99 m pertechnetate scan are performed. A blood sample needs to be checked to determine thyroid function levels by testing TSH and fT4 levels. The blood is also tested for antibodies against TPO and Tg, and the blood sedimentation rate is preferably checked. In addition, C-reactive protein and Tg levels can be helpful (Pearce et al. 2003).

In the following subsections, all thyroid syndromes but drug-induced thyroiditis will be presented. For drug-induced thyroid changes, please refer to Sect. 6.9.

6.8.1 Chronic Lymphocytic Thyroiditis—Hashimoto's Thyroiditis

This form of thyroiditis is the most frequent and is characterized by chronic, autoimmune-mediated lymphocytic inflammation and destruction of the thyroid. Clinically, patients present with gradually progressive thyroid failure with or without goiter formation. Depending on the total amount of damaged thyroid tissue, patients can present with normal thyroid function over a long and highly variable time period. With the gradual loss of follicular cells, subclinical and then overt hypothyroidism can develop. As for all autoimmune diseases, the pathogenesis is still not clearly understood. The immune defect is thought to be the result of a genetic susceptibility combined with one or several environmental factors (Pyzik et al. 2015).

The serological feature of chronic thyroiditis is the detection of elevated antibody titers against thyroid proteins, receptors and enzymes. Most commonly, antibodies against TPO and Tg are present, for which laboratory tests are conducted to determine their presence. However, other antibodies against thyroid proteins and enzymes like the NIS, pendrin, colloid antigen and thyroid hormones occur. However, they are not routinely been tested for (Pearce et al. 2003). Also blocking antibodies against the TSH receptor have been found and are thought to play a role in the development of thyroid atrophy in some patients suffering from chronic thyroiditis (Furmaniak et al. 2013).

In the general population, the prevalence of anti-TPO antibodies is between 8 % and 27 %; however, in approximately 90 % of cases with autoimmune thyroiditis, increased anti-TPO antibody titers can be found. The general frequency of Hashimoto's thyroiditis is estimated to be about 5 % in Caucasians. The majority of patients who have elevated anti-TPO antibodies have normal thyroid function. From an epidemiological standpoint, it is estimated that approximately 10 % of women with increased anti-TPO antibodies also have subclinical hypothyroidism, while only about 0.5 % have overt hypothyroidism (Vanderpump et al. 1995; Pearce et al. 2003; Pyzik et al. 2015).

There is no explicit correlation between the antibody titer level and risk of thyroid failure or inflammatory activity. However, it has been demonstrated that for all patients presenting with subclinical hypothyroidism, the risk of developing overt hypothyroidism is 4.6 % per year, if anti-TPO antibody titers are elevated. In patients without increased anti-TPO antibody levels, the risk is only 2.6 % per year. In general, the risk for patients with normal thyroid function to develop overt hypothyroidism is between 2 % and 4 % per year if elevated antibodies against the thyroid are present (Vanderpump et al. 1995).

Hashimoto's thyroiditis occurs five to ten times more often in women than men, and the incidence increases with age. As for other autoimmune diseases, the occurrence of chronic thyroiditis is significantly more common in patients who already have another preexisting autoimmune disorder.

The breakdown of self-tolerance and start of autoimmune aggression are thought to commence with activation of thyroid antigen-specific helper T lymphocytes. One theory indicates this activation is caused by molecular mimicry of viral antigens and thyroid antigens. However, no clear correlation between a specific viral infection and the occurrence of Hashimoto's thyroiditis has yet been found. Another theory claims that thyroid follicular cells start to wrongly present their own intracellular proteins to T lymphocytes. The causes of this irregular presentation are still a matter of discussion. Iodine deficiency tends to decrease this mistaken presentation of intracellular proteins. After amelioration of the iodine deficiency, this incorrect presentation seems more likely to occur and correlate with the higher prevalence of autoimmune thyroiditis in iodine-sufficient countries, as compared with iodine-insufficient countries (Pyzik et al. 2015; Zimmermann and Boelaert 2015).

Activated T lymphocytes induce antigen-specific B lymphocytes to migrate into the thyroid and to produce anti-thyroid-antigen antibodies. These antibodies are polyclonal and are a consequence, not the primary cause, of auto-aggression. Usually, these antibodies are of subclass IgG1 and IgG3, but any other class or subclass may occur. The loss of follicular cells is then a consequence of cytotoxic T cells and pro-inflammatory cytokine-induced apoptosis and may also be related to complement system–induced apoptosis that occurs after the binding of anti-thyroid antibodies to wrongly presented intracellular antigens on the follicular cell surface. If present, blocking antibodies against the TSH receptor normally do not activate complement but have an antigrowth or even atrophic effect on thyroid follicles because of the lack of a TSH-mediated growth stimulus. This partially explains the different clinical presentations in affected patients with goitrous and non-goitrous chronic thyroiditis.

Because of a misguided complex autoimmune process, no causal therapy is yet available to treat Hashimoto's thyroiditis. The wide range of adverse effects of immunosuppressive drugs leads to a highly unfavorable risk–benefit profile for treating chronic thyroiditis with immunosuppression. The standard treatment is to substitute absent thyroid hormones. This is mainly achieved by administering an individualized, oral T4 dose. Clear evidence exists regarding the benefit of substitution therapy in patients experiencing overt hypothyroidism. For patients presenting with subclinical hypothyroidism only, the evidence is weak. However, it has been proven that normalizing TSH levels in patients with chronic thyroiditis significantly reduces anti-thyroid autoimmunity and at least partially restores self-tolerance (Pyzik et al. 2015; Pearce et al. 2003). Only weak evidence exists regarding any additional benefits with selenium therapy (Kohrle 2015).

6.8.2 Painless Thyroiditis

This form of thyroiditis is considered to be a variant of chronic autoimmune thyroiditis, Hashimoto's thyroiditis, but with a different clinical appearance. The term painless or silent thyroiditis is used if the autoimmunity toward the thyroid has a sudden onset and if a significant number of thyroid follicles have been destroyed in a short-time period. This leads to a transient hyperthyroid phase as is commonly seen in subacute granulomatous thyroiditis, but without local thyroid pain.

Sometimes, a mild diffuse goiter can be a consequence of inflammatory activity in the thyroid. It is estimated that painless thyroiditis accounts for 0.5 %, and up to 5 %, of cases with hyperthyroidism (Samuels 2012).

The hyperthyroid phase usually resolves within 6–8 weeks and is sometimes followed by subclinical or overt hypothyroidism. In some cases, the thyroid can recover and regain its ability of producing proper amounts of thyroid hormones. It is estimated that, in up to 50 % of cases, hypothyroidism persists and hence those patients need T4 replacement therapy (Nikolai et al. 1981). In patients who have recovered functionally, the probability of thyroid failure later during lifetime, however, is high. The cause is a similar chronic autoimmune process as it occurs in Hashimoto's thyroiditis.

Again, there is no specific anti-inflammatory treatment. As hyperthyroidism is a consequence of cellular damage, and not the result of hormone overproduction, antithyroid agents are of no use and are not indicated. If a patient presents with pronounced hyperthyroidism symptoms, unspecific beta-blockers have proven to be useful in reducing discomfort. If affected patients develop hypothyroidism, the treatment of choice is T4 replacement (Samuels 2012; Pearce et al. 2003).

6.8.3 Postpartum Thyroiditis

Postpartum thyroiditis has the same clinical presentation as painless thyroiditis. The only difference is that it occurs only in women within 1 year of parturition or abortion. The frequency of occurrence is indicated to vary between 1 and 17 % globally. However, a prospective study describes a mean prevalence of 7–8 % in women of a general population but also rates of up to 25 % in women suffering from type 1 diabetes (Nicholson et al. 2006).

Most affected women experience no or only weak symptoms; therefore, many cases may not be diagnosed and do not require any specific treatment. The chances of regaining a normal thyroid function within a year are estimated to be about 80 %. However, there is a 50 % probability of recurring autoimmune-mediated hypothyroidism and permanent hypothyroidism within 7 years after having postpartum thyroiditis (Pearce et al. 2003; Samuels 2012).

6.8.4 Subacute Granulomatous Thyroiditis—De Quervain Thyroiditis

This type of thyroiditis is usually self-limited, and the common clinical appearance is local thyroid pain. The cause is not clearly understood. It is theorized that subacute thyroiditis is either the effect of a viral thyroid infection or a postviral, misguided inflammatory response in the thyroid. This hypothesis is supported by the observed coincidental seasonal peak in frequency of subacute thyroiditis, along with enteroviral infections. Many patients report a preceding upper respiratory infection before developing thyroid pain.

Also associations with other viral infections have been drawn when clusters of subacute thyroiditis cases have occurred. Yet, a direct proof of a viral component within an affected thyroid gland has never been found. Therefore, the theory of a misguided immune system is still favored as the pathophysiologic concept.

There is an association between the probability of developing subacute thyroiditis and having the HLA-B35 subtype of the major histocompatibility complex (MHC) (Ohsako et al. 1995). Immunological responses against the thyroid in subacute thyroiditis are self-limiting, not chronic as in autoimmune thyroiditis. Histopathological inflammation is not dominated by lymphocytic inflammation, but granulomatous inflammation with giant cells.

As for other thyroiditis syndromes, functional thyroid changes depend on the amount of affected thyroid tissue. Up to 50 % of patients develop subclinical or overt hyperthyroidism. The cause is related to sudden follicular cell damage. Therefore, antithyroid agents are not helpful. The hyperthyroid phase can be followed by subclinical or overt hypothyroidism. It is estimated that up to 85 % of patients have normalized thyroid function within 6–12 months after onset of this thyroiditis (Fatourechi et al. 2003; Rizzo et al. 2014; Pearce et al. 2003).

For diagnosing subacute thyroiditis, the main laboratory finding is a noticeably elevated erythrocyte sedimentation rate.

To ease hyperthyroidism symptoms, unspecific beta-blockers can be used. Anti-inflammatory agents such as non-steroidal anti-inflammatory drugs or glucocorticoids effectively reduce local pain. If a patient is experiencing hypothyroidism, the therapy of choice is T4 replacement.

6.8.5 Infectious Thyroiditis

A thyroiditis caused by microorganisms is rare. The thyroid is generally resistant to infection because of its high blood flow, iodide concentration and good lymphatic drainage. It can be either an acute or smoldering chronic process.

Acute septic thyroiditis is usually caused by streptococci or staphylococci species, but many other bacteria have also been reported. Upon presentation, patients are usually acutely ill with local pain at the thyroid, along with the signs and symptoms of infection. Thyroid function usually remains normal, but hypothyroid and hyperthyroid phases have been described. Intravenous antibiotics are normally necessary to treat acute infections.

Chronic infectious thyroiditis is usually seen in immunocompromised patients and can be caused by fungi, mycobacteria and parasites. Specific treatment can be established after diagnosing which microorganism is causing the infection (Rizzo et al. 2014; Pearce et al. 2003).

6.8.6 Fibrous Thyroiditis—Riedel's Thyroiditis

This thyroiditis is a local manifestation involving a chronic, progressing fibrotic process. Fibrotic changes destroy the thyroid architecture that can lead to hypothyroidism depending on how quickly the destruction occurs. Moreover, fibrosis may affect regions surrounding the thyroid. This can also lead to hypoparathyroidism, a tight feeling in the neck, dysphagia, hoarseness and dyspnea. In up to two-thirds of patients, antithyroid antibodies can be detected, but their appearance might be an effect of thyroid destruction, not the cause of fibrotic changes.

The diagnosis is suspected if a patient presents with a rock-hard goiter. A biopsy demonstrating fibrotic changes establishes the diagnosis. Treatment often requires surgical removal of the gland, but there are reports of the successful use of glucocorticoids, methotrexate and tamoxifen (Pearce et al. 2003; Lo et al. 1998).

6.9 Drug-Induced Thyroid Changes

Many drugs can interfere with normal thyroid function by disturbing either iodine or thyroid hormone metabolism. Some drugs can cause inflammatory thyroid responses.

Finally, many drugs might interfere with the intestinal resorption of T4 in patients undergoing thyroid hormone replacement therapy.

The most potent drugs with an ability to cause thyroid dysfunction are presented as follows.

6.9.1 Amiodarone

Amiodarone is an antiarrhythmic drug that has structural biochemical similarity to T4 and contains iodine, leading to excessive iodine intake as a treatment result. The drug itself and the high iodine amount can cause both, either hypothyroidism or hyperthyroidism. Approximately 40 % of patients develop a functional disturbance of the thyroid during treatment. About the half of these patients (20 %) develops a hypothyroid state. This can partly be explained by the induction of autoimmunity against the thyroid, especially in patients who have a pretreated iodine deficiency. As previously mentioned, the amelioration of an iodine deficiency leads to a higher autoimmunity toward the thyroid gland.

The treatment of hypothyroidism during amiodarone therapy usually requires an inadequately high T4 dosage because amiodarone reduces deiodinase activity that is necessary to convert supplemented T4 into T3 in the body.

The other half of patients developing a thyroid malfunction during amiodarone therapy suffer from hyperthyroidism. There are two different mechanisms described leading to hyperthyroidism. One involves the iodine amount that enables autonomous follicles or nodules to increase thyroid hormone production. The other involves the induction of a destructive thyroiditis, causing uncontrolled preformed hormone release (Gassanov et al. 2010; Pearce et al. 2003).

6.9.2 Lithium

Lithium is used mainly for the treatment of bipolar disorder. The drug reduces thyroid hormone production and secretion in the thyroid gland and seems to reduce deiodinase activity. Lithium has the ability to increase antithyroid antibody titers and seems to enhance antithyroid autoimmune activity. This can lead to subclinical and overt hypothyroidism especially in patients who have preexisting autoimmune thyroiditis. Furthermore, lithium reduces thyroid hormone activity, thus leading to central feedback mechanisms with elevated TSH levels. It has also been reported that lithium can have a direct toxic effect on thyroid cells leading to transient hyperthyroidism and hypothyroidism at a later time (Kibirige et al. 2013; Pearce et al. 2003).

6.9.3 Interferon-alpha

This drug is used for the treatment of chronic hepatitis C virus infection and is able to induce or exacerbate preexisting autoimmunity against the thyroid. About 5–15 % of patients without antithyroid antibodies prior to starting treatment develop antibodies during interferon-alpha treatment but show no clinical signs of a thyroid disorder. It is estimated that another 5–10 % of patients develop clinically manifested Hashimoto's thyroiditis, painless thyroiditis or Grave's disease during treatment (Narciso-Schiavon and Schiavon Lde 2015; Pearce et al. 2003).

6.9.4 Interleukin-2 and Other Cytokines

Up to 32 % of patients develop hypothyroidism during immunotherapy that is transient for most cases and rarely requires T4 replacement therapy. However, about 2 % of patients receiving interleukin-2 treatment develop a thyroiditis syndrome similar to painless thyroiditis. It is hypothesized that cytokines induce autoimmunity against the thyroid by causing thyroid follicular cells to present intracellular proteins and antigens via the MHC to T lymphocytes. A consequence of induced thyroiditis can be either a (normally transient) hyperthyroidism state or hypothyroidism development depending on the amount of thyroid destruction (Pearce et al. 2003; Schwartzentruber et al. 1991).

6.9.5 Tyrosine Kinase Inhibitors

According to a retrospective analysis, up to 85 % of patients treated with tyrosine kinase inhibitors develop hypothyroidism during treatment. Sunitinib seems to bear the greatest risk for thyroid dysfunction, along with other tyrosine kinase inhibitors that decrease thyroid function. The longer the treatment is necessary, the higher the probability of developing hypothyroidism. The cause is still not clear. It is thought to be an interaction with iodine metabolism, blood flow and TSH-mediated signaling and also might lead to the induction of a destructive thyroiditis that is caused by inducing and amplifying follicular cell apoptosis. The more intense the destruction, the more probable it is for a patient to develop a transient hyperthyroid phase because stored thyroid hormones are released. Further, patients who have already been treated with T4 seem to require higher doses after tyrosine kinase inhibitor exposure than they did before treatment (Torino et al. 2009).

References

Abduljabbar MA, Afifi AM (2012) Congenital hypothyroidism. J Pediatr Endocrinol Metab: JPEM 25(1–2):13–29. doi:10.1515/jpem.2011.408

Biondi B, Wartofsky L (2014) Treatment with thyroid hormone. Endocr Rev 35(3):433–512. doi:10.1210/er.2013-1083

Bizhanova A, Kopp P (2009) Minireview: the sodium-iodide symporter NIS and pendrin in iodide homeostasis of the thyroid. Endocrinology 150(3):1084–1090. doi:10.1210/en.2008-1437

Braun EM, Windisch G, Wolf G, Hausleitner L, Anderhuber F (2007) The pyramidal lobe: clinical anatomy and its importance in thyroid surgery. Surg Radiol Anat: SRA 29(1):21–27. doi:10.1007/s00276-006-0165-1

Brix K, Fuhrer D, Biebermann H (2011) Molecules important for thyroid hormone synthesis and action – known facts and future perspectives. Thyroid Res 4 Suppl 1:S9. doi:10.1186/1756-6614-4-S1-S9

Dietrich JW, Brisseau K, Boehm BO (2008) Absorption, transport and bio-availability of iodothyronines. Dtsch Med Wochenschr 133(31–32):1644–1648. doi:10.1055/s-0028-1082780

Fatourechi V, Aniszewski JP, Fatourechi GZ, Atkinson EJ, Jacobsen SJ (2003) Clinical features and outcome of subacute thyroiditis in an incidence cohort: Olmsted County, Minnesota, study. J Clin Endocrinol Metab 88(5):2100–2105. doi:10.1210/jc.2002-021799

Furmaniak J, Sanders J, Rees Smith B (2013) Blocking type TSH receptor antibodies. Auto Immun Highlights 4(1):11–26. doi:10.1007/s13317-012-0028-1

Gaitonde DY, Rowley KD, Sweeney LB (2012) Hypothyroidism: an update. Am Fam Physician 86(3):244–251

Garmendia Madariaga A, Santos Palacios S, Guillen-Grima F, Galofre JC (2014) The incidence and prevalence of thyroid dysfunction in Europe: a meta-analysis. J Clin Endocrinol Metab 99(3):923–931. doi:10.1210/jc.2013-2409

Gassanov N, Dietlein M, Caglayan E, Erdmann E, Er F (2010) Amiodaron-induzierte Schilddrüsenfunktionsstörungen (Amiodarone-induced thyroid gland dysfunctions). Dtsch Med Wochenschr 135(16):807–811. doi:10.1055/s-0030-1251935

Hoermann R, Midgley JE, Larisch R, Dietrich JW (2015) Homeostatic control of the thyroid-pituitary axis: perspectives for diagnosis and treatment. Front Endocrinol 6:177. doi:10.3389/fendo.2015.00177

Indra R, Patil SS, Joshi R, Pai M, Kalantri SP (2004) Accuracy of physical examination in the diagnosis of hypothyroidism: a cross-sectional, double-blind study. J Postgrad Med 50(1):7–11; discussion 11

Khandelwal D, Tandon N (2012) Overt and subclinical hypothyroidism: who to treat and how. Drugs 72(1):17–33. doi:10.2165/11598070-000000000-00000

Kibirige D, Luzinda K, Ssekitoleko R (2013) Spectrum of lithium induced thyroid abnormalities: a current perspective. Thyroid Res 6(1):3. doi:10.1186/1756-6614-6-3

Klein I, Danzi S (2016) Thyroid disease and the heart. Curr Probl Cardiol 41(2):65–92. doi:10.1016/j.cpcardiol.2015.04.002

Kohrle J (2015) Selenium and the thyroid. Curr Opin Endocrinol Diabetes Obes 22(5):392–401. doi:10.1097/med.0000000000000190

Krude H, Kuhnen P, Biebermann H (2015) Treatment of congenital thyroid dysfunction: achievements and challenges. Best Pract Res Clin Endocrinol Metab 29(3):399–413. doi:10.1016/j.beem.2015.04.004

Lo JC, Loh KC, Rubin AL, Cha I, Greenspan FS (1998) Riedel's thyroiditis presenting with hypothyroidism and hypoparathyroidism: dramatic response to glucocorticoid and thyroxine therapy. Clin Endocrinol (Oxf) 48(6):815–818

Meller J, Zappel H, Conrad M, Roth C, Emrich D, Becker W (1998) 123I-scintigraphy and perchlorate depletion test in the diagnosis of congenital hypothyroidism. Nuklearmedizin Nuclear Med 37(1):7–11

Miller FR (2003) Surgical anatomy of the thyroid and parathyroid glands. Otolaryngol Clin North Am 36(1):1–7, vii

Moslinger D, Frisch H, Strobl W, Stockler-Ipsiroglu S (1997) Neonatal screening for congenital hypothyroidism. Acta Med Austriaca 24(4):162–164

Narciso-Schiavon JL, Schiavon Lde L (2015) Autoantibodies in chronic hepatitis C: a clinical perspective. World J Hepatol 7(8):1074–1085. doi:10.4254/wjh.v7.i8.1074

Nicholson WK, Robinson KA, Smallridge RC, Ladenson PW, Powe NR (2006) Prevalence of postpartum thyroid dysfunction: a quantitative review. Thyroid : Off J Am Thyroid Assoc 16(6):573–582. doi:10.1089/thy.2006.16.573

Nikolai TF, Coombs GJ, McKenzie AK (1981) Lymphocytic thyroiditis with spontaneously resolving hyperthyroidism and subacute thyroiditis. Long-term follow-up. Arch Intern Med 141(11):1455–1458

Ohsako N, Tamai H, Sudo T, Mukuta T, Tanaka H, Kuma K, Kimura A, Sasazuki T (1995) Clinical characteristics of subacute thyroiditis classified according to human leukocyte antigen typing. J Clin Endocrinol Metab 80(12):3653–3656. doi:10.1210/jcem.80.12.8530615

Pearce EN, Farwell AP, Braverman LE (2003) Thyroiditis. N Engl J Med 348(26):2646–2655. doi:10.1056/NEJMra021194

Persani L (2012) Clinical review: central hypothyroidism: pathogenic, diagnostic, and therapeutic challenges. J Clin Endocrinol Metab 97(9):3068–3078. doi:10.1210/jc.2012-1616

Pollak A, Kasper DC (2014) Austrian newborn screening program: a perspective of five decades. J Perinat Med 42(2):151–158. doi:10.1515/jpm-2013-0113

Pyzik A, Grywalska E, Matyjaszek-Matuszek B, Rolinski J (2015) Immune disorders in Hashimoto's thyroiditis: what do we know so far? J Immunol Res 2015:979167. doi:10.1155/2015/979167

Rizzo LF, Mana DL, Bruno OD (2014) Non-autoimmune thyroiditis. Medicina 74(6):481–492

Samuels MH (2012) Subacute, silent, and postpartum thyroiditis. Med Clin North Am 96(2):223–233. doi:10.1016/j.mcna.2012.01.003

Schwartzentruber DJ, White DE, Zweig MH, Weintraub BD, Rosenberg SA (1991) Thyroid dysfunction associated with immunotherapy for patients with cancer. Cancer 68(11):2384–2390

Schweizer U, Steegborn C (2015) New insights into the structure and mechanism of iodothyronine deiodinases. J Mol Endocrinol 55(3):R37–R52. doi:10.1530/JME-15-0156

Sfakianakis GN, Ezuddin SH, Sanchez JE, Eidson M, Cleveland W (1999) Pertechnetate scintigraphy in primary congenital hypothyroidism. J Nucl Med: Off Publ Soc Nucl Med 40(5):799–804

Torino F, Corsello SM, Longo R, Barnabei A, Gasparini G (2009) Hypothyroidism related to tyrosine kinase inhibitors: an emerging toxic effect of targeted therapy. Nat Rev Clin Oncol 6(4):219–228

Twyffels L, Massart C, Golstein PE, Raspe E, Van Sande J, Dumont JE, Beauwens R, Kruys V (2011) Pendrin: the thyrocyte apical membrane iodide transporter? Cell Physiol Biochem: Int J Exp Cell Physiol Biochem Pharmacol 28(3):491–496. doi:10.1159/000335110

Vanderpump MP, Tunbridge WM, French JM, Appleton D, Bates D, Clark F, Grimley Evans J, Hasan DM, Rodgers H, Tunbridge F et al (1995) The incidence of thyroid disorders in the community: a twenty-year follow-up of the Whickham Survey. Clin Endocrinol (Oxf) 43(1):55–68

WHO Guidelines Approved by the Guidelines Review Committee (2014) In: Guideline: fortification of food-grade salt with iodine for the prevention and control of iodine deficiency disorders. World Health Organization Copyright (c) World Health Organization 2014, Geneva

Zimmermann AW, Weber MM (2012) Thyroid diseases during pregnancy. J Gynäkol Endokrinol 22 (01):6–12

Zimmermann MB, Boelaert K (2015) Iodine deficiency and thyroid disorders. Lancet Diabetes Endocrinol 3(4):286–295. doi:10.1016/S2213-8587(14)70225-6

The Controversial Role of Pendrin in Thyroid Cell Function and in the Thyroid Phenotype in Pendred Syndrome

7

Peter Kopp, Aigerim Bizhanova, and Liuska Pesce

Abstract

Thyroid hormones are essential for normal development, growth and differentiation of numerous tissues, and metabolic regulation. Structurally, they are unique because they contain iodine. Their synthesis in thyroid follicles thus requires a sufficient nutritional iodide intake, transport into the thyroid cells, and efflux into the follicular lumen where the actual biosynthesis occurs. Historically, Pendred syndrome has been defined by the triad of sensorineural deafness/hearing impairment in combination with goiter and an abnormal organification of iodide. After the identification of the molecular basis of Pendred syndrome, which is caused by biallelic mutations in the *SLC26A4/PDS* gene, functional studies revealed that pendrin is a multifunctional anion exchanger with affinity, among others, for chloride, iodide, and bicarbonate. This observation, together with the demonstration of pendrin protein expression at the apical membrane of thyrocytes, led to the hypothesis that pendrin might be involved in the efflux of iodide into the follicular lumen. Several experimental observations do indeed support a potential role of pendrin in mediating iodide efflux. However, iodide efflux is also possible in the absence of pendrin, and *Slc26a4 −/−* knockout mice do not have a thyroidal phenotype. These findings indicate that other exchangers or channels have a redun-

P. Kopp (✉)
Division of Endocrinology, Metabolism and Molecular Medicine, Department of Internal Medicine, Feinberg School of Medicine, Northwestern University, Chicago, IL 60611, USA
e-mail: p-kopp@northwestern.edu

A. Bizhanova
Harold Washington College, 30 E Lake St, Chicago, IL 60601, USA
e-mail: abizhanova@ccc.edu

L. Pesce
Stead Family Department of Pediatrics, Division of Pediatric Endocrinology and Diabetes, University of Iowa Carver School of Medicine, Iowa City, IA 52246, USA
e-mail: liuska-pesce@uiowa.edu

S. Dossena, M. Paulmichl (eds.), *The Role of Pendrin in Health and Disease*,
DOI 10.1007/978-3-319-43287-8_7

dant or perhaps predominant function. A potential candidate is anoctamin 1 (ANO1/TMEM16A), a calcium-activated anion channel. Anoctamin is also expressed at the apical membrane of thyrocytes, and it has affinity for iodide.

Further studies are needed in order to define the relative physiological role of pendrin and anoctamin in mediating iodide efflux, to characterize their affinity for iodide, and to analyze their species-specific expression pattern.

7.1 Introduction

Thyroid hormones are essential for normal development, differentiation, and metabolism of the majority of organs. Their synthesis requires intact follicles, which form the functional units of the gland, several regulated biochemical steps, and an adequate nutritional iodide uptake (Kopp 2012; Pesce and Kopp 2014). At the basolateral membrane of thyroid follicular cells, iodide uptake is mediated by the sodium-iodide symporter (NIS) (Portulano et al. 2014). The function of NIS is dependent on the sodium gradient generated by the Na^+/K^+-ATPase (Fig. 7.1), and a constitutively active potassium channel consisting of the KCNQ1 and KCNE2 subunits, which promotes K^+ efflux (Frohlich et al. 2011; Roepke et al. 2009). At the apical membrane, iodide enters the follicular lumen where the actual thyroid hormone synthesis occurs. Within the follicular lumen, iodide is oxidized by the membrane-bound heme enzyme thyroid peroxidase. The oxidation requires the presence of hydrogen peroxide (H_2O_2), which is generated by the dual oxidase (DUOX) system (Moreno and Visser 2007). Subsequently, oxidized iodide is organified into defined tyrosyl residues of thyroglobulin, which serves as the scaffold for thyroid hormone synthesis. Thyroglobulin is a heavily glycosylated and very large protein (330 kDa) that forms dimers (Di Jeso and Arvan 2016). In a first step, referred to as *organification*, the iodination leads to the formation of mono- and diiodotyrosines (MIT, DIT). A donor and an acceptor iodotyrosine are then fused in the coupling reaction, which generates T4 or T3. Iodinated thyroglobulin is digested by several endopeptidases, both in the follicular lumen and after uptake into the cells through macro- and micropinocytosis (Di Jeso and Arvan 2016). T4 and T3 are secreted into the bloodstream at the basolateral membrane, at least in part by the monocarboxylate transporter MCT8 (Di Cosmo et al. 2010). Remarkably, MIT and DIT are sorted and then deiodinated by an intracellular iodotyrosine dehalogenase (DEHAL1), which permits to recycle the released iodide into the follicular lumen (Moreno et al. 2008; Kopp 2012).

7.2 How Does Iodide Enter the Follicular Lumen?

In electrophysiological studies performed more than two decades ago with inverted thyroid membrane vesicles, Golstein et al. suggested that iodide efflux is mediated by two distinct channels (Golstein et al. 1992). The first one, referred to as the

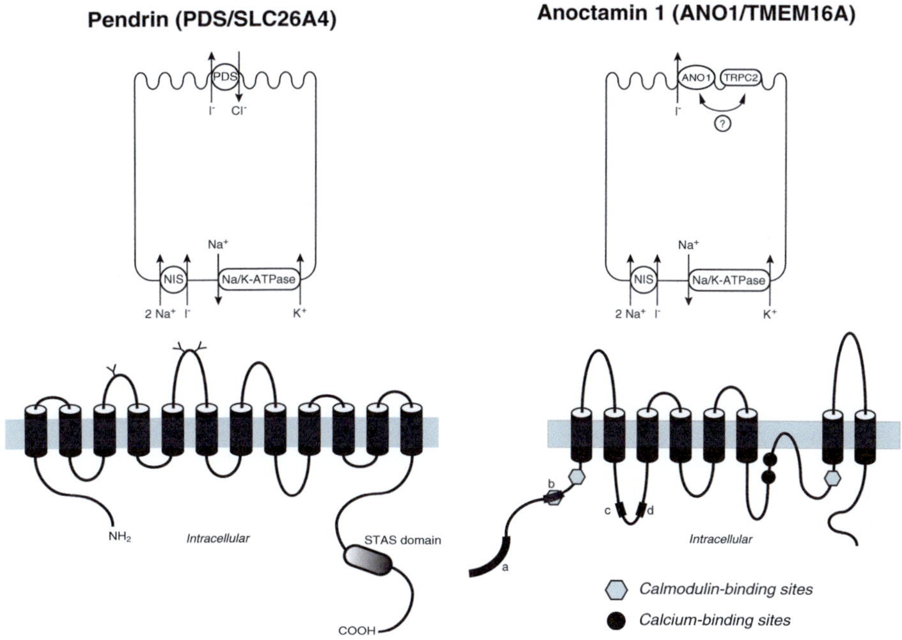

Fig. 7.1 Cellular localization (*top*) and putative structure (*bottom*) of pendrin (PDS/SLC26A4) and anoctamin 1 (ANO1/TMEM16A). Pendrin contains a carboxyterminal STAS (sulfate transporter and antisigma factor antagonist) domain that contains a protein kinase A phosphorylation site. The predicted calmodulin- and calcium-binding sites of ANO1 are indicated. The sodium-iodide symporter (NIS) transports iodide into the follicular cells at the basolateral membrane against the electrochemical gradient. NIS is dependent on the sodium gradient generated by the Na^+/K^+-ATPase. At the apical membrane, pendrin is thought to function as a iodide/chloride exchanger, and anoctamin is a channel that may function in conjunction with other channels such as the transient receptor potential channel 2 (TRPC2)

iodide channel, exhibited a high permeability and specificity for iodide with an approximate Km of 70 μM. The second channel was found to be about fourfold more permeable to iodide than chloride with a Km of about 33 mM. The authors postulated that the iodide channel is restricted to the apical membrane and transports iodide from the cytosol into the colloid space, whereas the second one may mediate predominantly chloride transport under physiological conditions. So far, the molecular identity of these two conductances, which could be anion channels or transporters, has however not been identified. Currently, two candidates have been considered: pendrin and anoctamin (Silveira and Kopp 2015) (Table 7.1 and Fig. 7.1). Chloride channels such as the cystic fibrosis transmembrane conductance regulator (CFTR) and the voltage-gated chloride channel 5 (CLCN5) are also expressed in thyroid follicular cells (Li et al. 2010; van den Hove et al. 2006). While they have affinity for iodide, they are not thought to play a role in mediating iodide efflux under physiological conditions. SLC5A8, a homologue of NIS, which was initially called human apical iodide transporter (hAIT) (Rodriguez et al. 2002), is

Table 7.1 Observations for and against a physiological role of pendrin and anoctamin as mediators of iodide efflux in thyroid follicular cells

Pendrin (SLC26A4)	Anoctamin 1 (ANO1, TMEM16A)
Findings supporting a physiological role of pendrin	*Findings supporting a physiological role of anoctamin*
Patients with Pendred syndrome have a partial iodide organification defect	ANO1 is abundantly expressed at the apical membrane of thyroid follicular cells
Patients with Pendred syndrome can develop a goiter	ANO1 expression is regulated by TSH
Patients with Pendred syndrome can present with congenital or acquired hypothyroidism	ANO1 can be stimulated by calcium
Pendrin is a multifunctional anion exchanger abundantly expressed at the apical membrane of thyroid follicular cells	ANO1 mediates efflux of iodide in transfected cells
Pendrin mediates iodide transport in transfected non-polarized cells	Inhibition or siRNA knockdown of ANO1 decreases iodide efflux from thyroid cells
Pendrin mediates vectorial iodide efflux in transfected polarized cells	ANO1 splice variants expressed in the thyroid have a higher affinity for iodide than chloride
Mutations of pendrin lose the ability to mediate iodide efflux	Studies in oocytes have demonstrated that ANO1 is a calcium-activated chloride channel with a preference for iodide over chloride
Pendrin has a higher affinity for iodide than for chloride	
TSH mediates rapid translocation of pendrin to the apical membrane and increases iodide efflux	
Membrane abundance and half-life increased under conditions of iodide excess	
Findings possibly questioning a primary physiological role of pendrin	*Findings possibly questioning a primary physiological role of ANO1*
TSH does not regulate *PDS/SLC26A4* gene expression	ANO1 may function in conjunction with other channels such as the transient receptor potential channel 2 (TRPC2)
Slc26a4 −/− mice do not display a thyroidal phenotype *This could, however, simply reflect a species difference*	In transfected HEK293 cells expressing pendrin, the iodide efflux is higher compared to cells expressing ANO1 (Fig. 7.2)
Pendrin has a distinct role as an exchanger of chloride and bicarbonate in other tissues such as the kidneys and inner ear	

siRNA small interfering ribonucleic acid, *HEK293* human embryonic kidney cell line 293

clearly not involved in apical iodide efflux as formally demonstrated in functional studies in oocytes and polarized MDCK cells (Paroder et al. 2006).

Thyroid-stimulating hormone (TSH) rapidly activates the efflux of iodide efflux at the apical membrane through the protein kinase A and C pathways (Weiss et al. 1984; Nilsson et al. 1990, 1992; Iosco et al. 2014). These pathways may be differentially regulated depending on the physiological status.

7.3 Pendrin

The classic phenotype of the autosomal recessive Pendred syndrome (OMIM #274600) consists of sensorineural deafness associated with inner ear malformations, especially enlarged vestibular aqueduct (EVA), goiter, and a partial iodide organification defect (PIOD) (Pendred 1896; Morgans and Trotter 1958; Bizhanova and Kopp 2010). Pendred syndrome is caused by biallelic (homozygous and compound heterozygous) mutations in the *PDS/SLC26A4* gene (Everett et al. 1997). Functionally, pendrin belongs to the SLC26 family of multifunctional anion transporters (Alper and Sharma 2013), and it was found to have affinity for iodide (Scott et al. 1999).

Because of the human phenotype with a partial iodide organification defect (PIOD), goiter, and in a subset of subjects congenital or acquired hypothyroidism (Gonzalez Trevino et al. 2001; Ladsous et al. 2014), the expression of pendrin at the apical membrane of thyroid follicular cells (Royaux et al. 2000) (Fig. 7.1), and its affinity for iodide (Scott et al. 1999), it is a plausible candidate for mediating iodide efflux at the apical membrane.

Initial studies in *Xenopus* oocytes demonstrated that pendrin is able to mediate uptake of anions such as chloride and iodide in a sodium-independent manner (Scott et al. 1999). In transfected unpolarized mammalian cells, pendrin was then shown to mediate iodide release (Yoshida et al. 2002). Next, studies performed in polarized cells expressing NIS at the basolateral membrane and pendrin at the apical membrane in a bicameral system demonstrated that pendrin can mediate vectorial iodide efflux at the apical membrane (Gillam et al. 2004). Importantly, more recent studies performed in oocytes have shown that pendrin functions as a coupled, electroneutral iodide/chloride, iodide/bicarbonate, or chloride/bicarbonate exchanger with a 1:1 stoichiometry, and that it has a preferential affinity for iodide, even in the presence of high chloride concentrations (Shcheynikov et al. 2008). Moreover, it has been shown that pendrin is expressed at the apical membrane of parotid gland ducts where it can mediate luminal iodide secretion (Shcheynikov et al. 2008).

Some of the more than hundred naturally occurring mutations have been tested functionally, mainly after transfection into heterologous mammalian cells. Disease-causing mutations result in a complete or partial loss in iodide efflux (Gillam et al. 2004; Taylor et al. 2002; Pera et al. 2008; Dossena et al. 2009, 2011), and many mutated proteins are retained in intracellular compartments such as the endoplasmic reticulum secondary to misfolding (Rotman-Pikielny et al. 2002). The human phenotype, which is characterized by goiter development under conditions of scarce iodide intake (Gonzalez Trevino et al. 2001), as well as the PIOD, is suggestive for a physiological role of pendrin in mediating or participating in iodide efflux in humans. In contrast, however, *Slc26a4 −/−* mice do not develop a goiter or abnormal thyroid hormone levels, even under conditions of iodine deficiency (Calebiro et al. 2011; Iwata et al. 2011). Although this has been used as an argument against a physiological role of pendrin in mediating iodide efflux in thyrocytes (Twyffels et al. 2011), it is currently unclear whether this simply reflects a species difference (Bizhanova and Kopp 2011).

Iodide efflux at the apical membrane is rapidly accelerated by TSH (Nilsson et al. 1990, 1992; Weiss et al. 1984). Stimulation of the protein kinase A pathway in rat thyroid cells results in a rapid increase in membrane insertion of pendrin and an increased iodide efflux (Bizhanova et al. 2011; Pesce et al. 2012). Stimulation with forskolin increases the membrane insertion of pendrin in PCCL3 rat thyroid cells, presumably through activation of a protein kinase A site in the intracellular carboxyterminus (Bizhanova et al. 2011). Deletion or targeted mutation of the protein kinase A site residing in the intracellular carboxyterminal tail containing the so-called STAS (sulfate transporter and antisigma factor antagonist) domain results in decreased basal function and membrane insertion of pendrin (Bizhanova et al. 2011). In addition, the protein kinase A site mutation (T717A) is partially functional, but it has a mitigated response to forskolin (Bizhanova et al. 2011). Trafficking of murine pendrin to the apical membrane is also activated by cAMP in microperfused mouse cortical collecting duct (CCD) and in polarized renal opossum kidney proximal tubule (OKP) cells via phosphorylation of S49 in the aminoterminal intracellular domain (Azroyan et al. 2012). In contrast to the rapid translocation mediated by the protein kinase A pathway, stimulation of the protein kinase C pathway in rat thyroid cells appears to result in a delayed translocation of pendrin to the plasma membrane (Muscella et al. 2008). Interestingly, rat thyroid cells incubated with supraphysiological amounts of iodide show an increased abundance of pendrin at the plasma membrane, the half-life of pendrin increases, and the efflux of iodide is enhanced (Calil-Silveira et al. 2016). These findings suggest that pendrin could have a role in mediating iodide efflux under conditions of iodide excess.

Under conditions of normal or abundant iodide intake, goiter development is unusual (Sato et al. 2001), but goitrous congenital and overt hypothyroidism developing later in life can be present in patients with Pendred syndrome (Gonzalez Trevino et al. 2001; Ladsous et al. 2014). In a study from Northern France, a region which has a normal to marginal iodide intake, Ladsous et al. characterized the thyroid phenotype in patients with Pendred syndrome and non-syndromic EVA (Ladsous et al. 2014). Fifteen out of the 19 patients with Pendred syndrome (79 %) presented with a goiter. Fifteen (79 %) subjects had hypothyroidism: 6/15 had congenital hypothyroidism, 5/15 had overt hypothyroidism, and 4/15 had subclinical hypothyroidism. Ten out of 16 (63 %) of these patients showed abnormal iodide organification as determined by a perchlorate test, a test that is unfortunately poorly standardized and subject to exogenous (e.g., iodine intake) and endogenous (e.g., autoimmune thyroid disease) modulators. The study by Ladsous et al. clearly demonstrates that there is a relatively wide spectrum in the thyroid phenotype among patients with Pendred syndrome, suggesting that it is influenced by genetic and environmental modifiers, including nutritional iodine intake (Ladsous et al. 2014). Intriguingly, biallelic mutations in the *SLC26A4* gene have also been identified in two patients with thyroid hypoplasia and congenital hypothyroidism from two unrelated families (Kühnen et al. 2014). The mutations found in these subjects have been previously identified in patients with the classical form of Pendred syndrome or familial EVA (Kühnen et al. 2014; Kopp 2014). The reasons why these two patients developed thyroid hypoplasia, rather than a goitrous phenotype, are unclear. It has been speculated that either the retained misfolded proteins or an increased production of free radicals in response to sustained stimulation by TSH could have a toxic effect leading to cell death, or that the hypoplastic phenotype requires the presence of additional modifying (genetic) factors (Kopp 2014; Kühnen et al. 2014).

7.4 Anoctamin

Three recent studies have suggested that anoctamin 1 (ANO1), also referred to as TMEM16A, could be involved in apical iodide efflux in thyroid cells (Viitanen et al. 2013; Iosco et al. 2014; Twyffels et al. 2014). ANO1 is a calcium-activated anion channel, which is expressed in numerous tissues, including thyroid follicular cells (Pedemonte and Galietta 2014; Ferrera et al. 2010). ANO1 is part of a family of ten paralogs (ANO1-10; TMEM16A-K) sharing a common transmembrane topology, but a wide spectrum of in part putative functional roles as ion channels, regulatory subunits of other channels or phospholipid scramblases, proteins responsible for the translocation of phospholipids between the two monolayers of the cell membrane lipid bilayer (Pedemonte and Galietta 2014; Picollo et al. 2015). The *ANO1* gene generates several splice variants, and most of them have a higher affinity for iodide than chloride (Ferrera et al. 2010). Human and rat thyroid cells predominantly express the so-called *abc* and the *ac* isoforms (Ferrera et al. 2009; Iosco et al. 2014), whereas the rat thyroid cell lines PCCL3 and FRTL-5 predominantly express the *ac* isoform, which is more sensitive to calcium (Ferrera et al. 2009, 2010). A functional study by Viitanen et al. performed in native FRTL-5 rat thyroid cells suggested that ANO1, in conjunction with the transient receptor potential channel 2 (TRPC2), mediates iodide release (Viitanen et al. 2013). Iosco et al. then demonstrated that the ANO1 protein is localized at the apical membrane of human thyrocytes (Fig. 7.1), and that its expression is more abundant in active cells (Iosco et al. 2014). Functional studies determining the intracellular iodide content revealed that iodide release can be stimulated by adenosine triphosphate (ATP) in a calcium-dependent manner from FRTL-5 cells, whereas treatment with inhibitors or siRNA knockdown decreased the iodide efflux. Similarly, iodide efflux was also increased in transfected mammalian cells expressing both NIS and ANO1, and iodide release could be further stimulated by calcium (Iosco et al. 2014). Twyffels et al. demonstrated that *Ano1* mRNA is stimulated by TSH and that the protein expression, which is relatively discrete under basal conditions, increases after stimulation by TSH (Twyffels et al. 2014). In unpolarized human embryonic kidney (HEK293) cells transfected with ANO1, the efflux of iodide is increased compared to untransfected cells. However, in HEK293 cells expressing pendrin, the iodide efflux is higher compared to ANO1 transfected cells, suggesting that pendrin is more efficient in mediating iodide efflux, at least in this model system (Fig. 7.2). Treatment with the calcium ionophore ionomycin was found to acutely stimulate the ANO1-mediated iodide

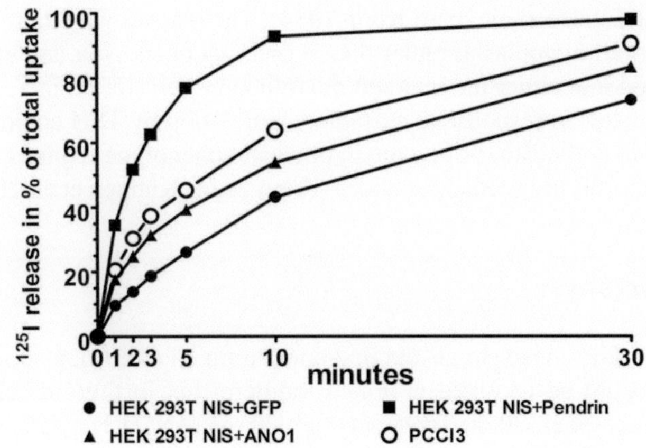

Fig. 7.2 Release of radioiodide from ^{125}I-preloaded PCCl3 cells (○) or HEK 293 T cells transfected to express Na^+/I^- symporter (*NIS*) in combination with *GFP* (●) or *ANO1* (▲) or *pendrin* (■). The data are representative of more than five independent experiments (From Ref. Twyffels et al. (2014). With permission). Note that the release of iodide is more efficient in these pendrin-transfected cells compared to ANO1-transfected cells

release (Twyffels et al. 2014). The mechanism by which calcium activates ANO1 seems to involve calmodulin as well as direct calcium binding to ANO1 (Pedemonte and Galietta 2014). Rat thyroid cell lines and human primary thyroid cells treated either with an ANO1 inhibitor (T16Ainh-A01) or siRNA show a decrease in iodide release (Twyffels et al. 2014). Studies in oocytes have demonstrated that ANO1 is a calcium-activated chloride channel with a preference for iodide over chloride (Schroeder et al. 2008; Yang et al. 2008) indirectly suggesting that it could be involved in mediating iodide efflux. ANO1 is able to mediate iodide efflux from FRTL-5 cells after ATP stimulation in the absence of chloride, which suggests that it functions independently of pendrin, which functions as an anion exchanger. In aggregate, these results indicate that ANO1, which has a preferential affinity for iodide over chloride (Schroeder et al. 2008; Yang et al. 2008), is able to mediate iodide release from thyroid cells.

7.5 Future Directions

In conclusion, the current body of data suggests that the multianion exchanger pendrin (PDS/SLC26A4) and the calcium-dependent channel anoctamin 1 (ANO1/TMEM16A) can mediate apical iodide efflux in thyroid cells and several model systems. It is conceivable that they are part of a redundant system. The exact physiological role of pendrin and ANO1 awaits further characterization, and it may be variable between basal conditions and conditions of thyroid dysfunction as illustrated by the regulation of ANO1 expression by TSH, and the differential regulation of pendrin trafficking by the protein kinase A and C pathways. As suggested by the absence of a thyroid phenotype in *Slc26a4 −/−* knockout mice, which differs with the human phenotype that can include goiter and congenital or acquired hypothyroidism, there may be relevant differences in the expression pattern and physiological roles of pendrin and ANO1 between species.

References

Alper SL, Sharma AK (2013) The SLC26 gene family of anion transporters and channels. Mol Aspects Med 34(2–3):494–515. doi:10.1016/j.mam.2012.07.009

Azroyan A, Morla L, Crambert G, Laghmani K, Ramakrishnan S, Edwards A, Doucet A (2012) Regulation of pendrin by cAMP: possible involvement in beta-adrenergic-dependent NaCl retention. Am J Physiol Renal Physiol 302(9):F1180–F1187. doi:10.1152/ajprenal.00403.2011

Bizhanova A, Kopp P (2010) Genetics and phenomics of Pendred syndrome. Mol Cell Endocrinol 322(1–2):83–90. doi:10.1016/j.mce.2010.03.006

Bizhanova A, Kopp P (2011) Controversies concerning the role of pendrin as an apical iodide transporter in thyroid follicular cells. Cell Physiol Biochem 28(3):485–490. doi:10.1159/000335103

Bizhanova A, Chew T, Khuon S, Kopp P (2011) Analysis of cellular localization and function of carboxy-terminal truncation mutants of pendrin. Cell Physiol Biochem 28:423–434

Calebiro D, Porazzi P, Bonomi M, Lisi S, Grindati A, De Nittis D, Fugazzola L, Marino M, Botta G, Persani L (2011) Absence of primary hypothyroidism and goiter in *Slc26a4 (−/−)* mice fed on a low iodine diet. J Endocrinol Invest 34:593–598. doi:10.3275/7262, 7262 [pii]

Calil-Silveira J, Serrano-Nascimento C, Kopp P, Nunes MT (2016) Iodide excess regulates its own efflux: a possible involvement of pendrin. Am J Physiol Cell Physiol ajpcell 00210:02015. doi:10.1152/ajpcell.00210.2015

Di Cosmo C, Liao XH, Dumitrescu AM, Philp NJ, Weiss RE, Refetoff S (2010) Mice deficient in MCT8 reveal a mechanism regulating thyroid hormone secretion. J Clin Invest 120(9):3377–3388. doi:10.1172/JCI42113, 42113 [pii]

Di Jeso B, Arvan P (2016) Thyroglobulin from molecular and cellular biology to clinical endocrinology. Endocr Rev 37(1):2–36. doi:10.1210/er.2015-1090

Dossena S, Rodighiero S, Vezzoli V, Nofziger C, Salvioni E, Boccazzi M, Grabmayer E, Botta G, Meyer G, Fugazzola L, Beck-Peccoz P, Paulmichl M (2009) Functional characterization of wild-type and mutated pendrin (SLC26A4), the anion transporter involved in Pendred syndrome. J Mol Endocrinol 43(3):93–103. doi:10.1677/JME-08-0175, JME-08-0175 [pii]

Dossena S, Bizhanova A, Nofziger C, Bernardinelli E, Ramsauer J, Kopp P, Paulmichl M (2011) Identification of allelic variants of pendrin (SLC26A4) with loss and gain of function. Cell Physiol Biochem 28:467–476

Everett LA, Glaser B, Beck JC, Idol JR, Buchs A, Heyman M, Adawi F, Hazani E, Nassir E, Baxevanis AD, Sheffield VC, Green ED (1997) Pendred syndrome is caused by mutations in a putative sulphate transporter gene *(PDS)*. Nat Genet 17(4):411–422. doi:10.1038/ng1297-411

Ferrera L, Caputo A, Ubby I, Bussani E, Zegarra-Moran O, Ravazzolo R, Pagani F, Galietta LJ (2009) Regulation of TMEM16A chloride channel properties by alternative splicing. J Biol Chem 284(48):33360–33368. doi:10.1074/jbc.M109.046607

Ferrera L, Caputo A, Galietta LJ (2010) TMEM16A protein: a new identity for Ca(2+)-dependent Cl(−) channels. Physiology (Bethesda) 25(6):357–363. doi:10.1152/physiol.00030.2010

Frohlich H, Boini KM, Seebohm G, Strutz-Seebohm N, Ureche ON, Foller M, Eichenmuller M, Shumilina E, Pathare G, Singh AK, Seidler U, Pfeifer KE, Lang F (2011) Hypothyroidism of gene-targeted mice lacking *Kcnq1*. Pflugers Arch 461(1):45–52. doi:10.1007/s00424-010-0890-5

Gillam MP, Sidhaye AR, Lee EJ, Rutishauser J, Stephan CW, Kopp P (2004) Functional characterization of pendrin in a polarized cell system. Evidence for pendrin-mediated apical iodide efflux. J Biol Chem 279(13):13004–13010. doi:10.1074/jbc.M313648200, M313648200 [pii]

Golstein P, Abramow M, Dumont JE, Beauwens R (1992) The iodide channel of the thyroid: a plasma membrane vesicle study. Am J Physiol 263(3 Pt 1):C590–C597

Gonzalez Trevino O, Karamanoglu Arseven O, Ceballos CJ, Vives VI, Ramirez RC, Gomez VV, Medeiros-Neto G, Kopp P (2001) Clinical and molecular analysis of three Mexican families with Pendred's syndrome. Eur J Endocrinol 144(6):585–593, doi:1440585 [pii]

Iosco C, Cosentino C, Sirna L, Romano R, Cursano S, Mongia A, Pompeo G, di Bernardo J, Ceccarelli C, Tallini G, Rhoden KJ (2014) Anoctamin 1 is apically expressed on thyroid follicular cells and contributes to ATP- and calcium-activated iodide efflux. Cell Physiol Biochem 34(3):966–980. doi:10.1159/000366313

Iwata T, Yoshida T, Teranishi M, Murata Y, Hayashi Y, Kanou Y, Griffith AJ, Nakashima T (2011) Influence of dietary iodine deficiency on the thyroid gland in *Slc26a4*-null mutant mice. Thyroid Res 4(1):1–6. doi:10.1186/1756-6614-4-10

Kopp P (2012) Thyroid hormone synthesis: thyroid iodine metabolism. In: Braverman L, Cooper D (eds) Wegner and Ingbar's the thyroid: a fundamental and clinical text, 10th edn. Lippincott, Williams & Wilkins, Philadelphia, pp 48–74

Kopp P (2014) Mutations in the Pendred Syndrome *(PDS/SLC26A)* gene: an increasingly complex phenotypic spectrum from goiter to thyroid hypoplasia. J Clin Endocrinol Metab 99(1):67–69. doi:10.1210/jc.2013-4319

Kühnen P, Turan S, Fröhler S, Güran T, Abali S, Biebermann H, Bereket A, Grüters A, Chen W, Krude H (2014) Identification of PENDRIN *(SLC26A4)* mutations in patients with congenital hypothyroidism and "apparent" thyroid dysgenesis. J Clin Endocrinol Metab 99(1):E169–E176. doi:10.1210/jc.2013-2619

Ladsous M, Vlaeminck-Guillem V, Dumur V, Vincent C, Dubrulle F, Dhaenens CM, Wemeau JL (2014) Analysis of the thyroid phenotype in 42 patients with Pendred syndrome and nonsyndromic enlargement of the vestibular aqueduct. Thyroid 24(4):639–648. doi:10.1089/thy.2013.0164

Li H, Ganta S, Fong P (2010) Altered ion transport by thyroid epithelia from *CFTR(−/−)* pigs suggests mechanisms for hypothyroidism in cystic fibrosis. Exp Physiol 95(12):1132–1144. doi:10.1113/expphysiol.2010.054700

Moreno JC, Visser TJ (2007) New phenotypes in thyroid dyshormonogenesis: hypothyroidism due to *DUOX2* mutations. Endocr Dev 10:99–117. doi:10.1159/0000106822

Moreno JC, Klootwijk W, van Toor H, Pinto G, D'Alessandro M, Leger A, Goudie D, Polak M, Gruters A, Visser TJ (2008) Mutations in the iodotyrosine deiodinase gene and hypothyroidism. N Engl J Med 358(17):1811–1818. doi:10.1056/NEJMoa0706819, 358/17/1811 [pii]

Morgans ME, Trotter WR (1958) Association of congenital deafness with goitre; the nature of the thyroid defect. Lancet 1(7021):607–609

Muscella A, Marsigliante S, Verri T, Urso L, Dimitri C, Botta G, Paulmichl M, Beck-Peccoz P, Fugazzola L, Storelli C (2008) PKC-epsilon-dependent cytosol-to-membrane translocation of pendrin in rat thyroid PC Cl3 cells. J Cell Physiol 217(1):103–112. doi:10.1002/jcp.21478

Nilsson M, Bjorkman U, Ekholm R, Ericson LE (1990) Iodide transport in primary cultured thyroid follicle cells: evidence of a TSH-regulated channel mediating iodide efflux selectively across the apical domain of the plasma membrane. Eur J Cell Biol 52(2):270–281

Nilsson M, Bjorkman U, Ekholm R, Ericson LE (1992) Polarized efflux of iodide in porcine thyrocytes occurs via a cAMP-regulated iodide channel in the apical plasma membrane. Acta Endocrinol 126(1):67–74

Paroder V, Spencer SR, Paroder M, Arango D, Schwartz S Jr, Mariadason JM, Augenlicht LH, Eskandari S, Carrasco N (2006) Na(+)/monocarboxylate transport (SMCT) protein expression correlates with survival in colon cancer: molecular characterization of SMCT. Proc Natl Acad Sci U S A 103(19):7270–7275. doi:10.1073/pnas.0602365103, 0602365103 [pii]

Pedemonte N, Galietta LJ (2014) Structure and function of TMEM16 proteins (anoctamins). Physiol Rev 94(2):419–459. doi:10.1152/physrev.00039.2011

Pendred V (1896) Deaf-mutism and goitre. Lancet ii:532

Pera A, Dossena S, Rodighiero S, Gandia M, Botta G, Meyer G, Moreno F, Nofziger C, Hernandez-Chico C, Paulmichl M (2008) Functional assessment of allelic variants in the *SLC26A4* gene involved in Pendred syndrome and nonsyndromic EVA. Proc Natl Acad Sci U S A 105(47):18608–18613. doi:10.1073/pnas.0805831105, 0805831105 [pii]

Pesce L, Kopp P (2014) Iodide transport: implications for health and disease. Int J Pediatr Endocrinol 8:1–12. doi:10.1186/1687-9856-2014-8

Pesce L, Bizhanova A, Caraballo JC, Westphal W, Butti ML, Comellas A, Kopp P (2012) TSH regulates pendrin membrane abundance and enhances iodide efflux in thyroid cells. Endocrinology 153(1):512–521. doi:10.1210/en.2011-1548

Picollo A, Malvezzi M, Accardi A (2015) TMEM16 proteins: unknown structure and confusing functions. J Mol Biol 427(1):94–105. doi:10.1016/j.jmb.2014.09.028

Portulano C, Paroder-Belenitsky M, Carrasco N (2014) The Na+/I- symporter (NIS): mechanism and medical impact. Endocr Rev 35(1):106–149. doi:10.1210/er.2012-1036

Rodriguez AM, Perron B, Lacroix L, Caillou B, Leblanc G, Schlumberger M, Bidart JM, Pourcher T (2002) Identification and characterization of a putative human iodide transporter located at the apical membrane of thyrocytes. J Clin Endocrinol Metab 87(7):3500–3503

Roepke TK, King EC, Reyna-Neyra A, Paroder M, Purtell K, Koba W, Fine E, Lerner DJ, Carrasco N, Abbott GW (2009) *Kcne2* deletion uncovers its crucial role in thyroid hormone biosynthesis. Nat Med 15(10):1186–1194. doi:10.1038/nm.2029

Rotman-Pikielny P, Hirschberg K, Maruvada P, Suzuki K, Royaux IE, Green ED, Kohn LD, Lippincott-Schwartz J, Yen PM (2002) Retention of pendrin in the endoplasmic reticulum is a major mechanism for Pendred syndrome. Hum Mol Genet 11(21):2625–2633

Royaux IE, Suzuki K, Mori A, Katoh R, Everett LA, Kohn LD, Green ED (2000) Pendrin, the protein encoded by the Pendred syndrome gene *(PDS)*, is an apical porter of iodide in the thyroid and is regulated by thyroglobulin in FRTL-5 cells. Endocrinology 141(2):839–845

Sato E, Nakashima T, Miura Y, Furuhashi A, Nakayama A, Mori N, Murakami H, Naganawa S, Tadokoro M (2001) Phenotypes associated with replacement of His by Arg in the Pendred syndrome gene. Eur J Endocrinol 145(6):697–703, doi:145697 [pii]

Schroeder BC, Cheng T, Jan YN, Jan LY (2008) Expression cloning of TMEM16A as a calcium-activated chloride channel subunit. Cell 134(6):1019–1029. doi:10.1016/j.cell.2008.09.003

Scott DA, Wang R, Kreman TM, Sheffield VC, Karniski LP (1999) The Pendred syndrome gene encodes a chloride-iodide transport protein. Nat Genet 21(4):440–443. doi:10.1038/7783

Shcheynikov N, Yang D, Wang Y, Zeng W, Karniski LP, So I, Wall SM, Muallem S (2008) The Slc26a4 transporter functions as an electroneutral Cl-/I-/HCO3- exchanger: role of Slc26a4 and Slc26a6 in I- and HCO3- secretion and in regulation of CFTR in the parotid duct. J Physiol 586(16):3813–3824. doi:10.1113/jphysiol.2008.154468

Silveira JC, Kopp PA (2015) Pendrin and anoctamin as mediators of apical iodide efflux in thyroid cells. Curr Opin Endocrinol Diabetes Obes 22(5):374–380. doi:10.1097/MED.0000000000000188

Taylor JP, Metcalfe RA, Watson PF, Weetman AP, Trembath RC (2002) Mutations of the PDS gene, encoding pendrin, are associated with protein mislocalization and loss of iodide efflux: implications for thyroid dysfunction in Pendred syndrome. J Clin Endocrinol Metab 87(4):1778–1784

Twyffels L, Massart C, Golstein PE, Raspe E, Van Sande J, Dumont JE, Beauwens R, Kruys V (2011) Pendrin: the thyrocyte apical membrane iodide transporter? Cell Physiol Biochem 28(3):491–496. doi:10.1159/000335110

Twyffels L, Strickaert A, Virreira M, Massart C, Van Sande J, Wauquier C, Beauwens R, Dumont JE, Galietta LJ, Boom A, Kruys V (2014) Anoctamin-1/TMEM16A is the major apical iodide channel of the thyrocyte. Am J Physiol Cell Physiol 307(12):C1102–C1112. doi:10.1152/ajpcell.00126.2014

van den Hove MF, Croizet-Berger K, Jouret F, Guggino SE, Guggino WB, Devuyst O, Courtoy PJ (2006) The loss of the chloride channel, ClC-5, delays apical iodide efflux and induces a euthyroid goiter in the mouse thyroid gland. Endocrinology 147(3):1287–1296. doi:10.1210/en.2005-1149, en.2005-1149 [pii]

Viitanen TM, Sukumaran P, Lof C, Tornquist K (2013) Functional coupling of TRPC2 cation channels and the calcium-activated anion channels in rat thyroid cells: implications for iodide homeostasis. J Cell Physiol 228(4):814–823. doi:10.1002/jcp.24230

Weiss SJ, Philp NJ, Grollman EF (1984) Effect of thyrotropin on iodide efflux in FRTL-5 cells mediated by Ca2+. Endocrinology 114(4):1108–1113

Yang YD, Cho H, Koo JY, Tak MH, Cho Y, Shim WS, Park SP, Lee J, Lee B, Kim BM, Raouf R, Shin YK, Oh U (2008) TMEM16A confers receptor-activated calcium-dependent chloride conductance. Nature 455(7217):1210–1215. doi:10.1038/nature07313

Yoshida A, Taniguchi S, Hisatome I, Royaux IE, Green ED, Kohn LD, Suzuki K (2002) Pendrin is an iodide-specific apical porter responsible for iodide efflux from thyroid cells. J Clin Endocrinol Metab 87(7):3356–3361

Part III

The Role of Pendrin in the Kidney

Pendrin and Its Partners in the Kidney: Roles in Vascular Volume and Acid Base Regulation

8

Manoocher Soleimani and Giovanna Valenti

Abstract

The Cl^-/HCO_3^- exchanger pendrin (SLC26A4, PDS) is located on the apical membrane of B-intercalated cells in the kidney cortical collecting duct (CCD) and the connecting tubules (CNTs) and mediates the secretion of bicarbonate and the reabsorption of chloride. Recent studies demonstrate the coordinated interaction of pendrin with several ion transporters and/or channels in the kidney distal nephron, resulting in systemic electrolyte and vascular volume homeostasis and acid base regulation. In this chapter, we will discuss the latest developments on the role of pendrin and its interacting partners in salt and water absorption in the distal nephron and their relevance to pathophysiologic states.

8.1 Introduction

The solute carrier (SLC) 26 family of anion transporters comprises both electrogenic and electroneutral chloride/base exchangers that are expressed in various epithelia and polarized cells (Sindić et al. 2007; Soleimani 2013). This family is genetically distinct from the SLC4 anion exchangers (AE1, AE2, AE3, and AE4) and comprises at least ten distinct members (SLC26A1-11) (Sindić et al. 2007; Soleimani 2013; Alper and Sharma 2013). Modes of transport mediated by SLC26 members include the exchange of chloride for bicarbonate, hydroxyl, sulfate, formate, iodide, or oxalate with variable

M. Soleimani, MD (✉)
Department of Medicine, University of Cincinnati, Cincinnati, OH, USA
e-mail: soleimm@ucmail.uc.edu; manoocher.soleimani@uc.edu

G. Valenti, PhD (✉)
Department of Biosciences, Biotechnologies and Biopharmaceutics, University of Bari, Bari, Italy
e-mail: giovanna.valenti@uniba.it

S. Dossena, M. Paulmichl (eds.), *The Role of Pendrin in Health and Disease*,
DOI 10.1007/978-3-319-43287-8_8

specificity (Melvin et al. 1999; Soleimani et al. 2001; Jiang et al. 2002; Xie et al. 2002; Chernova et al. 2005; Wang et al. 2002; Schweinfest et al. 2006; Petrovic et al. 2003; Xu et al. 2005, 2008, 2011). Several SLC26 family members can specifically function as Cl^-/HCO_3^- exchangers. These include SLC26A3 (DRA), SLC26A4 (pendrin), SLC26A6 (PAT1 or CFEX), SLC26A7, SLC26A9, and SLC26A11 (Melvin et al. 1999; Soleimani et al. 2001; Jiang et al. 2002; Xie et al. 2002; Chernova et al. 2005; Wang et al. 2002; Schweinfest et al. 2006; Petrovic et al. 2003; Xu et al. 2005, 2008, 2011). In addition to mediating chloride/base exchange, SLC26A7, SLC26A9, and SLC26A11 can also function as chloride channels (Bertrand et al. 2009; Kim et al. 2005; Dorwart et al. 2007).

Mutations in several SLC26 isoforms are linked to autosomal recessive genetic disorders, with SLC26A2 associated with chondrodysplasias (Hastbacka et al. 1994), SLC26A3 linked to chloride-losing diarrhea (Hoglund et al. 1996), and SLC26A4 connected to Pendred syndrome and nonsyndromic hereditary deafness (Everett et al. 1997). These findings confirm the intriguing and important roles of SLC26 isoforms in normal physiology and human pathophysiology. The generation of genetically engineered mouse models has significantly increased our understanding of the pathophysiology of genetic disorders linked to SLC26 isoforms. These models have further established the important role of SLC26 isoforms in the normal physiology of various organs, including the kidney.

The *SLC26A4* gene was first discovered by positional cloning studies in patients with Pendred syndrome (Everett et al. 1997), an autosomal recessive inherited disorder characterized by congenital deafness and thyroid goiter (Bizhanova and Kopp 2010). Pendrin is abundantly expressed in the thyroid and inner ear (Everett et al. 1997), with lower levels in the whole kidney (Bizhanova and Kopp 2010). Pendrin can function in Cl^-/HCO_3^- exchange, Cl^-/OH^- exchange, and Cl^-/I^- exchange modes (Soleimani et al. 2001; Royaux et al. 2001); however, its dominant role in the kidney is the mediation of Cl^-/HCO_3^- exchange. Immunofluorescence studies localized pendrin to the apical membrane of B and non-A-, non-B-intercalated cells in the connecting tubule (CNT) and the cortical collecting duct (CCD) (Royaux et al. 2001; Kim et al. 2002; Wall et al. 2003). The schematic diagram in Fig. 8.1 depicts the major salt-absorbing transporters in the distal nephron, including the Cl^-/HCO_3^- exchanger pendrin. This chapter focuses on the latest advances on the role of pendrin in the kidney physiology, as derived from studies on genetically engineered mouse models, humans with Pendred syndrome, and functional and localization experiments.

8.2 Interaction of Pendrin with the Sodium Chloride Cotransporter NCC

8.2.1 Pendrin and Salt Absorption

Genetic deletion of pendrin does not cause any significant salt wasting or excessive diuresis in mutant mice under basal conditions (Royaux et al. 2001; Verlander et al. 2003; Amlal et al. 2010). Kidney functions, including sodium

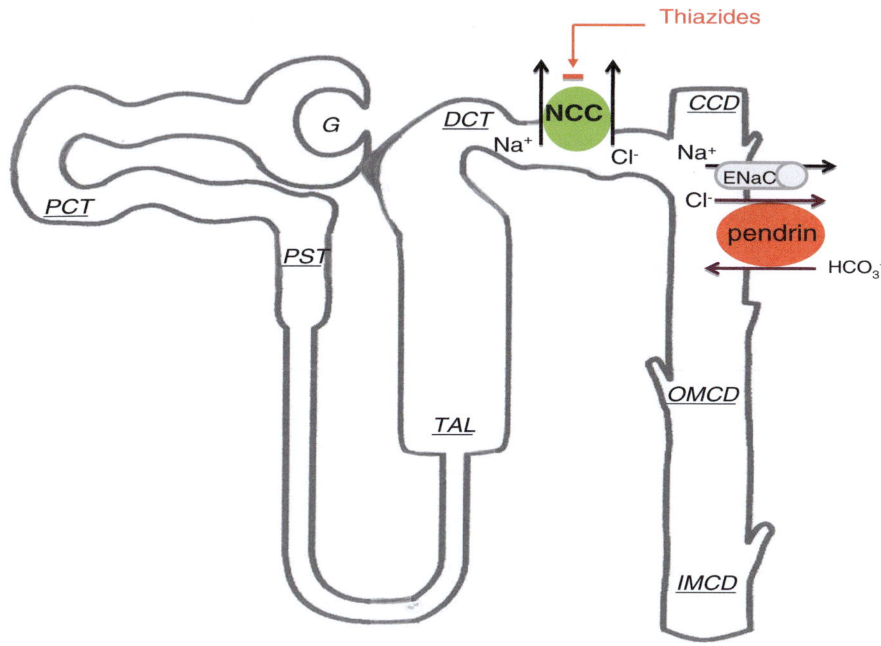

Fig. 8.1 A schematic diagram depicting the localization and function of pendrin, NCC, and ENaC in kidney tubules. Pendrin is located on the apical membrane of B-intercalated and non-A-, non-B-intercalated cells. *CCD* cortical collecting duct, *DCT* distal convoluted tubule, *ENaC* epithelial sodium channel, *G* glomerulus, *IMCD* inner medullary collecting duct, *NCC* sodium chloride cotransporter, *OMCD* outer medullary collecting duct, *PCT* proximal convoluted tubule, *PST* proximal straight tubule, *TAL* tick ascending limb

and chloride excretion, urine output, and blood urea nitrogen levels, are all normal in pendrin knockout (KO) mice (Royaux et al. 2001; Verlander et al. 2003; Amlal et al. 2010). Indeed, deafness is the only major phenotypic presentation in pendrin-deficient mice and in humans with Pendred syndrome, confirming an essential role for pendrin in the development of inner ear hearing structures in mammalians (Wangemann 2011) (see also Chaps. 2 and 3). An acidic urine pH is the only noticeable kidney phenotype in pendrin KO mice relative to wild-type (WT) animals under baseline conditions (Amlal et al. 2010). Mice with pendrin ablation, however, display impaired ability to conserve sodium when subjected to salt restriction, and develop volume depletion, metabolic alkalosis, and hypotension (Wall et al. 2004). These results clearly support an important role for pendrin in salt absorption and bicarbonate secretion in volume-depleted states. In addition, pendrin expression is increased, and its activity is enhanced in response to increased concentration of aldosterone, suggesting that pendrin plays an important role in the pathogenesis of hypertension in primary aldosteronism (Verlander et al. 2003).

8.2.2 Compensatory Role of Pendrin in Salt Reabsorption in the Distal Nephron in the Setting of NCC Inactivation

Based on its dual function of bicarbonate secretion and chloride reabsorption, it was originally thought that pendrin plays important roles in acid base regulation and electrolyte homeostasis under basal conditions. As noted above, the absence of any significant salt wasting in mice or human lacking a functional pendrin suggested that this transporter does not play an important role in salt absorption under baseline states. The upregulation of pendrin in response to salt restriction or increased aldosterone led to the conclusion that this transporter plays an important role in salt absorption during salt restriction or in response to aldosterone excess (Verlander et al. 2003; Wall et al. 2004). Of interest was a recent report indicating the development of metabolic alkalosis in a patient with Pendred syndrome during vascular volume depletion (Kandasamy et al. 2011), confirming the reports on the important role of pendrin in salt absorption during salt restriction.

The thiazide-sensitive NaCl cotransporter (NCC) is the main sodium-absorbing transporter in the distal convoluted tubule (DCT), whereas the epithelial sodium channel ENaC is the main molecule responsible for the absorption of sodium in the CNT and the CCD (Gamba 2005; Ellison 2003; Delpire et al. 1996; Câmpean et al. 2001; Delpire and Mount 2002; Ellison et al. 1987, 1989; Schultheis et al. 1998; Loffing et al. 2004). Chloride reabsorption in the DCT is coupled to the sodium absorption via NCC. Chloride reabsorption in the CNT and CCD is mediated via both paracellular and transcellular pathways, with the transcellular pathway being mediated primarily via pendrin, which in conjunction with ENaC is responsible for the bulk of sodium and chloride reabsorption in these two segments (Wall and Pech 2008; Wall and Weinstein 2013).

8.2.2.1 Pendrin Expression Is Significantly Enhanced in Kidneys of NCC KO Mice

Northern hybridization and immunofluorescence labeling indicated that the expression of pendrin is significantly increased in kidneys of NCC KO mice (Vallet et al. 2006; Patel-Chamberlin et al. 2016). In addition to the pendrin working in tandem with ENaC, the sodium-dependent chloride/bicarbonate exchanger (NDCBE) shows expression in the CCD and may participate in sodium absorption (Leviel et al. 2010; Eladari and Hübner 2011).

Similar to pendrin-deficient mice, genetically engineered mice lacking NCC do not display excessive salt wasting under baseline conditions (Schultheis et al. 1998). NCC-deficient mice display enhanced calcium absorption (hypocalciuria) along with magnesium wasting (Schultheis et al. 1998; Loffing et al. 2004); the latter is caused by the downregulation of magnesium-absorbing channel (transient receptor potential ion channel TRPM6) in the DCT and will result in low magnesium concentration in blood (hypomagnesium), recapitulating the common phenotypic presentations in patients with Gitelman syndrome.

8.2.2.2 Double Knockout of Pendrin and NCC Causes Severe Salt Wasting under Baseline Conditions

Very recent reports have unmasked the basis of incongruity between the mild phenotype in NCC KO or pendrin KO mice, and the role of pendrin and NCC as important players in salt reabsorption in the distal tubule. In these studies, mice with single deletion of NCC and pendrin were cross mated in order to generate mice with double deletion of pendrin and NCC (Soleimani et al. 2012). The pendrin/NCC double KO mice displayed significant salt wasting and developed severe volume depletion, kidney hypoperfusion, and metabolic alkalosis (Soleimani et al. 2012). These studies demonstrate that pendrin and NCC cross compensate for the loss of each other, therefore masking the role that each transporter plays in salt reabsorption under baseline conditions (Soleimani et al. 2012).

8.2.3 Compensatory Role of ENaC in Salt Absorption in the Distal Nephron in the Setting of NCC Inactivation

As previously mentioned (see Sect. 8.2.2), the sodium absorption in the CNT and CCD is mediated via the epithelial sodium channel ENaC in principal cells, whereas chloride is absorbed primarily via pendrin (Royaux et al. 2001; Verlander et al. 2003; Amlal et al. 2010; Costanzo 1984; Costanzo and Windhager 1978; Eaton et al. 2001; Roy et al. 2015).

8.2.3.1 Expression of ENaC Is Enhanced in Kidneys of NCC KO Mice

The northern hybridizations and Western blots showed significant upregulation of mRNA encoding ENaC α, β, and γ subunits in kidneys of NCC KO mice (Patel-Chamberlin et al. 2016). The analysis of Western blots by densitometric scanning demonstrated the upregulation of cleaved form (70 kDa) of ENaC γ subunit in kidneys of NCC KO mice, with the cleaved band *increasing by ~420% in NCC KO mice* (Patel-Chamberlin et al. 2016).

8.2.3.2 Effect of Amiloride on Salt Excretion in NCC KO Mice

To ascertain whether the upregulation of ENaC is associated with increased ability to conserve sodium, NCC-deficient mice were placed in metabolic cages and treated with daily injection of amiloride for 6 days. Urine output almost doubled in NCC KO mice after amiloride injection, whereas it did not change significantly in WT mice (Patel-Chamberlin et al. 2016). The increase in urine output was associated with enhanced sodium excretion in NCC KO mice. It is important to note that, as previously mentioned, sodium excretion in WT littermates was not significantly affected after amiloride injection (Patel-Chamberlin et al. 2016).

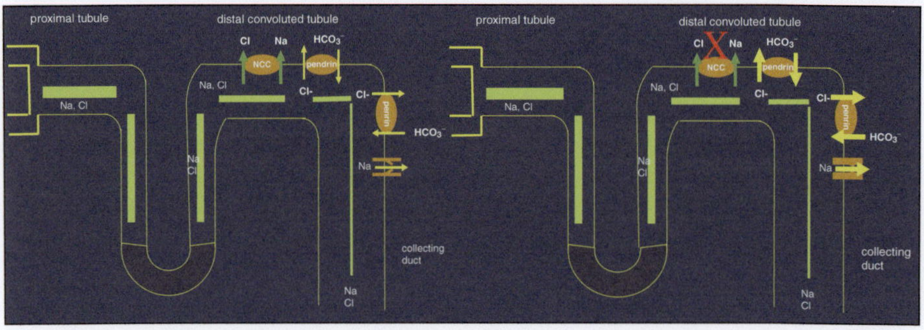

Fig. 8.2 A schematic diagram depicting the compensatory roles of pendrin and ENaC in salt absorption in the setting of NCC inactivation or inhibition. (*Left*) Salt absorption in the DCT and collecting duct: role of NCC, pendrin, and ENaC. (*Right*) NCC KO mice do not show significant salt wasting under baseline conditions due to the upregulation of pendrin and ENaC. The ablation of pendrin in the setting of NCC inactivation causes massive salt wasting and results in volume depletion and kidney hypoperfusion (Soleimani et al. 2012). *NCC* sodium chloride cotransporter

8.2.4 Conclusions

Pendrin and ENaC show significant activation in kidneys of NCC-deficient mice, and their downregulation or inhibition results in significant salt wasting, strongly suggesting that pendrin collaborates with ENaC to prevent salt wasting in the setting of NCC inactivation. The schematic diagram in Fig. 8.2 depicts the compensatory roles of pendrin and ENaC in salt absorption in the collecting duct in the setting of NCC inactivation or inhibition.

8.3 Impact of Pendrin and NCC Downregulation on Salt, Calcium, and Water Absorption in the Collecting Duct

8.3.1 Role of the Interplay of CaSR Signaling and Aquaporin-2

As previously described (see Sect. 8.2.2.2), the generation of mice with a genetic deletion of pendrin and NCC has showed several features revealing that under basal conditions, these two transporters compensate each other for their function. In fact, pendrin or NCC single KO mice did not show any salt wasting or volume depletion under basal conditions.

Interestingly, pendrin/NCC double KO mice display calcium (and phosphate) wasting (Soleimani et al. 2012). This feature is actually observed also in pendrin single KO mice, which produce a very acidic urine as a consequence of the lack of the apical Cl^-/HCO_3^- transporter in the CCD (Amlal et al. 2010).

In the kidney, the majority of calcium is reabsorbed through the paracellular pathway in the proximal tubule and in the thick ascending limb of Henle's loop (Hoenderop et al. 2005; Blanchard et al. 2001). In the distal nephron, calcium is reabsorbed through an active transcellular process that involves the concerted action of the apical epithelial calcium channel TRPV5, the Na^+/Ca^{2+} exchanger expressed on the basolateral membrane and calbindin (Hoenderop et al. 2002). Downregulation of TRPV5 is associated with profound calcium wasting (Gkika et al. 2006). Consistent with the critical role of pendrin in the pH regulation, it has been reported that calcium-absorbing pathway molecule (apical TRPV5 and basolateral Na^+/Ca^{2+} exchanger) is downregulated in pendrin KO mice. These changes were associated with a significant renal calcium wasting. Several lines of evidence demonstrate that hypercalciuria is associated with higher water excretion by the kidney. In this respect, we have previously provided evidence that high concentrations of luminal calcium in the renal collecting duct counteract vasopressin action impairing the trafficking of the vasopressin-sensitive water channel aquaporin 2 (AQP2). This effect is mediated by the activation of the calcium-sensing receptor (CaSR) (Procino et al. 2004, 2008). The data supporting this conclusion were obtained both from *in vitro* experiments using AQP2-expressing mouse collecting duct cells (MCD4 cells) and from hypercalciuric patients (Procino et al. 2012).

Additional evidence for the inverse relationship between calciuria and water metabolism was also provided in a bed-rest study in humans showing that bed-rest induced an increase in blood hematocrit (reflecting water loss), which coincided to a reduction of urinary AQP2 likely paralleled by an increase in urinary calcium due to bone demineralization (Tamma et al. 2014a). Of note, high external calcium has been found to reduce AQP2 expression both in the collecting duct cell line mpkCCD and in hypercalciuric rats (Bustamante et al. 2008; Sands et al. 1998). These data support a direct effect of luminal calcium on AQP2 expression in collecting duct principal cells and point to a role of calcium in regulating both AQP2 trafficking and expression. More recently, we have shown that CaSR activation in renal cells modulates AQP2 trafficking and/or expression by alteration of its phosphorylation state (Ranieri et al. 2015).

AQP2 phosphorylation event is required to increase the water permeability and water reabsorption of the collecting duct cells. Vasopressin-induced translocation of AQP2 coincides with AQP2 phosphorylation at serine (S) 256, S264, and threonine (T) 269 and dephosphorylation at S261. Specifically, we found that selective activation of CaSR expressed in the collecting duct prevented the cAMP-dependent increase in AQP2 phosphorylation at S256 (AQP2-pS256) and water reabsorption and reduced the basal AQP2-pS256 levels, thus having an opposite effect with respect to vasopressin action (Fig. 8.3). These data point to a crucial role of CaSR signaling in impairing the short-term vasopressin response resulting in water loss with urine (Ranieri et al. 2015). Of note, pendrin/NCC double KO mice have impaired kidney response to exogenous vasopressin and display low urine osmolality, despite severe volume depletion associated with downregulation of AQP2 in medullary collecting duct (MCD) despite increased circulating vasopressin levels (Soleimani et al. 2012). Pendrin/NCC double KO mice failed to increase their urine osmolality in response to desmopressin (dDAVP, a vasopressin analog), consistent with the impaired response to vasopressin (Soleimani et al. 2012).

Given that the inhibitory effect of CaSR signaling on AQP2 trafficking to the plasma membrane in response to vasopressin is mainly due to modulation of AQP2

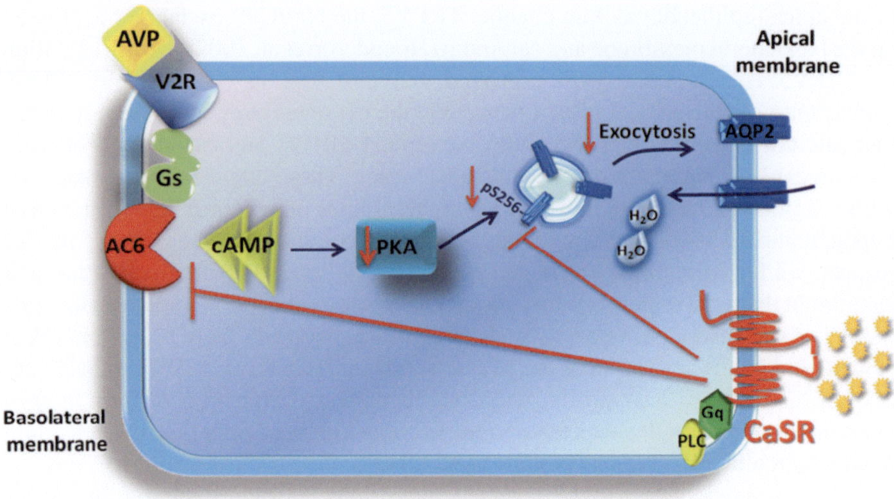

Fig. 8.3 A schematic diagram depicting the role of CaSR in regulating AQP2 trafficking *via* modulation of its phosphorylation state. Activation of CaSR expressed in apical membrane of the collecting duct principal cells by urinary calcium impairs vasopressin-stimulated water reabsorption through inhibition of the calcium inhibitable AC6 and reduction of cAMP generation, resulting in decreased PKA activity and AQP2-pS256 levels. This negative feedback from CaSR signaling to AQP2 sensitizes vasopressin to urinary calcium, representing an internal renal defense to mitigate the effect of hypercalciuria on the risk of calcium saturation during antidiuresis. *AC6* adenylyl cyclase6, *AQP2* aquaporin 2, *AVP* arginine vasopressin, *cAMP* cyclic AMP, *CaSR* calcium-sensing receptor, *Gq* G protein q, *Gs* stimulatory G protein, *PKA* protein kinase A, *PLC* phospholipase C, *V2R* arginine vasopressin receptor 2

phopshorylation, we hypothesize that in pendrin/NCC double KO mice, the high calcium delivery in the collecting duct activates a signaling mediated by CaSR counteracting vasopressin response and leading to reduced water absorption and volume depletion.

Quite interestingly, preliminary results suggest that pendrin/NCC double KO mice have significantly higher levels of AQP2-pS261 under basal conditions, consistent with an expected reduced response to vasopressin (Valenti et al., unpublished).

Consistent with that, data from literature showed that S261 in AQP2 is highly phosphorylated when unstimulated, but is dephosphorylated upon treatment with vasopressin (Hoffert et al. 2006). Moreover, a recent work showed that increased expression of AQP2-pS261 was concomitant with the increase in urine output in rat kidney inner medullary collecting duct (Trepiccione et al. 2014). The higher expression of AQP2-pS261 in pendrin/NCC double KO mice may reflect a higher AQP2 degradation rate. Quite interestingly, preliminary ex vivo experiments show that specific inhibition of CaSR in pendrin/NCC double KO mice results in 100 % increase in basal levels of AQP2-pS256, strongly supporting the hypothesis of a critical role of CaSR in impairing vasopressin response by reducing AQP2 phosphorylation at S256 and increasing AQP2 phosphorylation at S261, thus resulting in reduced AQP2 trafficking and volume depletion (Valenti et al., unpublished). The schematic diagram in Fig. 8.3 depicts the role of CaSR in regulating AQP2 trafficking.

8.4 Interaction of Pendrin with CFTR and Other Chloride Transporters

8.4.1 Pendrin and Cystic Fibrosis

Cystic fibrosis (CF) is the most common life-threatening genetic disease in the United States. It is caused by mutations in the cystic fibrosis transmembrane conductance regulator (CFTR), a cAMP/PKA, and ATP-regulated apical chloride channel, which is widely expressed in epithelial tissues (Frizzell and Hanrahan 2012; Kunzelmann et al. 2001). Due to its widespread distribution in epithelial cells, patients with CFTR mutation develop damage or injury in multiple organs, including lung injury and/or infection, pancreatic inflammation (pancreatitis), and/or intestinal obstruction (Derichs 2013; Hoffman and Ramsey 2013; Pier 2012). However, no apparent kidney phenotype has been identified in CF patients, despite abundant expression of CFTR in the kidney tubule, including intercalated cells in the collecting duct (Todd-Turla et al. 1996).

Patients with CF are prone to the development of metabolic alkalosis, which is manifested by an elevation of arterial blood pH and bicarbonate and produces a physiologic diminution of respiratory effort that is detrimental in CF patients (Sweetser et al. 2005; Baird et al. 2002; Al-Ghimlas et al. 2012; Yalçin et al. 2005). The pathogenesis of metabolic alkalosis in CF remains unknown. One major factor contributing to the generation of metabolic alkalosis in human is the impaired ability of the kidney to excrete the excess bicarbonate. The major transporter responsible for the excretion of bicarbonate in the kidney is pendrin (Sindić et al. 2007; Soleimani 2013; Alper and Sharma 2013; Royaux et al. 2001). It is therefore plausible that the inactivation of pendrin could impair the ability of the kidney to eliminate the excess bicarbonate, specifically in the setting of volume depletion which requires an active and functional pendrin.

CFTR plays a vital role in HCO_3^- secretion in multiple tissues, including the pancreatic duct and duodenum, by activating the apical Cl^-/HCO_3^- exchangers from SLC26 family (Ko et al. 2004; Bertrand et al. 2009). Recent studies have demonstrated that pendrin and CFTR activate each other and that CFTR regulates pendrin-mediated Cl^-/HCO_3^- exchange (Bertrand et al. 2009). These studies showed that a functional CFTR markedly activates pendrin-mediated Cl^-/HCO_3^- exchange, whereas a mutant CFTR does not (Shcheynikov et al. 2008). The interaction of CFTR and pendrin has also been shown in tracheal epithelial and thyroid follicular cells (Garnett et al. 2011). This phenomenon might have ramifications for patients with CF, who on many occasions present with metabolic alkalosis, specifically at a young age (Sweetser et al. 2005; Baird et al. 2002; Al-Ghimlas et al. 2012; Yalçin et al. 2005). While the generation of alkalosis has been attributed to volume depletion subsequent to the loss of electrolytes through the skin, it is plausible that pendrin, which plays an important role in chloride absorption and bicarbonate secretion in volume-depleted states, remains inactive in patients with cystic fibrosis, resulting in further loss of chloride and retention of HCO_3^- by the kidney.

Our recent studies have shown that pendrin is significantly downregulated in kidneys of a mouse model of cystic fibrosis, as shown by a profound reduction in apical expression and increased cytoplasmic and basolateral localization of pendrin in B-intercalated cells in CFTR KO mice (Varasteh Kia et al. Submitted for publication). As a result, CF mice developed metabolic alkalosis (as verified by increased arterial blood pH and bicarbonate) when subjected to excess exogenous bicarbonate load, subsequent to the impaired ability of their kidneys to eliminate excess bicarbonate (Varasteh Kia et al. Submitted for publication). It is noteworthy to mention that baseline parameters, including acid–base status and urine pH, were comparable in CF mice *vs.* WT mice. However, it was following a 3-day oral bicarbonate load that CF mice showed significant elevation in their blood bicaronate and pH (metabolic alkalosis) and impaired ability to excrete excess bicarbonate (low urine pH).

8.4.1.1 Conclusion

We propose that patients with cystic fibrosis are prone to the development of metabolic alkalosis in large part due to the impairment of pendrin, specifically during volume depletion, which is a common occurrence in CF patients.

8.4.2 Pendrin and Other Chloride Transporters

Is the epithelial Na^+ channel ENaC, working in conjunction with the paracellular Cl^- absorption, the major path for salt absorption in the collecting duct? Based on recent studies, the answer is negative. In addition to the Cl^-/HCO_3^- exchanger pendrin, studies over the last 5 years have identified several new players in transcellular chloride and/or sodium reabsorption in the collecting duct. These include the Na^+-dependent Cl^-/HCO_3^- exchanger Slc4a8 (NDCBE), and the chloride transporter/channel Slc26a11 (Kidney Brain Anion Transporter; KBAT) (Xu et al. 2011; Leviel et al. 2010). Aside from cross compensation between NCC and pendrin and their crucial role in salt absorption in the DCT and CCD, respectively, no transcellular chloride-absorbing pathway has been identified in medullary collecting duct (MCD) despite the presence of a sodium-absorbing molecule (ENaC) in the same nephron segment. Recent studies demonstrate that Slc26a11 (KBAT) is expressed on the apical membrane of A-intercalated cells in the CCD, outer and inner MCD, and mediates chloride transport (Xu et al. 2011). These studies further indicate that KBAT may play an important role in compensatory salt absorption in the collecting duct in disease states. The schematic diagram in Fig. 8.4 depicts the interaction of KBAT and NDCBE with pendrin and other ion transporters in the CCD.

The ambiguities with regard to the role of KBAT in salt reabsorption may be due to the lack of availability of pertinent mutant mouse models and the paucity of information on its collaborating transporters. We have generated mice with

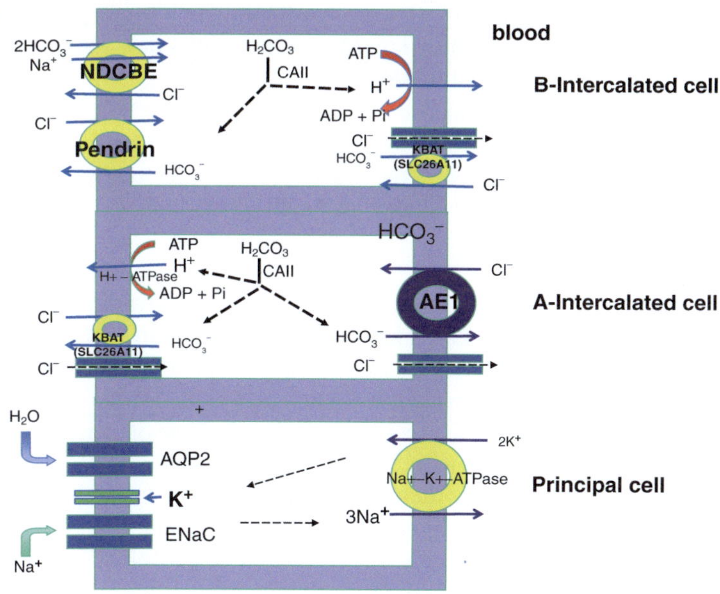

Fig. 8.4 Interaction of KBAT and NDCBE with pendrin and other ion transporters in the CCD. Slc26a11 (KBAT) can function as a chloride-absorbing transporter/channel and regulate systemic vascular volume. The Slc4a8 (NDCBE) can function in tandem with pendrin and facilitate the absorption of sodium and chloride. *AE1* anion exchanger 1, *AQP2* aquaporin 2, *CAII* carbonic anhydrase II, *ENaC* epithelial sodium channel, *KBAT* kidney brain anion transporter, *NDCBE* sodium-dependent chloride/bicarbonate exchanger

kidney-specific ablation of KBAT, which show significant salt wasting in response to furosemide or hydrochlorothiazide (HCTZ) treatment and following increased dietary salt intake (unpublished observation). Whether KBAT works in tandem with ENaC and/or NDCBE in the collecting duct to absorb sodium and chloride remains to be determined.

8.5 Regulation of Pendrin by the Carbonic Anhydrase II and Effect of Carbonic Anhydrase Inhibitors

Carbonic anhydrases play important roles in acid base transport in the proximal tubule and the collecting duct (Purkerson and Schwartz 2007; Schwartz 2002). In addition to facilitating the generation of intracellular bicarbonate in the proximal tubule cell, carbonic anhydrase II (CAII) is abundantly expressed in A- and B-intercalated cells and plays an important role in H^+ (proton) and HCO_3^- secretion in A- and B-intercalated cells, respectively. Inhibition of carbonic anhydrase activity in the proximal tubule by acetazolamide (ACTZ) blocks the apical Na^+/H^+ exchanger activity and decreases sodium and bicarbonate reabsorption (Puschett 1994). In addition to the inhibition of sodium and bicarbonate absorption in the proximal tubule, short-term (1 or 2 weeks) inhibition of carbonic anhydrases causes significant remodeling of cellular profile in the collecting duct, with a specific reduction in B-intercalated cells (Bagnis et al. 2001). The expression of pendrin was also significantly downregulated in a mouse of model CAII deficiency (Sun et al. 2008; Xu et al. 2013) consequent to a huge reduction in the number of intercalated cells, including B-intercalated cells. These results support the view that CAII is essential for the viability of B-intercalated cells in the kidney.

8.5.1 Carbonic Anhydrase Inhibitor Acetazolamide Downregulates the Expression of Pendrin in the Kidney

Recent studies have demonstrated that treatment with the carbonic anhydrase inhibitor (CAI) acetazolamide downregulates the expression of pendrin in rat kidney, presumably secondary to the reduction of B-intercalated cells (Zahedi et al. 2013).

Hydrochlorothiazide is the most widely used diuretic in the world for the treatment of mild and moderate hypertension (Ellison 2003; Câmpean et al. 2001; Ellison et al. 1987). Despite being a specific inhibitor of NCC in the DCT, hydrochlorothiazide causes a very mild diuretic response (Ellison 2003; Câmpean et al. 2001; Ellison et al. 1987). This observation is in agreement with studies indicating that NCC deletion in mouse causes very little salt wasting under basal conditions (Schultheis et al. 1998; Loffing et al. 2004). Based on published studies demonstrating that the double deletion of NCC and pendrin causes significant salt wasting (Soleimani et al. 2012), the effect of combination of acetazolamide and hydrochlorothiazide on salt excretion was tested.

8.5.2 Synergistic Effects of Acetazolamide and Hydrochlorothiazide on Salt Excretion, Vascular Volume, and Kidney Function

Carbonic anhydrase inhibitors (CAIs) are mild diuretics (Puschett 1994; Zahedi et al. 2013), hence not widely used in fluid overloaded states. Thiazides, specific inhibitors of NCC, are mild agents and the most widely used diuretics in the world for control of mild hypertension (Ellison 2003; Câmpean et al. 2001; Ellison et al. 1987).

It was postulated that carbonic anhydrase inhibitors downregulate pendrin, therefore leaving NCC as the major salt-absorbing transporter in the distal nephron. If that assumption is true, then the addition of hydrochlorothiazide in animals pretreated with acetazolamide should result in salt wasting.

To test the above hypothesis, rats were treated with acetazolamide (ACTZ), a known CAI, or hydrochlorothiazide (HCTZ) for 6 days. The results showed that ACTZ treatment caused mild diuresis, whereas daily treatment with HCTZ caused hardly any diuresis (Zahedi et al. 2013). However, treatment of rats that were pretreated with ACTZ for 6 days with a combination of ACTZ plus HCTZ for 4 additional days increased the urine output by greater than twofold ($p<0.001$, $n=5$) compared to ACTZ-treated animals (Zahedi et al. 2013). Sodium excretion increased by 80 % in the ACTZ plus HCTZ group, and animals developed significant volume depletion, metabolic alkalosis, and pre-renal failure.

These results support the conclusion that carbonic anhydrase inhibition by ACTZ downregulates pendrin (Zahedi et al. 2013), therefore leaving NCC as the only major salt-absorbing transporter in the distal nephron. As such, it was postulated that the addition of HCTZ, which inhibits NCC, should cause profound diuresis, subsequent to the inactivation of pendrin and NCC in the face of increased delivery of salt from proximal tubule. It was further proposed that patients that are treated with ACTZ for *pseudotumor cerebri* (idiopathic intracranial hypertension) or other non-kidney conditions, such as glaucoma, should avoid taking HCTZ for hypertension due to profound diuretic effect of the combination therapy.

References

Al-Ghimlas F, Faughnan ME, Tullis E (2012) Metabolic alkalosis in adults with stable cystic fibrosis. Open Respir Med J 6:59–62. doi:10.2174/1874306401206010059

Alper SL, Sharma AK (2013) The *SLC26* gene family of anion transporters and channels. Mol Aspects Med 34:494–515

Amlal H, Petrovic S, Xu J, Wang Z, Sun X, Barone S, Soleimani M (2010) Deletion of the anion exchanger Slc26a4 (pendrin) decreases apical Cl(−)/HCO3(−) exchanger activity and impairs bicarbonate secretion in kidney collecting duct. Am J Physiol Cell Physiol 299(1):C33–C41

Bagnis C, Marshansky V, Breton S, Brown D (2001) Remodeling the cellular profile of collecting ducts by chronic carbonic anhydrase inhibition. Am J Physiol Renal Physiol 280(3):F437–F448

Baird JS, Walker P, Urban A et al (2002) Metabolic alkalosis and cystic fibrosis. Chest 122:755–756. doi:10.1378/chest.122.2.755

Bertrand CA, Zhang R, Pilewski JM, Frizzell RA (2009) SLC26A9 is a constitutively active, CFTR-regulated anion conductance in human bronchial epithelia. J Gen Physiol 133(4):421–438. doi:10.1085/jgp.200810097

Bizhanova A, Kopp P (2010) Genetics and phenomics of Pendred syndrome. Mol Cell Endocrinol 322(1–2):83–90, Review

Blanchard A, Jeunemaitre X, Coudol P et al (2001) Paracellin-1 is critical for magnesium and calcium reabsorption in the human thick ascending limb of Henle. Kidney Int 59:2206–2215

Bustamante M, Hasler U, Leroy V, de Seigneux S, Dimitrov M, Mordasini D, Rousselot M, Martin P-Y, Feraille E (2008) Calcium-sensing receptor attenuates AVP-induced aquaporin-2 expression via a calmodulin dependent mechanism. J Am Soc Nephrol 19:109–116

Câmpean V, Kricke J, Ellison D et al (2001) Localization of thiazide sensitive Na(+)-Cl(−) cotransport and associated gene products in mouse DCT. Am J Physiol Renal Physiol 281(6): F1028–F1035

Chernova MN, Jiang L, Friedman DJ, Darman RB, Lohi H, Kere J, Vandorpe DH, Alper SL (2005) Functional comparison of mouse slc26a6 anion exchanger with human SLC26A6 polypeptide variants: differences in anion selectivity, regulation, and electrogenicity. J Biol Chem 280(9):8564–8580

Costanzo LS (1984) Comparison of calcium and sodium transport in early and late rat distal tubules: effect of amiloride. Am J Physiol Ren Fluid Electrolyte Physiol 246:F937–F945

Costanzo LS, Windhager EE (1978) Calcium and sodium transport by the distal convoluted tubule of the rat. Am J Physiol Ren Fluid Electrolyte Physiol 235:F492–F506

Delpire E, Mount DB (2002) Human and murine phenotypes associated with defects in cation-chloride cotransport. Annu Rev Physiol 64:803–843

Delpire E, Kaplan MR, Plotkin MD, Hebert SC (1996) The Na-(K)-Cl cotransporter family in the mammalian kidney: molecular identification and function(s). Nephrol Dial Transplant 11(10):1967–1973, Review

Derichs N (2013) Targeting a genetic defect: cystic fibrosis transmembrane conductance regulator modulators in cystic fibrosis. Eur Respir Rev 127(22):58–65. doi:10.1183/09059180.00008412, Review

Dorwart MR, Shcheynikov N, Wang Y, Stippec S, Muallem S (2007) SLC26A9 is a Cl^- channel regulated by the WNK kinases. J Physiol 584:333–345

Eaton DC, Malik B, Saxena NC, Al-Khalili OK, Yue G (2001) Mechanisms of aldosterone's action on epithelial Na^+ transport. J Membr Biol 184:313–319

Eladari D, Hübner CA (2011) Novel mechanisms for NaCl reabsorption in the collecting duct. Curr Opin Nephrol Hypertens 20(5):506–511, Review

Ellison DH (2003) The thiazide-sensitive Na-Cl cotransporter and human disease: reemergence of an old player. J Am Soc Nephrol 14:538–540, Review

Ellison DH, Velazquez H, Wright FS (1987) Thiazide-sensitive sodium chloride cotransport in early distal tubule. Am J Physiol Ren Fluid Electrolyte Physiol 253:F546–F554

Ellison DH, Velazquez H, Wright FS (1989) Adaptation of the distal convoluted tubule of the rat. Structural and functional effects of dietary salt intake and chronic diuretic infusion. J Clin Invest 83:113–126

Everett LA, Glaser B, Beck JC, Idol JR, Buchs A, Heyman M, Adawi F, Hazani E, Nassir E, Baxevanis AD, Sheffield VC, Green ED (1997) Pendred syndrome is caused by mutations in a putative sulphate transporter gene (*PDS*). Nat Genet 17:411–422

Frizzell RA, Hanrahan JW (2012) Physiology of epithelial chloride and fluid secretion. Cold Spring Harb Perspect Med 2(6):a009563. doi:10.1101/cshperspect.a009563

Gamba G (2005) Molecular physiology and pathophysiology of electroneutral cation-chloride cotransporters. Physiol Rev 85(2):423–493, Review

Garnett JP, Hickman E, Burrows R et al (2011) Novel role for pendrin in orchestrating bicarbonate secretion in cystic fibrosis transmembrane conductance regulator (CFTR)-expressing airway serous cells. J Biol Chem 286:41069–41082. doi:10.1074/jbc.M111.266734

Gkika D, Hsu YJ, van der Kemp AW et al (2006) Critical role of the epithelial Ca2+ channel TRPV5 in active Ca2+ reabsorption as revealed by TRPV5/calbindin-D28K knockout mice. J Am Soc Nephrol 17:3020–3027

Hastbacka J, de la Chapelle A, Mahtani MM, Clines G, Reeve-Daly MP, Daly M, Hamilton BA, Kusumi K, Trivedi B, Weaver A, Coloma A, Lovett M, Buckler A, Kaitila I, Lander ES (1994) The diastrophic dysplasia gene encodes a novel sulfate transporter: positional cloning by fine-structure linkage disequilibrium mapping. Cell 78:1073–1087

Hoenderop JG, Nilius B, Bindels RJ (2002) Molecular mechanism of active Ca2+ reabsorption in the distal nephron. Annu Rev Physiol 64:529–549

Hoenderop JG, Nilius B, Bindels RJ (2005) Calcium absorption across epithelia. Physiol Rev 85:373–422

Hoffert JD, Pisitkun T, Wang G, Shen RF, Knepper MA (2006) Quantitative phosphoproteomics of vasopressin-sensitive renal cells: regulation of aquaporin-2 phosphorylation at two sites. Proc Natl Acad Sci U S A 103:7159–7164

Hoffman LR, Ramsey BW (2013) Cystic fibrosis therapeutics: the road ahead. Chest 143(1):207–213. doi:10.1378/chest.12-1639, Review

Hoglund P, Haila S, Socha J, Tomaszewski L, Saarialho-Kere U, Karjalainen-Lindsberg M-L, Airola K, Holmberg C, de la Chapelle A, Kere J (1996) Mutations of the Down-regulated in adenoma (*DRA*) gene cause congenital chloride diarrhea. Nat Genet 14:316–319

Jiang Z, Grichtchenko II, Boron WF, Aronson PS (2002) Specificity of anion exchange mediated by mouse Slc26a6. J Biol Chem 277(37):33963–33967

Kandasamy N, Fugazzola L, Evans M, Chatterjee K, Karet F (2011) Life-threatening metabolic alkalosis in Pendred syndrome. Eur J Endocrinol 165(1):167–170

Kim YH, Kwon TH, Frische S, Kim J, Tisher CC, Madsen KM, Nielsen S (2002) Immunocytochemical localization of pendrin in intercalated cell subtypes in rat and mouse kidney. Am J Physiol Renal Physiol 283(4):F744–F754

Kim KH, Shcheynikov N, Wang Y, Muallem S (2005) SLC26A7 is a Cl^- channel regulated by intracellular pH. J Biol Chem 280:6463–6470

Ko SB, Zeng W, Dorwart MR, Luo X, Kim KH, Millen L, Goto H, Naruse S, Soyombo A, Thomas PJ, Muallem S (2004) Gating of CFTR by the STAS domain of SLC26 transporters. Nat Cell Biol 6(4):343–350. doi:10.1038/ncb1115

Kunzelmann K, Schreiber R, Boucherot A (2001) Mechanisms of the inhibition of epithelial Na^+ channels by CFTR and purinergic stimulation. Kidney Int 60:455–461. doi:10.1046/j.1523-1755.2001.060002455.x

Leviel F, Hübner CA, Houillier P et al (2010) The Na^+ –dependent chloride-bicarbonate exchanger SLC4A8 mediates an electroneutral Na^+ reabsorption process in the renal cortical collecting ducts of mice. J Clin Invest 120(5):1627–1635

Loffing J, Vallon V, Loffing-Cueni D et al (2004) Altered renal distal tubule structure and renal Na(+) and Ca(2+) handling in a mouse model for Gitelman's syndrome. J Am Soc Nephrol 15(9):2276–2288

Melvin JE, Park K, Richardson L, Schultheis PJ, Shull GE (1999) Mouse Down-regulated in Adenoma (DRA) is an intestinal Cl(−)/HCO(3)(−) exchanger and is upregulated in colon of mice lacking the NHE-3 Na(+)/H(+) exchanger. J Biol Chem 274:22855–22861

Patel-Chamberlin M, Varasteh Kia M, Xu J, Barone S, Zahedi K, Soleimani M (2016) The role of epithelial sodium channel ENaC and the apical Cl^-/HCO_3^- exchanger pendrin in compensatory salt reabsorption in the setting of Na-Cl cotransporter (NCC) inactivation. PLoS One 11(3):e0150918

Petrovic S, Ju X, Barone S, Seidler U, Alper SL, Lohi H, Kere J, Soleimani M (2003) Identification of a basolateral Cl-/HCO3- exchanger specific to gastric parietal cells. Am J Physiol Gastrointest Liver Physiol 284(6):G1093–G1103

Pier GB (2012) The challenges and promises of new therapies for cystic fibrosis. J Exp Med 209(7):1235–1239. doi:10.1084/jem.20121248, Review

Procino G, Carmosino M, Tamma G, Gouraud S, Laera A, Riccardi D, Svelto M, Valenti G (2004) Extracellular calcium antagonizes forskolininduced aquaporin 2 trafficking in collecting duct cells. Kidney Int 66:2245–2255

Procino G, Mastrofrancesco L, Mira A, Tamma G, Carmosino M, Emma F, Svelto M, Valenti G (2008) Aquaporin 2 and apical calcium-sensing receptor: new players in polyuric disorders associated with hypercalciuria. Semin Nephrol 28:297–305

Procino G, Mastrofrancesco L, Tamma G, Lasorsa DR, Ranieri M, Stringini G, Emma F, Svelto M, Valenti G (2012) Calcium-sensing receptor and aquaporin 2 interplay in hypercalciuria-associated renal concentrating defect in humans. An in vivo and in vitro study. PLoS ONE 7:e33145

Purkerson JM, Schwartz GJ (2007) The role of carbonic anhydrases in renal physiology. Kidney Int 71(2):103–115, Review

Puschett JB (1994) Pharmacological classification and renal actions of diuretics. Cardiology 84(Suppl 2):4–13, Review

Ranieri M, Tamma G, Di Mise A, Russo A, Centrone M, Svelto M, Calamita G, Valenti G (2015) Negative feedback from CaSR signaling to aquaporin-2 sensitizes vasopressin to extracellular Ca2. J Cell Sci 128(13):2350–2360

Roy A, Al-Bataineh MM, Pastor-Soler NM (2015) Collecting duct intercalated cell function and regulation. Clin J Am Soc Nephrol 10(2):305–324

Royaux IE, Wall SM, Karniski LP, Everett LA, Suzuki K, Knepper MA, Green ED (2001) Pendrin, encoded by the Pendred syndrome gene, resides in the apical region of renal intercalated cells and mediates bicarbonate secretion. Proc Natl Acad Sci U S A 98:4221–4226

Sands JM, Flores FX, Kato A, Baum MA, Brown EM, Ward DT, Hebert SC, Harris HW (1998) Vasopressin-elicited water and urea permeabilities are altered in IMCD in hypercalcemic rats. Am J Physiol 274:F978–F985

Schultheis PJ, Lorenz JN, Meneton P et al (1998) Phenotype resembling Gitelman's syndrome in mice lacking the apical Na^+ $-Cl^-$ cotransporter of the distal convoluted tubule. J Biol Chem 273(44):29150–29155
Schwartz GJ (2002) Physiology and molecular biology of renal carbonic anhydrase. J Nephrol 15(Suppl 5):S61–S74, Review
Schweinfest CW, Spyropoulos DD, Henderson KW, Kim JH, Chapman JM, Barone S, Worrell RT, Wang Z, Soleimani M (2006) Slc26a3 (dra)-deficient mice display chloride-losing diarrhea, enhanced colonic proliferation, and distinct up-regulation of ion transporters in the colon. J Biol Chem 281:37962–37971
Shcheynikov N, Yang D, Wang Y et al (2008) The Slc26a4 transporter functions as an electroneutral Cl-/I-/HCO3- exchanger: role of Slc26a4 and Slc26a6 in I- and HCO3- secretion and in regulation of CFTR in the parotid duct. J Physiol 586:3813–3824. doi:10.1113/jphysiol.2008.154468
Sindić A, Chang MH, Mount DB et al (2007) Renal physiology of SLC26 anion exchangers. Curr Opin Nephrol Hypertens 16:484–490
Soleimani M (2013) SLC26 Cl-/HCO3- exchangers in the kidney: roles in health and disease. Kidney Int 84:657–666
Soleimani M, Greeley T, Petrovic S, Wang Z, Amlal H, Kopp P, Burnham CE (2001) Pendrin: an apical Cl-/OH-/HCO3- exchanger in the kidney cortex. Am J Physiol Renal Physiol 280:F356–F364
Soleimani M, Barone S, Xu J, Shull GE, Siddiqui F, Zahedi K, Amlal H (2012) Double knockout of pendrin and Na-Cl cotransporter (NCC) causes severe salt wasting, volume depletion, and renal failure. Proc Natl Acad Sci U S A 109(33):13368–13373
Sun X, Soleimani M, Petrovic S (2008) Decreased expression of Slc26a4 (Pendrin) and Slc26a7 in the kidneys of carbonic anhydrase II-deficient mice. Cell Physiol Biochem 21(1–3):95–108. doi:10.1159/000113751
Sweetser L, Douglas J, Riha L, Bell S (2005) Clinical presentation of metabolic alkalosis in an adult patient with cystic fibrosis. Respirology 10:254–256. doi:10.1111/j.1440-1843.2005.00650.x
Tamma G, Di Mise A, Ranieri M, Svelto M, Pisot R, Bilancio G, Cavallo P, De Santo NG, Cirillo M, Valenti G (2014) A decrease in aquaporin 2 excretion is associated with bed rest induced high calciuria. J Transl Med 12:133
Todd-Turla KM, Rusvai ERZSEBET, Naray-Fejes-Toth A, Fejes-Toth G (1996) CFTR expression in cortical collecting duct cells. Am J Physiol Ren Physiol 270(1):F237–F244
Trepiccione F, Pisitkun T, Hoffert JD, Poulsen SB, Capasso G, Nielsen S, Knepper MA, Fenton RA, Christensen BM (2014) Early targets of lithium in rat kidney inner medullary collecting duct include p38 and ERK1/2. Kidney Int 86(4):757–767
Vallet M, Picard N, Loffing-Cueni D et al (2006) Pendrin regulation in mouse kidney primarily is chloride-dependent. J Am Soc Nephrol 17(8):2153–2163
Varasteh Kia M, Barone S, McDonough AA et al Dysregulation of Acid Base (H^+/HCO_3^-) transport machinery in the kidney collecting duct in cystic fibrosis: role in the pathogenesis of metabolic alkalosis. Submitted for publication
Verlander JW, Hassell KA, Royaux IE, Glapion DM, Wang ME, Everett LA, Green ED, Wall SM (2003) Deoxycorticosterone upregulates PDS (Slc26a4) in mouse kidney: role of pendrin in mineralocorticoid-induced hypertension. Hypertension 179:356–362
Wall SM, Pech V (2008) The interaction of pendrin and the epithelial sodium channel in blood pressure regulation. Curr Opin Nephrol Hypertens 17(1):18–24, Review
Wall SM, Weinstein AM (2013) Cortical distal nephron Cl(−) transport in volume homeostasis and blood pressure regulation. Am J Physiol Renal Physiol 305(4):F427–F438, Review
Wall SM, Hassell KA, Royaux IE, Green ED, Chang JY, Shipley GL, Verlander JW (2003) Localization of pendrin in mouse kidney. Am J Physiol Renal Physiol 284(1):F229–F241
Wall SM, Kim YH, Stanley L, Glapion DM, Everett LA, Green ED, Verlander JW (2004) NaCl restriction upregulates renal Slc26a4 through subcellular redistribution: role in Cl- conservation. Hypertension 44(6):982, −7, 2004
Wang Z, Petrovic S, Mann E, Soleimani M (2002) Identification of an apical Cl(−)/HCO3(−) exchanger in the small intestine. Am J Physiol Gastrointest Liver Physiol 282(3):G573–G579

Wangemann P (2011) The role of pendrin in the development of the murine inner ear. Cell Physiol Biochem 28:527–534, Review

Xie Q, Welch R, Mercado A, Romero MF, Mount DB (2002) Molecular characterization of the murine Slc26a6 anion exchanger: functional comparison with Slc26a1. Am J Physiol Renal Physiol 283:F826–F838

Xu J, Henriksnäs J, Barone S, Witte D, Shull GE, Forte JG, Holm L, Soleimani M (2005) SLC26A9 is expressed in gastric surface epithelial cells, mediates Cl-/HCO3- exchange, and is inhibited by NH4+. Am J Physiol Cell Physiol 289(2):C493–C505

Xu J, Song P, Miller ML, Borgese F, Barone S, Riederer B, Wang Z, Alper SL, Forte JG, Shull GE, Ehrenfeld J, Seideler U, Soleimani M (2008) Deletion of the chloride transporter Slc26a9 causes loss of tubulovesicles in parietal cells and impairs acid secretion in the stomach. Proc Natl Acad Sci U S A 105(46):17955–17960

Xu J, Barone S, Li H, Holiday S, Zahedi K, Soleimani M (2011) Slc26a11, a chloride transporter, localizes with the vacuolar H(+)-ATPase of A-intercalated cells of the kidney. Kidney Int 80(9):926–937

Xu J, Barone S, Brooks MB, Soleimani M (2013) Double knockout of carbonic anhydrase II (CAII) and Na(+)-Cl(−) cotransporter (NCC) causes salt wasting and volume depletion. Cell Physiol Biochem 32(7):173–183

Yalçin E, Kiper N, Doğru D, Ozçelik U, Aslan AT (2005) Clinical features and treatment approaches in cystic fibrosis with pseudo-Bartter syndrome. Ann Trop Paediatr 25(2):119–124

Zahedi K, Barone S, Xu J, Soleimani M (2013) Potentiation of the effect of thiazide derivatives by carbonic anhydrase inhibitors: molecular mechanisms and potential clinical implications. PLoS One 8(11):e79327

Part IV

The Role of Pendrin in the Airways

The Role of Pendrin in the Airways: Links with Asthma and COPD

9

Kenji Izuhara, Shoichi Suzuki, Charity Nofziger, Masahiro Ogawa, Shoichiro Ohta, Yasuhiro Nanri, Yasutaka Mitamura, Tomohito Yoshihara, Nicoletta Pedemonte, Luis J.V. Galietta, and Markus Paulmichl

Abstract

Interleukin (IL)-4 and IL-13 are related cytokines correlated with type 2 immunity involved in the pathogenesis of bronchial asthma. Pendrin is induced by IL-4 or IL-13 in airway epithelial cells and is highly expressed in the lung tissues of asthma model mice or asthma patients. The signal transducer and activator of transcription (STAT) 6, the critical transcriptional factor for IL-4 or IL-13 signals, is required for IL-4– or IL-13–induced pendrin expression. Although the pathological roles of pendrin have been confirmed by the analyses of model mice, the underlying mechanisms are still unclear. Furthermore, pendrin has a potential to be correlated with other pulmonary diseases—chronic obstructive pulmonary disease, environmental chemical compound–

K. Izuhara (✉) • M. Ogawa • Y. Nanri • Y. Mitamura • T. Yoshihara • S. Ohta
Division of Medical Biochemistry, Department of Biomolecular Sciences,
Saga Medical School, 5-1-1, Nabeshima, Saga 849-8501, Japan
e-mail: kizuhara@cc.saga-u.ac.jp

S. Suzuki
Infectious Disease Unit, Asia International Institute of Infectious Disease Control,
Teikyo University, Tokyo, Japan

C. Nofziger • M. Paulmichl
Institute of Pharmacology and Toxicology, Paracelsus Medical University,
Salzburg, Austria

N. Pedemonte • L.J.V. Galietta
Istituto Giannina Gaslini, Genoa, Italy

S. Dossena, M. Paulmichl (eds.), *The Role of Pendrin in Health and Disease*,
DOI 10.1007/978-3-319-43287-8_9

exposed diseases or infectious diseases. In addition to these detrimental effects in pathological settings, the physiological role of pendrin in the lung, such as promoting antimicrobial activity, may be protective. Thus, whereas pendrin inhibition appears as a promising therapeutic strategy to treat asthma and other chronic respiratory diseases, it will be important to evaluate the effect of this inhibition in the lungs and other organs.

9.1 Introduction

Bronchial asthma is a common and chronic respiratory disease characterized by variable symptoms such as wheezing, shortness of breath and cough, and by variable expiratory airflow limitation (Global Initiative for Asthma 2015), and is estimated to affect at least 300 million people worldwide, therefore representing a significant medical and social problem. Inhaled corticosteroids (ICSs) are used as the first-line drugs for bronchial asthma, and furthermore, various molecularly targeted drugs mainly against type 2 immunity–associated molecules are now underway of development (Izuhara et al. 2015). Although ICSs are very effective agents for bronchial asthma, 5–10 % of asthma patients are resistant or hyporesponsive to ICS, accounting for about 50 % of the total costs for treating asthma (Adcock and Lane 2003). Molecularly targeted drugs for bronchial asthma are expected as alternative drugs in ICS-resistant patients; however, since asthma patients are heterogenous and molecularly targeted drugs are efficacious only in specific patient subgroups, stratification of asthma patients is required for their successful application (Izuhara et al. 2015). Furthermore, molecularly targeted drugs, mostly biologics, are expensive and cause economic burden. Therefore, the elucidation of the molecular mechanisms underlying the pathogenesis of bronchial asthma and development of novel therapeutic agents based on such mechanisms is still needed.

Chronic obstructive pulmonary disease (COPD) is another chronic pulmonary disease characterized by airway limitation. COPD, now the fourth leading cause of death in the world, will become the third by 2020 (Global Initiative for Chronic Obstructive Lung Disease 2015). Inhaled cigarette smoke and other noxious particles are major risk factors for lung inflammation underlying the pathogenesis of COPD. Although COPD is a disease that can be prevented by reducing exposure to risk factors and can be pharmacologically treated with bronchodilators, corticosteroids and phosphodiesterase-4 inhibitors (Global Initiative for Chronic Obstructive Lung Disease 2015), it is important to elucidate the underlying pathogenic mechanisms to develop novel and specific therapeutic agents.

In this chapter, while we describe the discovery of pendrin as a respiratory disease–associated molecule and focus on its roles in bronchial asthma and COPD, we also present some evidences supporting a beneficial role of this protein in the airways.

9.2 Overview of the Signal Transduction of Interleukin-4 and Interleukin-13 and Their Significance in Bronchial Asthma

Interleukin (IL)-4 and IL-13 are related cytokines correlated with type 2 immunity (Izuhara et al. 2002, 2006). These cytokines share their receptors and signal pathways. Two IL-4 receptors (IL-4Rs) are known: type I and type II IL-4Rs. Type I IL-4R is composed of the IL-4R α chain (IL-4Rα) and the common γ chain (γc), whereas type II IL-4R is composed of IL-4Rα and the IL-13R α1 chain (IL-13Rα1), *i.e.,* type I and type II IL-4Rs share IL-4Rα. Type II IL-4R also functions as the only IL-13R that can transduce IL-13 signals. The Janus kinase (JAK)–signal transducer and activator of transcription (STAT) pathway is critical for most cytokine signals. In case of IL-4 and IL-13, JAK1, JAK3 and TYK2 associate with IL-4Rα, γc and IL-13Rα1, respectively, followed by activation of STAT6. This means that all signals *via* IL-4R and IL-13R lead to STAT6 activation, explaining the common biological activities of IL-4 and IL-13.

IL-4 and IL-13 are secreted from T_H2 cells, mast cells, basophil and eosinophil granulocytes and group 2 innate lymphoid cells (only IL-13) during the type 2 immune response following allergic inflammation or parasite infection (Izuhara 2003; Lambrecht and Hammad 2013). Although IL-4 and IL-13 share their receptor and signaling pathways, the IL-4 actions are more hematological or immune cell (e.g., T_H2 cells, eosinophils and mast cells)–oriented, whereas the IL-13 actions are more resident cell (e.g., epithelial cells, fibroblasts and smooth muscle cells)–oriented due to the different distribution of these cytokines and their receptors (Izuhara 2003). The significance of IL-4 and/or IL-13, particularly the latter, in the pathogenesis of bronchial asthma was established in the 1990s mainly based on the analyses of model mice. Inhalation of IL-13 alone causes asthma-like phenotypes, whereas blockage of IL-13 signals diminishes those phenotypes (Grünig et al. 1998; Wills-Karp et al. 1998). High expression of type 2 cytokines including IL-4 and IL-13 in the bronchial tissues of asthma patients, as well as genetic associations of single-nucleotide polymorphisms in IL-4– or IL-13–related genes such as IL-13, IL-4Rα and STAT6 with bronchial asthma supported the significance of IL-4 and IL-13 in the pathogenesis of asthma (Kotsimbos et al. 1996; Bodey et al. 1999; Shirakawa et al. 2000). At this moment, the population of adult asthma patients showing type 2–dominant immunity is estimated to be 50–70% (Woodruff et al. 2009; Peters et al. 2014). Based on these findings, several IL-4 or IL-13 signaling antagonists—anti–IL-13 (Lebrikizumab and Tralokinumab) and IL-4Rα (Dupilumab) antibodies—are now under clinical development as anti-asthma drugs (Corren et al. 2011; Brightling et al. 2015; Wenzel et al. 2013).

9.3 Identification of Pendrin as an IL-4– or IL-13–Inducible Molecule and Its Involvement in Allergic Respiratory Inflammation

Substantial attention has been paid on how IL-13 causes asthma phenotypes acting on resident cells, particularly airway epithelial cells, because the analyses of genetically manipulated mice showed the significance of IL-13 actions on airway

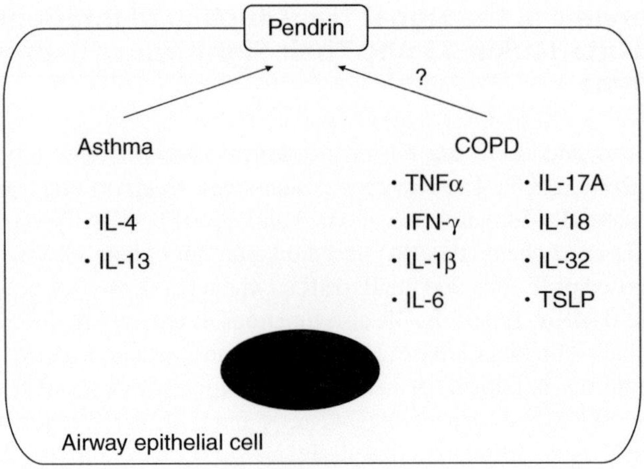

Fig. 9.1 Relationships between the triggers that increase pendrin expression and pathological pulmonary conditions. The triggers that induce pendrin expression and the pulmonary pathological conditions correlated with these triggers are depicted. The *question mark* indicates that the role of the COPD-related molecules in increasing pendrin expression is currently hypothetic

epithelial cells for the pathogenesis of bronchial asthma (Zhu et al. 1999; Kuperman et al. 2002). Moreover, it was shown that IL-13 causes in vitro differentiation of airway epithelial cells into mucus-producing cells, a key feature of bronchial asthma (Kondo et al. 2002; Zhen et al. 2007; Nakao et al. 2008). While both IL-4 and IL-13 have been shown to alter the expression of a multitude of ion transport and channel proteins (Nofziger et al. 2011a), we and others found that the *SLC26A4* gene was induced by IL-13 in human airway epithelial cells with microarray and in vitro cell-culture methods (Fig. 9.1, Nakao et al. 2008; Zhen et al. 2007; Kuperman et al. 2005). Interestingly, we found that IL-4 increased thiocyanate (SCN^-) transport in human airway epithelial cells independently of the cystic fibrosis transmembrane conductance regulator (CFTR) and tried to identify the transporter involved (Pedemonte et al. 2007). We examined expression of SLC26 transporters and other related channels in IL-4–stimulated airway epithelial cells using microarray, and found that only the *SLC26A4* gene was significantly induced by IL-4 among the investigated transporters and that pendrin/SLC26A4 is responsible for the SCN^-/Cl^- exchange using an in vitro system. Thus, pendrin has appeared as an IL-4– or IL-13–inducible molecule.

In accordance with the *in vitro* experiments, we and others demonstrated that pendrin expression is upregulated in lungs of asthma-mimicking mice such as ovalbumin- or IL-13–inhaled mice, as well as IL-13–transgenic mice (Nakao et al. 2008; Kuperman et al. 2005; Zhen et al. 2007; Di Valentin et al. 2009). Additionally, we showed that pendrin is expressed in the apical side of airway epithelial cells in ovalbumin-inhaled mice (Nakao et al. 2008). The finding that upregulation of pendrin expression is observed in the IL-13–overexpressing mice, in which STAT6 is expressed only in nonciliated airway epithelial cells, suggests that these cells are the pendrin-expressing cells upon stimulation of IL-4/IL-13 *in vivo* (Kuperman et al. 2005). It is of note that pendrin expression is enhanced in both acute and chronic asthma model mice (Di Valentin et al. 2009).

Several studies have been performed to examine the expression of pendrin in lung tissues of asthma patients. An initial study showed no difference in pendrin expression between asthmatic and control subjects (Kuperman et al. 2005). However, it was later reported that pendrin expression was enhanced in asthma patients compared to control subjects (Yick et al. 2013). The reason of the discrepancy between these studies is unknown; the different methods used to obtain the experimental material (endobronchial biopsy *vs.* brushing) or the different background of asthma patients (steroid-free atopic asthma patients in the study of Yick et al.) may account for it. We recently found that pendrin expression was up-regulated in bronchial epithelial cells of some asthma patients, suggesting that pendrin may be involved in the pathogenesis of at least some asthma patients (Suzuki, in press). Moreover, Ishida and collaborators demonstrated that pendrin is expressed in the nasal mucosa of the patients with allergic rhinitis or chronic rhinosinusitis with nasal polyps (Ishida et al. 2012). Allergic rhinitis is another type 2 immunity–dominant disease (Khan 2014), and chronic rhinosinusitis with nasal polyps is tightly associated with bronchial asthma (Okano et al. 2015). These results suggest the involvement of pendrin in both upper and lower respiratory allergic inflammation and the common characteristics in their pathogenesis.

9.4 Regulation of the *SLC26A4* Gene by IL-4/IL-13

As described earlier, STAT6 is a transcriptional factor critical for both the IL-4 and IL-13 signals (Izuhara et al. 2002, 2006). Since the *SLC26A4* gene is induced by IL-4 or IL-13 in airway epithelial cells, it was reasonable to think that STAT6 acts on the induction of the *SLC26A4* gene. Accordingly, we found two consensus sequences for the STAT6 binding site (TTC(N_4)GAA) at −3472 to −3463 (motif 1) and −1812 to −1803 (motif 2) of the 5'-flanking region of the *SLC26A4* gene and examined whether these regions were critical for cytokine-induced expression (Nofziger et al. 2011b; Vanoni et al. 2013). We found that while STAT6 is able to bind both consensus sequences following IL-4 exposure, only motif 2 is essential for IL-4– or IL-13–inducible pendrin expression using HEK-Blue cells stably overexpressing STAT6. Thus, it has been confirmed that the 5'-flanking region of the *SLC26A4* gene has a *cis*-acting sequence for STAT6.

9.5 Pathological Roles of Pendrin/SLC26A4 in Bronchial Asthma

Two papers have thus far demonstrated the pathological roles of pendrin in bronchial asthma using model mice. We, in one of the papers, showed the effects of pendrin overexpression in bronchial tissues, whereas Nakagami and collaborators

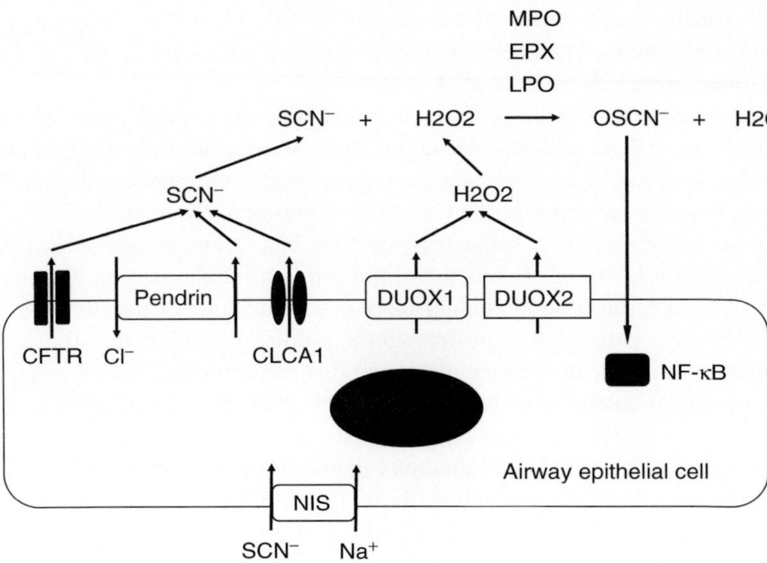

Fig. 9.2 Hypothetic schematic model of the pendrin/DUOX/peroxidase system in airway epithelial cells. The production pathway of $OSCN^-$ (hypothiocyanite) and its hypothetic detrimental effects linked to the activation of NF-κB are depicted

demonstrated the effects of genetic deficiency of pendrin on allergen-inducing inflammation (Nakao et al. 2008; Nakagami et al. 2008). Overexpression of pendrin in bronchial tissues causes mucus hyper-production, enhanced airway reactivity, neutrophil-dominant inflammation and upregulated expression of chemokines for recruitment of neutrophils. Reciprocally, pendrin deficiency decreases airway reactivity and infiltration of inflammatory cells in bronchoalveolar lavage fluid. Accordingly, systemic IgE production, mucus production and type 2 cytokines production are unaffected in allergen-challenged, pendrin-deficient mice. These results became formal proof showing a pathological role of pendrin in bronchial asthma.

The molecular mechanism of how pendrin exerts detrimental effects in bronchial asthma still remains unclear. Airway surface is covered by a thin layer of fluid, the airway surface liquid (ASL), whose composition and volume are critical to ensure proper mucociliary clearance and innate defense systems. Nakagami and collaborators showed that ASL is thickened in the pendrin-deficient tracheal cells stimulated by IL-13, probably because of dysregulated anion transport, suggesting the possibility that the thickened ASL can enhance mucus clearance and improve airway function (Nakagami et al. 2008). These authors concluded that pendrin-increased expression following allergens challenge may lead to ASL dehydration and airway inflammation and obstruction, therefore representing an important contributor of asthma exacerbations. We have another hypothesis to explain the harmful effects of pendrin in bronchial asthma (Fig. 9.2). SCN^- is actively transported into the bronchial lumen via the Na^+/I^- symporter (NIS)/SLC5A5 at the basal side and via several anion transporters—pendrin, CFTR and the chloride channel regulator

(CLCA)1/calcium-activated chloride channel (CaCC)—at the apical side in airway epithelial cells. SCN^-, together with H_2O_2 generated by dual oxidases (DUOX) 1 and 2, members of the NADPH oxidases family, is oxidized by peroxidases to $OSCN^-$ (hypothiocyanite). Three peroxidases—myeloperoxidase (MPO), eosinophil peroxidase (EPX) and lactoperoxidase (LPO)—are involved in this reaction in lung tissues. $OSCN^-$ serves as a potent innate defense system against microbes in lungs (Hawkins 2009; Barrett and Hawkins 2012). We recently found that inhibition of all peroxidase activities by panperoxidase inhibitors impared airway inflammation in the model mice and that $OSCN^-$ causes activation of NF-κB, a transcription factor critical for inflammatory responses or necrotic cell death in human airway epithelial cells (Suzuki in press and unpublished data). We assume that, in the asthmatic status, pendrin-induced overproduction of $OSCN^-$ leads to inflammatory responses in airway epithelial cells.

9.6 Pathological Roles of Pendrin/SLC26A4 in Other Respiratory Conditions

Although pendrin was first identified as a downstream molecule of IL-4/IL-13 signals or in the asthmatic status in the respiratory system, it was later found that various cytokines or environmental stimuli cause pendrin expression in lung tissues (Fig. 9.1). IL-1β (Pedemonte et al. 2007; Hogmalm et al. 2014), interferon-γ (IFN-γ; combined with rhinovirus, Nakagami et al. 2008) and IL-17A (Scanlon et al. 2014; Adams et al. 2014) have been listed as the cytokines that can induce pendrin expression. Moreover, several environmental chemical compounds such as silica (Sellamuthu et al. 2013), welding fumes (Oh et al. 2012), C_{60} fullerene (Fujita et al. 2010), single-wall carbon nanotubes (Fujita et al. 2015) as well as pathogenic microbes or microbe-derived molecules such as rhinovirus (combined with IFN-γ, Nakagami et al. 2008) and pertussis toxin (Scanlon et al. 2014; Connelly et al. 2012) have been shown to have the same ability.

We examined pendrin expression at mRNA level in the lung tissues of esterase-inhaled mice mimicking COPD (Nakao et al. 2008). Pendrin expression was enhanced 1 week after the inhalation, followed by expression of Muc5ac and Muc5b. Pendrin was localized at the apical side of epithelial cells as well as in that of asthma model mice. These results suggest a possibility that pendrin is involved in the pathogenesis of COPD, although thus far there is no report showing pendrin expression in COPD patients. The pathogenesis of COPD is complex and still undetermined; several pro-inflammatory cytokines—tumor necrosis factor α (TNFα), IFN-γ, IL-1β, IL-6, IL-17A, IL-18, IL-32 and thymic stromal lymphopoietin (TSLP)—have potentials to contribute to the pathogenesis of COPD (Caramori et al. 2014; Halwani et al. 2013; Rogliani et al. 2015). Any of these cytokines may be responsible for pendrin expression in the COPD condition (Fig. 9.1).

Carbonetti and collaborators have elegantly shown the pathological role of pendrin in pertussis disease, characterized by severe and prolonged cough and progression to a critical stage with pulmonary inflammation and death in young infants

(Connelly et al. 2012; Scanlon et al. 2014). They generated a *Bordetella pertussis* strain in which pertussis toxin is defected and found that its loading in mice reduces *Bordetella pertussis*–inducible pendrin expression while improving pulmonary inflammation. They demonstrated that pendrin expression induced by *Bordetella pertussis* is not STAT6-dependent, but IL-17A–dependent. Furthermore, they showed that pendrin deficiency improves *Bordetella pertussis*–inducible inflammation, but does not affect bacteria colonization. These results suggest the pathological roles of pendrin in pertussis disease, as well as its potential significance in other pulmonary infectious diseases or respiratory failure statuses.

9.7 Pendrin in Lung: Detrimental or Beneficial?

Although the association between pendrin overexpression and pathogenesis of asthma and COPD seems to be clearly established, there is also a number of evidences supporting a beneficial role of pendrin in the airways. As previously mentioned, composition and volume of ASL are critical to ensure proper mucociliary clearance and innate defense systems. The balance between fluid secretion and absorption results from the coordinated activities of different ion channels and transporters, in particular the epithelial sodium channel (ENaC) and CFTR, the chloride channel that is mutated in cystic fibrosis (CF). In CF, defective CFTR leads not only to inappropriately low chloride transport and impaired airway surface hydration but also to enhanced sodium absorption and further dehydration of the epithelial surface. As a consequence, the periciliary fluid covering the airway surface becomes abnormally shallow thus causing the arrest of mucociliary transport (Boucher 2007). In addition to differences in liquid volume, it has been reported that ASL from CF cultures as well as secretions from CF nasal submucosal glands are acidic as compared to non-CF (Coakley et al. 2003; Song et al. 2006). ASL acidic pH has been shown to reduce ciliary beating frequency and bacterial killing by phagocytic cells (Allen et al. 1997; Simchowitz 1985). A very important molecule to control the pH of the luminal microenvironment in epithelial tissues is HCO_3^-, due to its buffering role. It has been demonstrated that in bronchial and tracheal epithelia from CF humans and pigs, the lack of CFTR activity is paralleled by a reduction in electrogenic HCO_3^- secretion (Smith and Welsh 1992; Ostedgaard et al. 2011). In addition, by studying fluid and anion secretion in polarized cultures of Calu-3 cells, a model of human tracheobronchial submucosal gland serous cells, several investigators also concluded that CFTR is the main (only?) route for apical HCO_3^- secretion (Illek et al. 1997; Devor et al. 1999; Krouse et al. 2004). In addition to its buffering role, HCO_3^- has been reported to be very important for mucus homeostasis. Indeed, mucus secretion depends on HCO_3^- secretion (Garcia et al. 2009; Quinton 2001), and HCO_3^- has been reported to regulate mucin viscosity as well as mucin expansion (Chen et al. 2010; Muchekehu and Quinton 2010). Consequently, in CF, defective CFTR also leads to the loss of pH control and reduced mucus solubilization due to a decrease in HCO_3^- secretion, therefore contributing to the worsening of the pathology. To further remark the importance of

HCO_3^- in the airways, it has been recently shown that this anion has an important antibacterial function (Pezzulo et al. 2012) and, hence, that defective HCO_3^- secretion through CFTR has multiple pathogenic consequences in CF. However, the fact that, in the airway epithelium, CFTR is the only route for HCO_3^- secretion is somewhat atypical. Indeed, in different HCO_3^- secreting epithelia, including salivary glands and gastrointestinal tracts (Melvin et al. 2005; Stewart et al. 2009; Singh et al. 2010), HCO_3^- secretion is usually mediated by coupled activities of CFTR and at least one anion exchanger belonging to the SLC26 family (possibly SLC26A3 and/or SLC26A6). In the airways, expression of SLC26A3 and SLC26A9 is well documented (Wheat et al. 2000; Lohi et al. 2002), and it was reported that SLC26A9 acts as a CFTR-regulated constitutively active Cl^- channel of the bronchial epithelium (Bertrand et al. 2009). Recently, by studying HCO_3^- secretion in polarized epithelia of Calu-3 serous cells, Gray and collaborators described the existence of a luminal cAMP/PKA-activated Cl^-/HCO_3^- exchange activity, whose pharmacological and functional profiles were consistent with those of pendrin (Garnett et al. 2011). These authors demonstrated that transepithelial liquid secretion rate and liquid pH were increased upon cAMP/PKA stimulation, that CFTR inhibition reduced the rate of liquid secretion (but not the pH), whereas decreasing pendrin activity lowered pH with little effect on volume (Garnett et al. 2011). The authors concluded that CFTR predominately controls the rate of liquid secretion, whereas pendrin regulates the composition of the secreted fluid, thus identifying a critical role for this anion exchanger in transcellular HCO_3^- secretion in airway serous cells (Garnett et al. 2011).

As discussed earlier, another possible beneficial role for pendrin in the airways is associated to the transport of SCN^-. SCN^- is an important physiological anion involved in innate defense of mucosal surfaces. SCN^- is oxidized by H_2O_2, a reaction catalyzed by LPO, to produce $OSCN^-$, a molecule with bactericidal or bacteriostatic activity (Fig. 9.2 and Ratner and Prince 2000; Hawkins 2009; Barrett and Hawkins 2012). It has been demonstrated that LPO is secreted on the airway surface (Gerson et al. 2000; Wijkstrom-Frei et al. 2003), that the airway epithelium has the ability to generate H_2O_2 at the apical membrane through DUOX1 and DUOX2 (Schwarzer et al. 2004; Forteza et al. 2005) and that polarized preparations of airway epithelium transport SCN^- from the basolateral to the apical side in a way that involves CFTR (Fragoso et al. 2004). Indeed, CF cells show a defective bacterial killing due to lack of SCN^- transport (Moskwa et al. 2007). Such observations suggest that the $LPO/H_2O_2/SCN^-$ antimicrobial system is present in the airways, where it could play an important role in innate defense function, and imply that CFTR dysfunction in CF could favor bacterial colonization also by causing a deficit in SCN^- transport. Other anion channels and transporters could also contribute to the $OSCN^-$ antimicrobial system by providing an alternate route for SCN^- transport. In this respect, we demonstrated that stimulation with IL-4 causes a dramatic upregulation of an electroneutral transport for Cl^- and SCN^-, which is mediated by pendrin (Pedemonte et al. 2007).

Summarizing, pendrin may be important as a protective molecule by representing a route for HCO_3^- and SCN^- secretion in physiological conditions. Both anions appear to be important as the source of antimicrobial activity (Moskwa et al. 2007; Pezzulo et al. 2012). Furthermore, secretion of HCO_3^- is required to clear mucus secretions (Hoegger et al. 2014). On the other hand, increased expression of pendrin in pathological settings has been shown to worsen asthmatic inflammation (Nakao et al. 2008; Nakagami et al. 2008), an effect that may involve $OSCN^-$-mediated activation of NF-kB (unpublished data). Future studies are needed to elucidate the specific role of pendrin in airway pathophysiology and the possible beneficial or undesired consequences of its pharmacological modulation.

9.8 Perspectives

After pendrin was identified as a bronchial asthma-related molecule downstream of IL-4 and IL-13 signals, our comprehension about its involvement in pulmonary dysfunction has been extended. Given its possible pro-inflammatory activity, pendrin inhibition appears as a promising therapeutic strategy to treat asthma and other chronic respiratory diseases. Therefore, development of highly specific and potent pharmacological inhibitors of pendrin is needed. However, since pendrin has been also shown to play a role in multiple functions, it will be important to evaluate the effect of its inhibition in the lungs and other organs.

References

Adams KM, Abraham V, Spielman D, Kolls JK, Rubenstein RC, Conner GE, Cohen NA, Kreindler JL (2014) IL-17A induces Pendrin expression and chloride-bicarbonate exchange in human bronchial epithelial cells. PLoS One 9:e103263

Adcock IM, Lane SJ (2003) Corticosteroid-insensitive asthma: molecular mechanisms. J Endocrinol 178:347–355

Allen DB, Maguire JJ, Mahdavian M, Wicke C, Marcocci L, Scheuenstuhl H, Chang M, Le AX, Hopf HW, Hunt TK (1997) Wound hypoxia and acidosis limit neutrophil bacterial killing mechanisms. Arch Surg 132:991–996

Barrett TJ, Hawkins CL (2012) Hypothiocyanous acid: benign or deadly? Chem Res Toxicol 25:263–273

Bertrand CA, Zhang R, Pilewski JM, Frizzell RA (2009) SLC26A9 is a constitutively active, CFTR-regulated anion conductance in human bronchial epithelia. J Gen Physiol 133:421–438

Bodey KJ, Semper AE, Redington AE, Madden J, Teran LM, Holgate ST, Frew AJ (1999) Cytokine profiles of BAL T cells and T-cell clones obtained from human asthmatic airways after local allergen challenge. Allergy 54:1083–1093

Boucher RC (2007) Evidence for airway surface dehydration as the initiating event in CF airway disease. J Intern Med 261:5–16

Brightling CE, Chanez P, Leigh R, O'Byrne PM, Korn S, She D, May RD, Streicher K, Ranade K, Piper E (2015) Efficacy and safety of tralokinumab in patients with severe uncontrolled asthma: a randomised, double-blind, placebo-controlled, phase 2b trial. Lancet Respir Med. doi:10.1016/s2213-2600(15)00197-6

Caramori G, Adcock IM, Di Stefano A, Chung KF (2014) Cytokine inhibition in the treatment of COPD. Int J Chron Obstruct Pulmon Dis 9:397–412

Chen EY, Yang N, Quinton PM, Chin WC (2010) A new role for bicarbonate in mucus formation. Am J Physiol 299:L542–L549
Coakley RD, Grubb BR, Paradiso AM, Gatzy JT, Johnson LG, Kreda SM, O'Neal WK, Boucher RC (2003) Abnormal surface liquid pH regulation by cultured cystic fibrosis bronchial epithelium. Proc Natl Acad Sci U S A 100:16083–16088
Connelly CE, Sun Y, Carbonetti NH (2012) Pertussis toxin exacerbates and prolongs airway inflammatory responses during *Bordetella pertussis* infection. Infect Immun 80:4317–4332
Corren J, Lemanske RF, Hanania NA, Korenblat PE, Parsey MV, Arron JR, Harris JM, Scheerens H, Wu LC, Su Z, Mosesova S, Eisner MD, Bohen SP, Matthews JG (2011) Lebrikizumab treatment in adults with asthma. N Engl J Med 365:1088–1098
Devor DC, Singh AK, Lambert LC, DeLuca A, Frizzell RA, Bridges RJ (1999) Bicarbonate and chloride secretion in Calu-3 human airway epithelial cells. J Gen Physiol 113:743–760
Di Valentin E, Crahay C, Garbacki N, Hennuy B, Gueders M, Noel A, Foidart JM, Grooten J, Colige A, Piette J, Cataldo D (2009) New asthma biomarkers: lessons from murine models of acute and chronic asthma. Am J Physiol Lung Cell Mol Physiol 296:L185–L197
Forteza R, Salathe M, Miot F, Forteza R, Conner GE (2005) Regulated hydrogen peroxide production by Duox in human airway epithelial cells. Am J Respir Cell Mol Biol 32:462–469
Fragoso MA, Fernandez V, Forteza R, Randell SH, Salathe M, Conner GE (2004) Transcellular thiocyanate transport by human airway epithelia. J Physiol 561:183–194
Fujita K, Morimoto Y, Endoh S, Uchida K, Fukui H, Ogami A, Tanaka I, Horie M, Yoshida Y, Iwahashi H, Nakanishi J (2010) Identification of potential biomarkers from gene expression profiles in rat lungs intratracheally instilled with C_{60} fullerenes. Toxicology 274:34–41
Fujita K, Fukuda M, Fukui H, Horie M, Endoh S, Uchida K, Shichiri M, Morimoto Y, Ogami A, Iwahashi H (2015) Intratracheal instillation of single-wall carbon nanotubes in the rat lung induces time-dependent changes in gene expression. Nanotoxicology 9:290–301
Garcia MA, Yang N, Quinton PM (2009) Normal mouse intestinal mucus release requires cystic fibrosis transmembrane regulator-dependent bicarbonate secretion. J Clin Invest 119: 2613–2622
Garnett JP, Hickman E, Burrows R, Hegyi P, Tiszlavicz L, Cuthbert AW, Fong P, Gray MA (2011) Novel role for pendrin in orchestrating bicarbonate secretion in cystic fibrosis transmembrane conductance regulator (CFTR)-expressing airway serous cells. J Biol Chem 286: 41069–41082
Gerson C, Sabater J, Scuri M, Torbati A, Coffey R, Abraham JW, Lauredo I, Forteza R, Wanner A, Salathe M, Abraham WM, Conner GE (2000) The lactoperoxidase system functions in bacterial clearance of airways. Am J Respir Cell Mol Biol 22:665–671
Global Initiative for Asthma (2015) Global strategy for asthma management and prevention. http://www.ginasthma.org/documents/4. Accessed 6 Nov 2015
Global Initiative for Chronic Obstructive Lung Disease (2015) Global strategy for the diagnosis, management, and prevention of chronic obstructive pulmonary disease. http://www.goldcopd.org/guidelines-global-strategy-for-diagnosis-management.html. Accessed 6 Nov 2015
Grünig G, Warnock M, Wakil AE, Venkayya R, Brombacher F, Rennick DM, Sheppard D, Mohrs M, Donaldson DD, Locksley RM, Corry DB (1998) Requirement for IL-13 independently of IL-4 in experimental asthma. Science 282:2261–2263
Halwani R, Al-Muhsen S, Hamid Q (2013) T helper 17 cells in airway diseases: from laboratory bench to bedside. Chest 143:494–501
Hawkins CL (2009) The role of hypothiocyanous acid (HOSCN) in biological systems. Free Radic Res 43:1147–1158
Hoegger MJ, Fischer AJ, McMenimen JD, Ostedgaard LS, Tucker AJ, Awadalla MA, Moninger TO, Michalski AS, Hoffman EA, Zabner J, Stoltz DA, Welsh MJ (2014) Impaired mucus detachment disrupts mucociliary transport in a piglet model of cystic fibrosis. Science 345:818–822
Hogmalm A, Bry M, Strandvik B, Bry K (2014) IL-1β expression in the distal lung epithelium disrupts lung morphogenesis and epithelial cell differentiation in fetal mice. Am J Physiol Lung Cell Mol Physiol 306:L23–L34

Illek B, Yankaskas JR, Machen TE (1997) cAMP and genistein stimulate HCO3− conductance through CFTR in human airway epithelia. Am J Physiol 272:L752–L761
Ishida A, Ohta N, Suzuki Y, Kakehata S, Okubo K, Ikeda H, Shiraishi H, Izuhara K (2012) Expression of pendrin and periostin in allergic rhinitis and chronic rhinosinusitis. Allergol Int 61:589–595
Izuhara K (2003) The role of interleukin-4 and interleukin-13 in the non-immunologic aspects of asthma pathogenesis. Clin Chem Lab Med 41:860–864
Izuhara K, Arima K, Yasunaga S (2002) IL-4 and IL-13: their pathological roles in allergic diseases and their potential in developing new therapies. Curr Drug Targets Inflamm Allergy 1:263–269
Izuhara K, Arima K, Kanaji S, Ohta S, Kanaji T (2006) IL-13: a promising therapeutic target for bronchial asthma. Curr Med Chem 13:2291–2298
Izuhara K, Matsumoto H, Ohta S, Ono J, Arima K, Ogawa M (2015) Recent developments regarding periostin in bronchial asthma. Allergol Int 64:S3–S10
Khan DA (2014) Allergic rhinitis and asthma: epidemiology and common pathophysiology. Allergy Asthma Proc 35:357–361
Kondo M, Tamaoki J, Takeyama K, Nakata J, Nagai A (2002) Interleukin-13 induces goblet cell differentiation in primary cell culture from guinea pig tracheal epithelium. Am J Respir Cell Mol Biol 27:536–541
Kotsimbos TC, Ernst P, Hamid QA (1996) Interleukin-13 and interleukin-4 are coexpressed in atopic asthma. Proc Assoc Am Physicians 108:368–373
Krouse ME, Talbott JF, Lee MM, Joo NS, Wine JJ (2004) Acid and base secretion in the Calu-3 model of human serous cells. Am J Physiol 287:L1274–L1283
Kuperman DA, Huang X, Koth LL, Chang GH, Dolganov GM, Zhu Z, Elias JA, Sheppard D, Erle DJ (2002) Direct effects of interleukin-13 on epithelial cells cause airway hyperreactivity and mucus overproduction in asthma. Nat Med 8:885–889
Kuperman DA, Lewis CC, Woodruff PG, Rodriguez MW, Yang YH, Dolganov GM, Fahy JV, Erle DJ (2005) Dissecting asthma using focused transgenic modeling and functional genomics. J Allergy Clin Immunol 116:305–311
Lambrecht BN, Hammad H (2013) Asthma: the importance of dysregulated barrier immunity. Eur J Immunol 43:3125–3137
Lohi H, Kujala M, Makela S, Lehtonen E, Kestila M, Saarialho-Kere U, Markovich D, Kere J (2002) Functional characterization of three novel tissue-specific anion exchangers SLC26A7, −A8, and -A9. J Biol Chem 277:14246–14254
Melvin JE, Yule D, Shuttleworth T, Begenisich T (2005) Regulation of fluid and electrolyte secretion in salivary gland acinar cells. Annu Rev Physiol 67:445–469
Moskwa P, Lorentzen D, Excoffon KJ, Zabner J, McCray PB Jr, Nauseef WM, Dupuy C, Bánfi B (2007) A novel host defense system of airways is defective in cystic fibrosis. Am J Respir Crit Care Med 175:174–183
Muchekehu RW, Quinton PM (2010) A new role for bicarbonate secretion in cervico-uterine mucus release. J Physiol 588:2329–2342
Nakagami Y, Favoreto S Jr, Zhen G, Park SW, Nguyenvu LT, Kuperman DA, Dolganov GM, Huang X, Boushey HA, Avila PC, Erle DJ (2008) The epithelial anion transporter pendrin is induced by allergy and rhinovirus infection, regulates airway surface liquid, and increases airway reactivity and inflammation in an asthma model. J Immunol 181:2203–2210
Nakao I, Kanaji S, Ohta S, Matsushita H, Arima K, Yuyama N, Yamaya M, Nakayama K, Kubo H, Watanabe M, Sagara H, Sugiyama K, Tanaka H, Toda S, Hayashi H, Inoue H, Hoshino T, Shiraki A, Inoue M, Suzuki K, Aizawa H, Okinami S, Nagai H, Hasegawa M, Fukuda T, Green ED, Izuhara K (2008) Identification of pendrin as a common mediator for mucus production in bronchial asthma and chronic obstructive pulmonary disease. J Immunol 180:6262–6269
Nofziger C, Dossena S, Suzuki S, Izuhara K, Paulmichl M (2011a) Pendrin function in airway epithelia. Cell Physiol Biochem 28:571–578
Nofziger C, Vezzoli V, Dossena S, Schonherr T, Studnicka J, Nofziger J, Vanoni S, Stephan S, Silva ME, Meyer G, Paulmichl M (2011b) STAT6 links IL-4/IL-13 stimulation with pendrin expression in asthma and chronic obstructive pulmonary disease. Clin Pharmacol Ther 90:399–405

Oh JH, Yang MJ, Heo JD, Yang YS, Park HJ, Park SM, Kwon MS, Song CW, Yoon S, Yu IJ (2012) Inflammatory response in rat lungs with recurrent exposure to welding fumes: a transcriptomic approach. Toxicol Ind Health 28:203–215

Okano M, Kariya S, Ohta N, Imoto Y, Fujieda S, Nishizaki K (2015) Association and management of eosinophilic inflammation in upper and lower airways. Allergol Int 64:131–138

Ostedgaard LS, Meyerholz DK, Chen JH, Pezzulo AA, Karp PH, Rokhlina T, Ernst SE, Hanfland RA, Reznikov LR, Ludwig PS, Rogan MP, Davis GJ, Dohrn CL, Wohlford-Lenane C, Taft PJ, Rector MV, Hornick E, Nassar BS, Samuel M, Zhang Y, Richter SS, Uc A, Shilyansky J, Prather RS, McCray PB Jr, Zabner J, Welsh MJ, Stoltz DA (2011) The ΔF508 mutation causes CFTR misprocessing and cystic fibrosis-like disease in pigs. Sci Transl Med 3:74ra24

Pedemonte N, Caci E, Sondo E, Caputo A, Rhoden K, Pfeffer U, Di Candia M, Bandettini R, Ravazzolo R, Zegarra-Moran O, Galietta LJ (2007) Thiocyanate transport in resting and IL-4-stimulated human bronchial epithelial cells: role of pendrin and anion channels. J Immunol 178:5144–5153

Peters MC, Mekonnen ZK, Yuan S, Bhakta NR, Woodruff PG, Fahy JV (2014) Measures of gene expression in sputum cells can identify T_H2-high and T_H2-low subtypes of asthma. J Allergy Clin Immunol 133:388–394

Pezzulo AA, Tang XX, Hoegger MJ, Alaiwa MH, Ramachandran S, Moninger TO, Karp PH, Wohlford-Lenane CL, Haagsman HP, van Eijk M, Bánfi B, Horswill AR, Stoltz DA, McCray PB Jr, Welsh MJ, Zabner J (2012) Reduced airway surface pH impairs bacterial killing in the porcine cystic fibrosis lung. Nature 487:109–113

Quinton PM (2001) The neglected ion: HCO_3^-. Nat Med 7:292–293

Ratner AJ, Prince A (2000) Lactoperoxidase: new recognition of an "old" enzyme in airway defenses. Am J Respir Cell Mol Biol 22:642–644

Rogliani P, Calzetta L, Ora J, Matera MG (2015) Canakinumab for the treatment of chronic obstructive pulmonary disease. Pulm Pharmacol Ther 31:15–27

Scanlon KM, Gau Y, Zhu J, Skerry C, Wall SM, Soleimani M, Carbonetti NH (2014) Epithelial anion transporter pendrin contributes to inflammatory lung pathology in mouse models of *Bordetella pertussis* infection. Infect Immun 82:4212–4221

Schwarzer C, Machen TE, Illek B, Fischer H (2004) NADPH oxidase-dependent acid production in airway epithelial cells. J Biol Chem 279:36454–36461

Sellamuthu R, Umbright C, Roberts JR, Cumpston A, McKinney W, Chen BT, Frazer D, Li S, Kashon M, Joseph P (2013) Molecular insights into the progression of crystalline silica-induced pulmonary toxicity in rats. J Appl Toxicol 33:301–312

Shirakawa T, Deichmann KA, Izuhara K, Mao I, Adra CN, Hopkin JM (2000) Atopy and asthma: genetic variants of IL-4 and IL-13 signalling. Immunol Today 21:60–64

Simchowitz L (1985) Intracellular pH modulates the generation of superoxide radicals by human neutrophils. J Clin Invest 76:1079–1089

Singh AK, Riederer B, Chen M, Xiao F, Krabbenhöft A, Engelhardt R, Nylander O, Soleimani M, Seidler U (2010) The switch of intestinal Slc26 exchangers from anion absorptive to HCO_3^- secretory mode is dependent on CFTR anion channel function. Am J Physiol 298:C1057–C1065

Smith JJ, Welsh MJ (1992) cAMP stimulates bicarbonate secretion across normal, but not cystic fibrosis airway epithelia. J Clin Invest 89:1148–1153

Song Y, Salinas D, Nielson DW, Verkman AS (2006) Hyperacidity of secreted fluid from submucosal glands in early cystic fibrosis. Am J Physiol 290:C741–C749

Stewart AK, Yamamoto A, Nakakuki M, Kondo T, Alper SL, Ishiguro H (2009) Functional coupling of apical Cl^-/HCO_3^- exchange with CFTR in stimulated HCO_3^- secretion by guinea pig interlobular pancreatic duct. Am J Physiol 296:G1307–G1317

Suzuki S, Ogawa M, Ohta S, Arima K, Nunomura S, Nanri Y, Mitamura Y, Yoshihara T, Nakamura Y, Yamauchi K, Chibana K, Ishii Y, Lee JJ, Aratani Y, Kakuta S, Kubo S, Iwakura Y, Yoshida H, Izuhara K (in press) The potential for repositioning anti-thyroid agents as anti-asthma drugs. J Allergy Clin Immunol

Vanoni S, Nofziger C, Dossena S, Soyal SM, Patsch W, Plevani P, Duschl A, Paulmichl M (2013) The human pendrin promoter contains two N_4 GAS motifs with different functional relevance. Cell Physiol Biochem 32:238–248

Wenzel S, Ford L, Pearlman D, Spector S, Sher L, Skobieranda F, Wang L, Kirkesseli S, Rocklin R, Bock B, Hamilton J, Ming JE, Radin A, Stahl N, Yancopoulos GD, Graham N, Pirozzi G (2013) Dupilumab in persistent asthma with elevated eosinophil levels. N Engl J Med 368:2455–2466

Wheat VJ, Shumaker H, Burnham C, Shull GE, Yankaskas JR, Soleimani M (2000) CFTR induces the expression of DRA along with Cl^-/HCO_3^- exchange activity in tracheal epithelial cells. Am J Physiol 279:C62–C71

Wijkstrom-Frei C, El-Chemaly S, Ali-Rachedi R, Gerson C, Cobas MA, Forteza R, Salathe M, Conner GE (2003) Lactoperoxidase and human airway host defense. Am J Respir Cell Mol Biol 29:206–212

Wills-Karp M, Luyimbazi J, Xu X, Schofield B, Neben TY, Karp CL, Donaldson DD (1998) Interleukin-13: central mediator of allergic asthma. Science 282:2258–2261

Woodruff PG, Modrek B, Choy DF, Jia G, Abbas AR, Ellwanger A, Koth LL, Arron JR, Fahy JV (2009) T-helper type 2-driven inflammation defines major subphenotypes of asthma. Am J Respir Crit Care Med 180:388–395

Yick CY, Zwinderman AH, Kunst PW, Grunberg K, Mauad T, Dijkhuis A, Bel EH, Baas F, Lutter R, Sterk PJ (2013) Transcriptome sequencing (RNA-Seq) of human endobronchial biopsies: asthma *versus* controls. Eur Respir J 42:662–670

Zhen G, Park SW, Nguyenvu LT, Rodriguez MW, Barbeau R, Paquet AC, Erle DJ (2007) IL-13 and epidermal growth factor receptor have critical but distinct roles in epithelial cell mucin production. Am J Respir Cell Mol Biol 36:244–253

Zhu Z, Homer RJ, Wang Z, Chen Q, Geba GP, Wang J, Zhang Y, Elias JA (1999) Pulmonary expression of interleukin-13 causes inflammation, mucus hypersecretion, subepithelial fibrosis, physiologic abnormalities, and eotaxin production. J Clin Invest 103:779–788

Part V

The Pendrin Gene and Protein

Transcriptional Regulation and Epigenetics of Pendrin

10

Julia Rozenfeld, Charity Nofziger, Orly Kladnitsky, Seth L. Alper, and Israel Zelikovic

Abstract

The *SLC26A4* gene's 21 exons on chromosome 7q31.1 encode the pendrin (SLC26A4) Cl^-/anion exchanger. *SLC26A4* is highly expressed in the epithelial cells of kidney, thyroid, inner ear and airways, in which the pendrin polypeptide is essential for normal chloride reabsorption, bicarbonate secretion and iodide accumulation. These activities regulate systemic volume and acid-base status, airway surface liquid composition, thyroid hormone synthesis and endolymph ion balance. *SLC26A4* biallelic mutations underlie autosomal recessive Pendred syndrome, a major cause of congenital deafness variably accompanied by euthyroid goiter.

Recent studies of transcriptional regulation of pendrin expression have revealed that *SLC26A4* promoter binding sites for thyroid transcription factors 1 and 2 (TTF-1/2) and forkhead box transcription factor I1 (FOXI1) are essential for cell-specific promoter activity in thyroid and inner ear, respectively. Systemic

J. Rozenfeld • O. Kladnitsky
Laboratory of Developmental Nephrology, Department of Physiology and Biophysics, Faculty of Medicine, Technion-Israel Institute of Technology, Haifa, Israel

C. Nofziger
Institute of Pharmacology and Toxicology, Paracelsus Medical University, Salzburg, Austria

S.L. Alper
Division of Nephrology and Center for Vascular Biology Research, Beth Israel Deaconess Medical Center, Beth, Israel

Harvard Medical School, Boston, MA, USA

I. Zelikovic (✉)
Laboratory of Developmental Nephrology, Department of Physiology and Biophysics, Faculty of Medicine, Technion-Israel Institute of Technology, Haifa, Israel

Laboratory of Developmental Nephrology, Rappaport Faculty of Medicine, Technion – Israel Institute of Technology, P.O. Box 9649, Haifa 3109601, Israel
e-mail: israelz@tx.technion.ac.il

S. Dossena, M. Paulmichl (eds.), *The Role of Pendrin in Health and Disease*, DOI 10.1007/978-3-319-43287-8_10 pH, Cl^- concentration, aldosterone, uroguanylin (UGN) and interleukins 4 (IL-4) and 13 (IL-13) each modulate pendrin transcription through distinct regulatory elements. The *SLC26A4* promoter elements regulated by pH, Cl^- and aldosterone remain undefined, but defined binding sites for heat shock factor 1 (HSF1) and signal transducer and activator of transcription protein 6 (STAT6), respectively, mediate transcriptional effects of UGN and IL-4/IL-13 on the *SLC26A4* promoter. Recent findings also suggest epigenetic regulation of pendrin gene expression.

Although the detailed molecular mechanisms of pendrin transcriptional and epigenetic regulation remain incomplete, they will provide insight into pendrin roles in acid-base, electrolyte and water homeostasis, blood pressure regulation and airway function. They will, in addition, influence our understanding and potential treatment of hypertension, heart failure, hepatic failure and other fluid retention syndromes, and airway diseases such as asthma and chronic obstructive lung disease.

10.1 Introduction

10.1.1 The Pendrin Gene *(SLC26A4/PDS)* and Its Product, SLC26A4/Pendrin Protein

Everett et al. (1997) identified the *SLC26A4/PDS* gene as the locus of mutations causing Pendred syndrome, an autosomal recessive disorder characterized by sensorineural hearing loss and enlargement of the thyroid gland (Campbell et al. 2001; Reardon and Trembath 1996). The 21 exons *SLC26A4* gene resides immediately 3' from the closely homologous *SLC26A3/DRA* gene on chromosome 7q31.1 (Haila et al. 1998), suggesting an ancient gene duplication (Soleimani and Xu 2006). The ~5 kb *SLC26A4* transcript is most highly expressed in thyroid, kidney, inner ear and inflamed respiratory epithelium (Everett et al. 1997, 1999; Wall et al. 2003; Pedemonte et al. 2007), but is detected also in trophoblasts (Bidart et al. 2000a), endometrium (Suzuki et al. 2002), testis (Lacroix et al. 2006), *vas deferens* (Carlin et al. 2006), mammary gland (Rillema and Hill 2003) and the adrenal medulla (Fernandez-Lazo et al. 2015). The *SLC26A4* transcript encodes the 780 amino acid pendrin protein, sharing with other solute carrier (SLC) 26 superfamily proteins the predicted topographic disposition of a short N-terminal cytoplasmic domain followed by ~14 transmembrane spanning segments and a C-terminal cytoplasmic sulfate transporter and anti-sigma-factor-antagonist (STAS) domain (Everett et al. 1997; Alper and Sharma 2013; Shelden et al. 2010; Dossena et al. 2009; Gorbunov et al. 2014). Functional studies in heterologous expression systems have revealed that pendrin functions as an electroneutral plasmalemmal anion exchanger (Shcheynikov et al. 2008; Reimold et al. 2011). Pendrin functions in the renal cortical collecting duct (CCD) in Cl^-/HCO_3^- (Royaux et al. 2001; Verlander et al. 2003) and Cl^-/I^- exchange (Kim et al. 2009). Pendrin is also believed to mediate Cl^-/

HCO_3^- exchange in the inner ear and respiratory epithelium, and Cl^-/I^- exchange in the thyroid (Nakaya et al. 2007; Garnett et al. 2011; Yoshida et al. 2004).

10.1.2 Pendred Syndrome

Pendred syndrome is the most common form of syndromic deafness and accounts for up to 10% of hereditary deafness (Kopp 2000; Park et al. 2003; Sheffield et al. 1996). The profound hearing loss in Pendred syndrome is particularly pronounced at high frequency, and occasionally accompanied by impaired vestibular function (Kopp 2000). It is associated with radiologically detectable enlargement of the vestibular aqueduct (EVA) and with temporal bone abnormalities (Campbell et al. 2001; Johnsen et al. 1987). Pendred syndrome is further characterized by a thyroidal iodide organification defect variably accompanied by euthyroid goiter (Campbell et al. 2001; Reardon and Trembath 1996). Pendrin mutations without thyroid dysfunction are linked to recessive non-syndromic deafness (DFNB4) (Anwar et al. 2009) with EVA (Choi et al. 2009a). More than 300 different *SLC26A4* missense mutations have been identified in deaf patients, and many of the protein variants encoded by these mutations have been functionally evaluated ((Choi et al. 2009a, b; Dai et al. 2009), among many others) by several methodological approaches (see Chap. 11 for more information).

10.2 Expression and Function of Pendrin

10.2.1 Pendrin in the Kidney

The CCD plays a vital role in acid-base homeostasis and electrolyte balance, and includes principal cells and intercalated cells (types A, B and non-A non-B (Alpern et al. 2000; Schwartz 2001)). Pendrin is located at the luminal membrane of B and non-A non-B intercalated cells of the CCD (Royaux et al. 2001). Pendrin contributes to acid-base balance by secreting HCO_3^- into the tubular lumen in exchange for luminal Cl^- (Royaux et al. 2001), and that same Cl^- reabsorption regulates body fluid homeostasis and blood pressure (Verlander et al. 2003; Quentin et al. 2004; Vallet et al. 2006; Wall and Lazo-Fernandez 2015). Pendrin protein expression in the apical membrane of B intercalated cells is increased by systemic HCO_3^- loading and decreased by acid loading (Frische et al. 2003; Petrovic et al. 2003; Wagner et al. 2002). Systemic and tubule lumen Cl^- concentrations and/or Cl^- loads also regulate pendrin protein levels and activity (Quentin et al. 2004; Vallet et al. 2006; Verlander et al. 2006).

The importance of intercalated cells (and of pendrin, in particular) to systemic salt balance and blood pressure regulation has been clearly shown by recent animal studies demonstrating: (1) the role of pendrin in the pathogenesis of mineralocorticoid and/or angiotensin II-induced hypertension (Verlander et al. 2003, 2011; Pech et al. 2007); (2) the functional link between pendrin and Na^+ transport mechanisms in the CCD, including the amiloride-sensitive epithelial Na^+ channel of the principal cell, ENaC (Kim et al. 2007; Pech et al. 2010) and the Na^+-driven Cl^-/HCO_3^- exchanger of the B intercalated cell, NDCBE/SLC4A8 (Leviel et al. 2010); (3) Cl^--sensitive hypertension in mice overexpressing pendrin in intercalated cells (Jacques et al. 2013); (4) severe salt wasting and volume depletion in pendrin/NaCl cotransporter (NCC) double knockout mice (Soleimani et al. 2012); and (5) the major role of pendrin in the integrated compensatory NaCl reabsorption network activated in the kidney of SPAK kinase-deficient (and hence, NCC phosphorylation-absent) mouse (Grimm et al. 2015).

Our deletion analysis of the 5'-flanking region of the *SLC26A4* gene defined both positive and negative regulatory elements in the promoter and proposed a major role for these control elements in the epithelial cell-specific, regulated expression of this gene in the kidney as well as in the inner ear and thyroid (Adler et al. 2008). We also showed that pendrin is transcriptionally regulated by systemic pH and aldosterone (Adler et al. 2008; Rozenfeld et al. 2011), by extracellular Cl^- concentration (Efrati et al. 2007) as well as by the "intestinal natriuretic hormone" uroguanylin (UGN) (Rozenfeld et al. 2012, 2013) in renal epithelial cells (see below), further demonstrating the important role of pendrin in electrolyte balance and blood pressure regulation.

It should be noted that in contrast to the impaired inner ear and thyroid function in patients with Pendred syndrome, no overt acid-base disturbances such as metabolic alkalosis or electrolyte abnormalities have been detected in these patients under unstressed basal conditions. However, two Pendred syndrome patients with intercurrent illnesses exhibited unusually severe metabolic alkalosis (Pela et al. 2008; Kandasamy et al. 2011).

10.2.2 Pendrin in the Inner Ear

In the inner ear, pendrin is expressed in several locations of the cochlea and the vestibular apparatus of the mouse inner ear, most abundantly in endolymphatic duct and sac (Dou et al. 2004; Choi et al. 2011), whose pendrin-positive cells morphologically resemble the pendrin-expressing non-A non-B intercalated cells of the renal CCD (Dou et al. 2004). Pendrin is essential for normal auditory development and function, likely by maintaining endolymph volume, pH and ionic composition in the endolymphatic sac (Dou et al. 2004; Everett et al. 1999) during a critical time window of embryonic development (Wangemann 2011).

There is evidence that the transcription factor forkhead box I1 (FOXI1), which is necessary for pendrin expression in type B intercalated cells of the mouse CCD (Blomqvist et al. 2004), is also an upstream regulator of the pendrin gene during mouse inner ear development (Hulander et al. 2003). A consensus FOXI1 binding site is located between nt −889 and −873 of the *SLC26A4* promoter (Fig. 10.1), but its contributions to pendrin transcriptional regulation remain to be tested. The *SLC26A4* 5'-UTR (untranslated region) polymorphism c.-103 T>C has been found as a compound heterozygous variant in *trans* with a missense loss-of-function

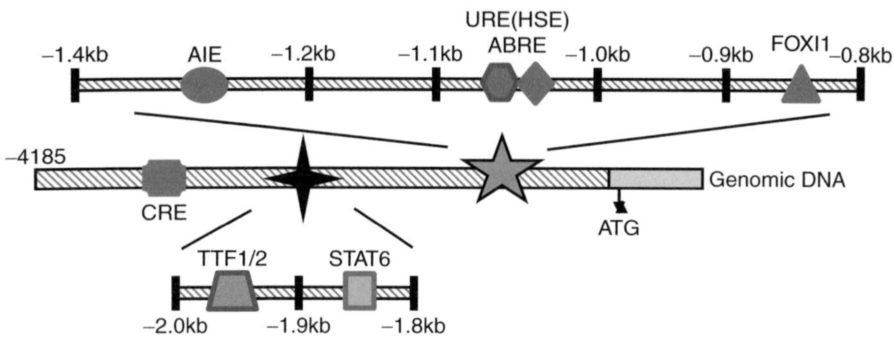

Fig. 10.1 A schematic model of the 5'-flanking region of the human pendrin gene with details of the *SLC26A4* promoter region of 600 bp between −0.8 and −1.4 kb and the 200 bp region between −1.8 and −2.0 kb. These regions contain functionally significant elements that appear to contribute to gene regulation through various modulators and transcription factors, such as the acid-base response element (*ABRE*), aldosterone-induced element (*AIE*), uroguanylin response element (*URE*), heat shock element (*HSE*), forkhead box transcription factor I1 (*FOXI1*), signal transducer and activator of transcription protein 6 (*STAT6*), thyroid transcription factors 1 and 2 (*TTF-1/2*). Also depicted is the putative chloride response element (*CRE*), likely residing between −2 and −4 kb on the promoter. See text for details (Adapted from Rozenfeld et al. (2011) with permission)

mutation (Yang et al. 2007). However, the interpretation of and evidence supporting the functional importance of this proposed FOXI1 binding site located within exon 1 of the *SLC26A4* gene have been questioned (Choi et al. 2009a). We have provided evidence that pendrin is transcriptionally regulated by systemic pH in inner ear cells (Adler et al. 2008) (see Sect. 10.3.1). However, pendrin exact role in inner ear function, the molecular mechanisms controlling pendrin activity in the inner ear and how pendrin defects lead to the anatomical and functional abnormalities observed in Pendred syndrome remain under investigation.

10.2.3 Pendrin in the Thyroid Gland

In the thyroid, pendrin resides on the apical membrane of thyrocytes and, by its activity as a Cl^-/I^- exchanger, contributes a variable and sometimes major proportion of I^- efflux from the thyroid epithelial cell and accumulation in the follicular lumen (Royaux et al. 2000; Scott et al. 1999; Yoshida et al. 2002). Pendred syndrome-associated loss-of-function mutations in the pendrin protein impair the ability of the thyroid gland to accumulate iodide in the follicular lumen which leads to insufficient thyroid hormone synthesis and compensatory goiter in the minority of pendrin mutation-associated deaf patients who develop Pendred syndrome (Campbell et al. 2001).

Thyroid stimulating hormone (TSH) increased *SLC26A4* mRNA expression, and pendrin expression and localization were regulated by insulin and influenced by PKC-ε–dependent intracellular pathway in the rat-differentiated thyroid cell line PC

C13 (Muscella et al. 2008). Both thyroglobulin (Tg) (Royaux et al. 2000) and thyroid transcription factor 1 (TTF-1) (Dentice et al. 2005), a regulator of thyroid differentiation *in vivo*, were found to regulate the expression of pendrin in rat thyroid FRTL-5 cells. Our deletion analysis of the 5'-flanking sequence of the human pendrin gene demonstrated that transcription factor 1 and 2 (TTF-1 and TTF-2) binding sites located between nucleotides (nt) −1946 and −1938 and between nt −1942 and −1933, respectively, on the *SLC26A4* promoter are essential for the activity of this promoter in human thyroid follicular carcinoma (LA2; WRO) cells (Adler et al. 2008) (Fig. 10.1).

10.2.4 Pendrin in the Airway Epithelium

Pendrin is highly expressed and its activity is upregulated at the apical membrane of bronchial epithelial cells following cytokine or antigen exposure, or in models of asthma or chronic obstructive lung disease (Pedemonte et al. 2007; Rozenfeld et al. 2013; Nakagami et al. 2008; Nakao et al. 2008; Ishida et al. 2012). The stimulated Cl^-/HCO_3^- exchange activity of pendrin results in increased production of mucus and increased viscosity of the airway surface liquid (ASL), thereby exacerbating airway disease (Nakagami et al. 2008; Nakao et al. 2008). Pertussis toxin was recently shown to contribute to pertussis pathology via the upregulation of pendrin, which promotes conditions favoring inflammatory pathology (Scanlon et al. 2014). The cytokines interleukin-4 (IL-4) and interleukin-13 (IL-13), known triggers of airway hyperreactivity and disease, have been found to stimulate *SLC26A4* mRNA levels in mouse and human airway epithelial cells (Nakao et al. 2008; Kuperman et al. 2005) and to upregulate pendrin activity by transcriptional activation of the pendrin gene promoter (Rozenfeld et al. 2011; Nofziger et al. 2011; Vanoni et al. 2013) (see below). Similarly, interleukin-17A (IL-17A) induced pendrin expression and pendrin-mediated Cl^-/HCO_3^- exchange in human bronchial epithelial cells (Adams et al. 2014). A thick ASL volume and weak mucin expression were demonstrated in airway epithelial cells from patients with *SLC26A4* mutations, and both phenomena were refractory to stimulation with IL-13 (Lee et al. 2015).

Of note are studies investigating HCO_3^- secretion in human airway submucosal gland cell line, Calu-3, which yielded conflicting results; some studies (Garnett et al. 2011) showed an important role for pendrin in HCO_3^- secretion in this cell line, whereas other studies (Shan et al. 2012; Huang et al. 2012; Kim et al. 2014) provided no evidence for pendrin's involvement in this transport process.

Pendrin also functions in the bronchial epithelial cell as a SCN^-/Cl^- exchanger (Pedemonte et al. 2007). Since SCN^- is oxidized by leukocytes to the potent antimicrobial hypothiocyanite, pendrin-mediated secretion of SCN^- into the lumen (in exchange for Cl^- entering the cell) likely contributes greatly to innate immune defense of the mucosal surface (Pedemonte et al. 2007; Adams et al. 2013).

Interestingly, however, evidence of resistance to respiratory disease has not been reported in individuals with pendrin mutations (Madeo et al. 2009). Further information regarding the role of pendrin in the airway epithelium can be found in Chap. 9.

10.3 Transcriptional Regulation of Pendrin Activity

10.3.1 Effect of Systemic pH

Ambient pH modulates pendrin mRNA and protein expression in the CCD (Frische et al. 2003; Petrovic et al. 2003; Wagner et al. 2002), providing evidence for the involvement of pendrin in acid-base homeostasis. Pendrin also plays an important role in controlling the acid-base status of the endolymphatic sac (Dou et al. 2004; Everett et al. 1999; Griffith and Wangemann 2011). To explore further the molecular mechanisms of the regulatory influences of pH on pendrin, we examined the direct effect of ambient pH on activity of the *SLC26A4* promoter (Adler et al. 2008). For this purpose, human embryonic kidney (HEK)293, mouse embryo inner ear ventral otocyst-derived (VOT36) and human thyroid follicular carcinoma (LA2; WRO) cells were transiently transfected with reporter vectors expressing luciferase under the upstream control of *SLC26A4* promoter fragments of decreasing length. Transfected cells were exposed to acidic pH (7.0–7.1), normal pH (7.35–7.45) or alkaline pH (7.6–7.7). Ambient acid pH decreased and alkaline pH increased *SLC26A4* promoter activity in renal and inner ear cells but not in thyroid cells. These results suggest that ambient pH likely plays a role in modulating pendrin activity at the transcriptional level in kidney and inner ear, but is without effect in the thyroid.

Further fine deletion analysis of the *SLC26A4* promoter provided evidence for the presence of a putative acid-base response element (ABRE) within the 96 bp between nt −1140 and −1044 on the promoter in the kidney and the inner ear (Figs. 10.1 and 10.2a, b). The findings are in agreement with the modulatory effects of acidosis and alkalosis on pendrin protein activity observed in the rodent kidney at the tubular and cellular levels (Verlander et al. 2003; Frische et al. 2003; Petrovic et al. 2003; Wagner et al. 2002; Pech et al. 2007) and demonstrate, for the first time, that systemic pH may directly regulate at the transcriptional level pendrin-mediated, renal tubular Cl^- reabsorption and HCO_3^- secretion. These data further suggest that systemic pH may modulate endolymph pH by directly regulating pendrin gene transcription.

The 96 bp stretch of the *SLC26A4* promoter harboring the putative pH-response element (ABRE) is a region without previously known pH-sensitive elements, but contains binding sites for transcription factors known to be regulated by mitogen-activated protein kinase (MAPK) signaling pathways, including Elk-1, c-Jun, activating transcription factor-2 (ATF-2), cAMP response element-binding (CREB), nuclear factor of activated T cells-4 (NFAT-4), c-Myc and N-Myc, among others. Thus, the MAPK pathway may be involved in pH-induced modulation of the

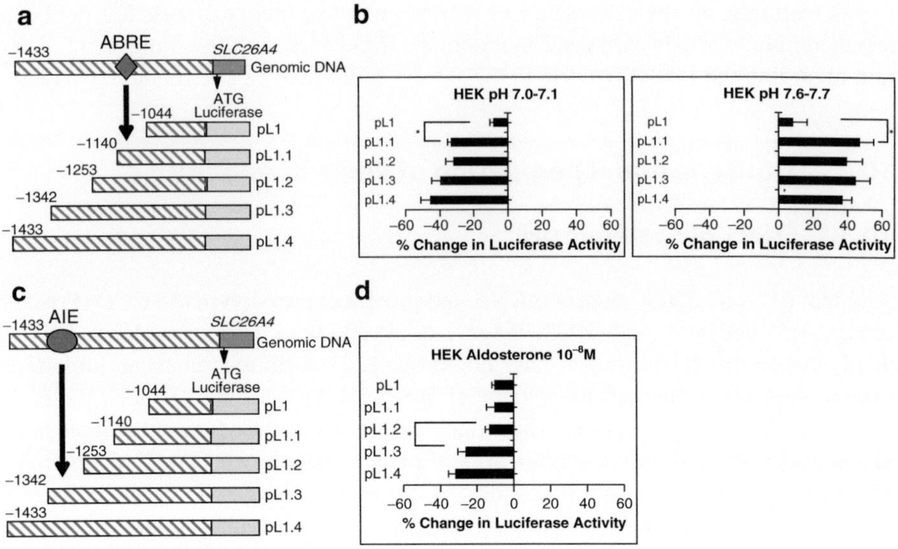

Fig. 10.2 Effect of ambient pH (**a**, **b**) and of aldosterone (**c**, **d**) on the human *SLC26A4* promoter in HEK293 cells. (**a**, **c**) schematic representation of the DNA constructs of *SLC26A4* promoter in the luciferase reporter vector, pGL3-basic, used in these experiments. ABRE, putative acid-base response element; AIE, putative aldosterone-induced element. (**b**, **d**) cells were transfected with pGL3-basic or pGL3-basic containing segments of decreasing size corresponding to the 5'-flanking region of *SLC26A4* (pL1.4, pL1.3, pL1.2, pL1.1 and pL1). Cells were exposed to acidic pH (7.0–7.1) or alkaline pH (7.6–7.7) (**b**) or to 10^{-8} M aldosterone (**d**) for 24 h. Subsequently, luciferase activity was measured. To control for transfection efficiency, cells were cotransfected with the LacZ-containing vector, pCH110, and luciferase activity was normalized to β-galactosidase activity. Data represent the % change in luciferase activity in cells exposed to experimental media (with acidic pH, alkaline pH or aldosterone) relative to cells exposed to control medium (pH 7.35–7.45 without aldosterone). Values are mean ± standard error of the mean of three to five independent quadruplicate experiments. Acidic pH-induced inhibition and alkaline pH-induced stimulation of luciferase activity were evident in HEK293 cells, which markedly decreased when the fragment size was shortened from 1.1 to 1 kb (**b**). Aldosterone-induced inhibition of luciferase activity was demonstrated in HEK293 cells, which markedly diminished when the fragment size was truncated from 1.3 to 1.2 kb (**d**). $^*P<0.01$ (Adapted from Rozenfeld et al. (2011) with permission)

pendrin gene promoter, just as it has been implicated in regulation of other H^+ and HCO_3^- transporting pathways (Li et al. 2004; Gluck 2004; Espiritu et al. 2002). Alternatively, a direct effect of ambient pH-mediated alterations in intracellular pH on this pH-response element residing on the *SLC26A4* promoter cannot be excluded. It should be noted that the ABRE overlaps with the heat shock element (HSE)-containing hypothetical uroguanylin response element (URE) on the *SLC26A4* promoter (see below; Fig. 10.1). It is, therefore, of great interest to explore whether the heat shock system is involved in pH-induced modulation of the pendrin gene. Further studies are needed to elucidate the exact location and the molecular nature and regulation of this novel pH-response element.

Our findings on the effects of ambient pH on the *SLC26A4* promoter in transfected renal cells were further supported by our experiments demonstrating acid-induced inhibition and alkali-induced stimulation of endogenous *SLC26A4* mRNA levels in renal HEK293 cells (Adler et al. 2008).

10.3.2 Effect of Aldosterone

Aldosterone modulates pendrin protein and mRNA expression in the CCD (Verlander et al. 2003; Mohebbi et al. 2013). To explore the molecular mechanisms of this aldosterone-induced effect on pendrin activity, we examined the direct effect of aldosterone on the *SLC26A4* promoter (Adler et al. 2008). To this end, HEK293 as well as LA2 (WRO) and VOT36 cells were transiently transfected with *SLC26A4* promoter fragments of different lengths, as above, and exposed to 10^{-8} M aldosterone. Aldosterone caused a marked decrease in *SLC26A4* promoter activity in HEK293 cells without altering pH of the experimental medium (7.35–7.45). Aldosterone had no effect on *SLC26A4* promoter activity in transfected LA2 and VOT36 cells. Taken together, these results suggest that aldosterone may directly affect the *SLC26A4* promoter to transcriptionally modulate pendrin activity in the kidney, but not in thyroid and inner ear.

Further fine deletion analysis of the *SLC26A4* promoter provided evidence for the presence of a putative aldosterone response element (AIE) likely residing within the 89 bp between nt −1342 and −1253 on this promoter (Figs. 10.1 and 10.2c, d). This promoter region also contains a mineralocorticoid receptor (MR) binding site between positions −1293 and −1287 (Rozenfeld et al. 2011). Additional aldosterone-sensitive sites within this 89 bp sequence remain undefined.

Reverse transcription PCR analysis detected MR mRNA in the aldosterone-responsive HEK293 cells, but not in the aldosterone-unresponsive cell lines LA2 and VOT36 (Adler et al. 2008). The association of aldosterone-regulated promoter activity with the presence of MR in HEK293 kidney cells suggests that aldosterone influences pendrin gene activity directly *via* the genomic pathway, and this may reflect the biology of the intact kidney. In line with this kidney-specific inhibitory effect of aldosterone on *SLC26A4* promoter activity, was our finding demonstrating an aldosterone-induced decrease in endogenous pendrin mRNA levels in renal HEK293 cells (Adler et al. 2008).

Taken together, these findings demonstrated for the first time that aldosterone may regulate pendrin-mediated HCO_3^- secretion at the transcriptional level. Moreover, our finding of isolated, aldosterone-induced direct inhibition of the pendrin gene promoter – as opposed to the complex aldosterone/alkalosis-induced stimulation of pendrin activity demonstrated *in vivo* (Verlander et al. 2003) – is in concert with the net H^+ secretion and HCO_3^- reabsorption associated with aldosterone action (Gennari and Maddox 2000; Hamm and Nakhoul 2008).

10.3.3 Effect of Ambient Cl^-

10.3.3.1 Effect in Kidney

Recent studies indicate that Cl^-, in addition to Na^+, has an important specific role in determining the extracellular fluid volume and blood pressure, and that pendrin-driven Cl^- reabsorption in β intercalated cells of the CCD is pivotal in mediating this physiological effect of Cl^- (Wall and Lazo-Fernandez 2015; Eladari et al. 2012; McCallum et al. 2015; Knepper 2015).

Systemic Cl^- depletion increases and Cl^- excess decreases pendrin protein abundance in rodent kidney (Quentin et al. 2004; Vallet et al. 2006; Verlander et al. 2006; Mohebbi et al. 2013), but little is known of the molecular mechanisms by which Cl^- regulates pendrin activity. Therefore, we investigated the effect of Cl^- concentration changes on pendrin gene expression and the direct effect of Cl^- on *SLC26A4* promoter activity in the kidney (Efrati et al. 2007). In a series of *in vivo* and *in vitro* experiments, we have demonstrated a Cl^--induced decrease in pendrin mRNA and protein expression in mouse kidney, a Cl^--dependent decrease in pendrin mRNA in HEK293 cells, as well as a high Cl^- concentration–induced decrease and a low Cl^- concentration–induced increase in *SLC26A4* promoter activity in transfected HEK293 cells. The findings suggested that a putative "chloride response element" (CRE) likely resides between −2 and −4 kb on the promoter (Fig. 10.1). A possible role for the Cl^- sensor WNK kinases (Piala et al. 2014) in regulating the apparent CRE of the pendrin promoter remains to be determined.

Taken together, these findings show that the response to Cl^- occurs along the entire pathway of pendrin biosynthesis: from changes in activity of its gene promoter, to changes in mRNA level and finally to changes at the level of pendrin protein abundance. The effect of ambient Cl^- on pendrin transcriptional activity demonstrated by us may play an important role in extracellular fluid homeostasis and blood pressure control in health and disease.

10.3.3.2 Effect in Thyroid

The modulators of acid-base and electrolyte balance used in our studies, namely, ambient pH and aldosterone, had no effect on the *SLC26A4* promoter activity in the thyroid cell line LA2 (see above). This finding is not surprising considering that acid-base conditions are not known to regulate thyroid hormone metabolism and that the thyroid gland does not contain aldosterone-responsive epithelial cells.

Pendrin functions in both thyroid gland (Royaux et al. 2000; Scott et al. 1999; Yoshida et al. 2002) and kidney (Reimold et al. 2011) as an exchanger of Cl^- and I^-. Previous studies showed that ambient iodide level regulates pendrin activity in the thyroid (Rozenfeld et al. 2011), but the effect of ambient Cl^- on this activity is unknown. Therefore, we have investigated whether ambient Cl^- influences thyroidal pendrin activity and whether this effect occurs at the level of transcription of the pendrin gene. We have demonstrated a high Cl^- concentration–induced decrease and a low Cl^- concentration-induced increase in *SLC26A4* promoter activity in reporter-transfected LA2 cells (unpublished data). Further studies examining the modulatory effect of Cl^- and I^- on pendrin activity and transcription in thyrocytes and collecting duct intercalated cells may improve our understanding of the molecular pathways of iodide transport, and may help to clarify the role of pendrin in iodide balance.

10.3.4 Effect of Uroguanylin

Guanylin (GN) and UGN are low-molecular weight peptide hormones produced mainly in the intestinal mucosa in response to oral salt load. GN and UGN resemble in structure and activity the secretory diarrhea-causing heat-stable enterotoxin (STa) of *Escherichia coli* (Forte et al. 1996) and induce secretion of electrolytes and water in both the intestine and kidney (Kita et al. 1999; Sindic and Schlatter 2006). GN and UGN are co-expressed along the intestinal tract with guanylyl cyclase C (GC-C), the principal GN receptor (Kita et al. 1994). Salt ingestion induces secretion of GN and UGN into the intestinal lumen, and both hormones activate GC-C-mediated intestinal secretion of electrolytes and water (Sindic and Schlatter 2006).

10.3.4.1 Effect in Kidney

GN and UGN play a major role in the regulatory link between the intestine and the kidney by increasing urinary NaCl and water excretion in response to enteral (but not intravenous) intake of NaCl, thereby serving as "intestinal natriuretic factors" (Kita et al. 1999; Sindic and Schlatter 2006; Fonteles et al. 1998). The effect of the GN peptides on renal electrolyte and water handling is achieved via both endocrine and paracrine/autocrine mechanisms (Sindic and Schlatter 2006, 2007; Fonteles and Falcao do Nascimento 2011; Moss et al. 2008; Qian et al. 2008; 2011; Carrithers et al. 2004; Mueller and Dieplinger 2012). The mode of action and signaling pathways for UGN differ according to nephron segment. UGN promotes natriuresis in the proximal tubule via GC-C-mediated, cGMP- and protein kinase G-dependent modulation of Na^+/H^+ exchange, K^+ channels and Na^+-K^+-ATPase. In contrast, the natriuretic effect of UGN on CCD principal cells involves PLA2-mediated inhibition of ROMK (Rozenfeld et al. 2013; Sindic and Schlatter 2006; Carrithers et al. 2004; Forte 2003; Potthast et al. 2001; Sindic et al. 2005). However, despite accumulating data on renal expression and function of GN peptides, the cellular and molecular pathways mediating UGN action in the CCD remain poorly understood.

Considering the major role of both GN peptides and pendrin in the regulation of total body NaCl content, maintenance of extracellular fluid volume and control of blood pressure, we investigated UGN modulation of pendrin expression and explored the molecular mechanisms responsible for this modulation (Rozenfeld et al. 2012). We first showed that injection of UGN into mice resulted in decreased renal expression of pendrin mRNA and protein. UGN also decreased endogenous pendrin mRNA levels in HEK293 cells (Rozenfeld et al. 2012). We next examined possible modulation by UGN of human *SLC26A4* transcription at the level of the *SLC26A4* promoter (Rozenfeld et al. 2012).

For this purpose, *SLC26A4* promoter fragments of decreasing length were cloned into luciferase reporter vectors and transiently transfected into HEK293 cells.

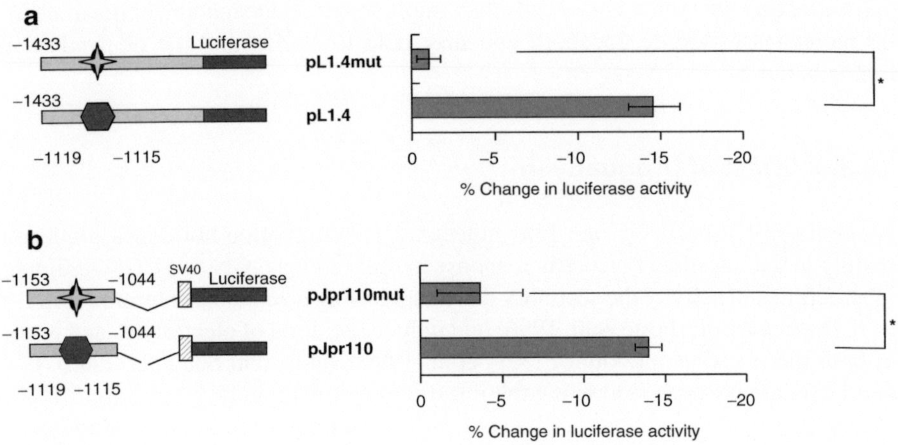

Fig. 10.3 Role of a defined heat shock element (HSE) in the effect of UGN on the *SLC26A4* promoter in HEK293 cells. (**a**) effect of UGN on the 1.4 kb 5'-flanking region. Cells were transfected with 0.3 μg pGL3-basic or 0.3 μg pGL3-basic containing the 1.4 kb 5'-flanking region of *SLC26A4* (pL1.4) or 0.3 μg pL1.4 harboring a point-mutated HSE site (pL1.4mut). Cells were exposed to 1 μM UGN or control medium for 24 h and luciferase activity was then measured. (**b**) effect of UGN on the 110 bp HSE-containing DNA region. Cells were transfected with 0.3 μg pGL3-SV40 minimal promoter vector or 0.3 μg pGL3-SV40 containing the 110 bp *SLC26A4* promoter fragment with a wild-type HSE site (pJpr110) or 0.3 μg pJpr110 harboring a point-mutated HSE site (pJpr110mut). Cells were exposed to UGN or control as in (**a**), and luciferase activity was measured. Luciferase activity in (**a**, **b**) was normalized to β-galactosidase activity. Data represent the % change in luciferase activity in cells exposed to experimental medium (with 1 μM UGN) relative to cells exposed to control medium (without UGN). Whereas UGN inhibited luciferase activity in cells transfected with the wild-type 1.4 kb promoter fragment (**a**) or the wild-type 110 bp fragment (**b**), mutating the HSE site in either of these fragments markedly diminished the UGN-induced effect. Values are mean ± standard error of the mean of three to five independent quadruplicate experiments. *$P<0.05$ (Adapted from Rozenfeld et al. (2012) with permission)

Exposure of transfected cells to UGN decreased *SLC26A4* promoter activity (Rozenfeld et al. 2012). Fine deletion analysis of the *SLC26A4* promoter provided evidence for the presence of a putative URE located within the 52 bp between nt −1153 and −1101 (Rozenfeld et al. 2012) (Fig. 10.1). Further truncation and mutation analyses of this promoter sequence suggested that UGN modulates *SLC26A4* activity by a mechanism that requires an HSE located between nt −1119 and −1115 on the promoter (Fig. 10.3a, b) (Rozenfeld et al. 2012).

In subsequent experiments, we showed the expression of mRNA encoding heat shock factor 1 (HSF1, a transcription factor that recognizes the HSE motif; see below) in HEK293 cells, and demonstrated that transfection of HSF1 small interfering RNA (siRNA) markedly reduced endogenous HSF1 mRNA levels in HEK293 cells (Rozenfeld et al. 2012). We then showed that a 30 % reduction in *SLC26A4* mRNA levels by UGN treatment of HEK293 cells transfected with control siRNA was completely abolished in HEK293 cells transfected with HSF1 siRNA

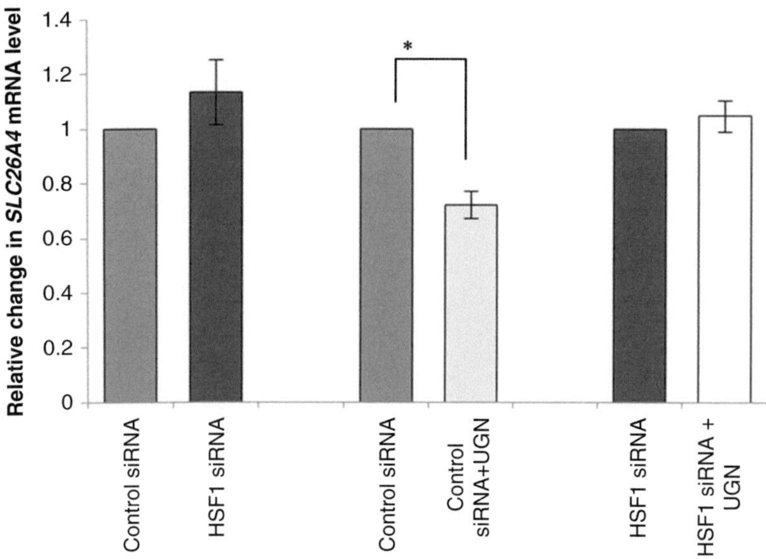

Fig. 10.4 Effect of heat shock factor-1 (HSF1) small interfering RNA (siRNA) on endogenous *SLC26A4* mRNA levels in UGN-treated HEK293 cells. HEK293 cells were transfected with HSF1 siRNA or control siRNA before treatment with medium containing or lacking UGN (1 μM). Subsequently, total RNA was extracted, and real-time PCR analysis of the *SLC26A4* mRNA was performed. Results represent *SLC26A4* mRNA in cells transfected with HSF1 siRNA compared with cells treated with control siRNA (*left*), cells transfected with control siRNA and exposed to UGN-containing medium relative to control siRNA-transfected cells exposed to medium without UGN (*middle*), or cells transfected with HSF1 siRNA and exposed to UGN-containing medium relative to HSF1 siRNA–transfected cells exposed to medium without UGN (*right*). Values were normalized to the housekeeping gene encoding the TATA Box Binding Protein (TBP). Values are mean ± standard error of the mean of four independent experiments, each performed in duplicate. HSF1 siRNA transfection reduced HSF1 mRNA expression by 76 % (not shown). HSF1 siRNA transfection had no effect on *SLC26A4* mRNA level (*left*), UGN reduced *SLC26A4* mRNA level by 30 % in cells transfected with control siRNA (*middle*), whereas *SLC26A4* mRNA level remained unchanged in UGN-treated cells transfected with HSF1 siRNA (*right*). $^*P<0.05$ (Adapted from Rozenfeld et al. (2012) with permission)

(Fig. 10.4). These findings provided strong evidence for involvement of HSF1 in regulation of the *SLC26A4* gene by UGN.

Heat shock factors (HSFs) comprise a group of transcription factors that regulate the heat shock response (HSR) by binding to specific HSEs which were first defined upstream of cytoprotective heat shock genes including heat shock protein (HSP)70, HSP90, HSP27 and other molecular chaperonins of the HSR network (Anckar and Sistonen 2007; Morimoto 2008; Morimoto and Santoro 1998). While the role of the HSF/HSE system in regulating the activity of HSPs mediating the HSR has been well established, the full range of biological target genes for the HSFs, particularly in the kidney, remains to be established. Our study is the first report of a mammalian kidney solute transporter transcriptionally regulated by the HSF/HSE system in a hormone-specific manner.

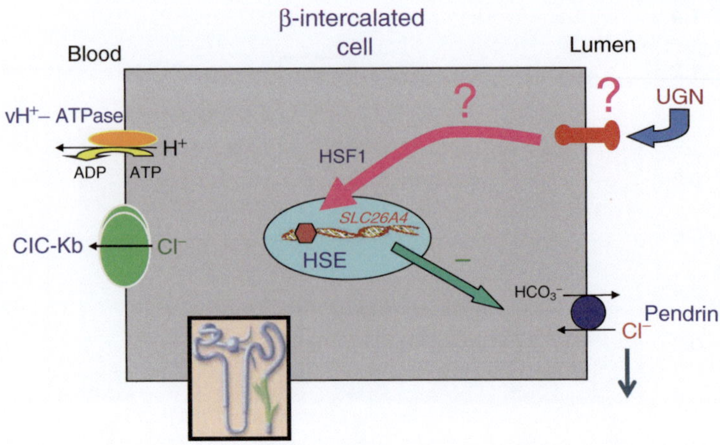

Fig. 10.5 A schematic model of the transcriptional inhibition of the pendrin gene (*SLC26A4*) promoter by uroguanylin (*UGN*) in β intercalated cells of the cortical collecting duct. This effect of UGN is achieved via heat shock factor 1 (*HSF1*) active at a defined heat shock element (*HSE*) site on the promoter. The membrane-associated receptor that binds UGN and the signal transduction pathway whereby UGN triggers HSF1 binding to HSE remain unidentified. vH^+-ATPase, vacuolar type H^+-ATPase (proton pump); ClC-Kb, voltage-gated Cl^- channel-Kb. Not shown are basolateral barttin and apical SLC4A8 (Adapted from Rozenfeld et al. (2013) with permission)

An evolutionary precedent of this transcriptional regulation is exhibited by the intestinal GN system of teleost fish, which plays an important role in seawater adaptation (Forte 2003; Comrie et al. 2001; Yuge et al. 2003). UGN mRNA is upregulated in the intestine and kidney of eels upon seawater exposure (Comrie et al. 2001; Yuge et al. 2003). The primary structure of UGN, conserved throughout vertebrate evolution, suggests that UGN-mediated regulation of systemic NaCl balance is the mammalian counterpart of UGN-mediated osmoregulation in teleost fishes of both fresh and salt water environments (Forte 2003). Our findings demonstrating the involvement of the stress-stimulated HSF/HSE axis in the UGN-induced modulation of pendrin gene transcription support this notion. These findings also raise the possibility of an adaptive chloriuretic response to osmotic or salt load stress mediated in the kidney by this novel UGN-HSR-pendrin pathway.

Taken together, our findings have identified the pendrin Cl^-/HCO_3^- exchanger of the β intercalated cells of the CCD as an important renal target of UGN, have defined a novel enterorenal link controlling electrolyte and water homeostasis and have provided a possible novel explanation for a significant part of UGN-induced chloriuresis (Fig. 10.5). However, the identity of the intercalated cell membrane receptor that binds UGN and the signaling pathway by which UGN triggers HSF1 binding to the HSE of the *SLC26A4* promoter to repress transcription both remain to be clarified.

10.3.4.2 Effect in Airways

The GN family of GC-C/cGMP-regulating peptides is expressed in airway epithelium and plays a role in airway function (Rozenfeld et al. 2013; Forte 2004; Cetin et al. 1995; Krause et al. 1990; Kulaksiz et al. 2002; Ohbayashi and Yamaki 2000; Zhang et al. 1998). GN/UGN and the GC-C receptor have been identified in the apical membrane of bronchiolar non-ciliated secretory (Clara) cells of rodents (Cetin et al. 1995; Krause et al. 1990). GN peptides operating *via* the GC-C/cGMP pathway were found to activate Cl^- conductance in human airway epithelium through both the CFTR (Kulaksiz et al. 2002) and other Cl^--conductive pathways (Zhang et al. 1998). GN peptides also exert beneficial effects on airways in a guinea pig model of asthma and relaxant effects on ovalbumin-induced bronchoconstriction and leukotriene C4-induced airway microvascular leakage (Ohbayashi and Yamaki 2000; Fruhbeck 2012; Ohbayashi et al. 1998). As discussed above, pendrin is highly expressed in inflamed airway epithelia and has been implicated in airway function and disease (Pedemonte et al. 2007; Rozenfeld et al. 2013; Nakagami et al. 2008; Nakao et al. 2008; Scanlon et al. 2014).

However, nothing is known about a regulatory or functional link between the UGN and pendrin systems in the airways.

10.3.5 Effect of IL-4 and IL-13

Upregulation of pendrin expression by IL-4 and IL-13 results in increased production of mucus and increased viscosity of the ASL (Nakagami et al. 2008; Nakao et al. 2008). To investigate the molecular mechanisms responsible for the effect of IL-4 and IL-13 on pendrin expression, we examined the effect of these cytokines on the *SLC26A4* promoter (Nofziger et al. 2011; Vanoni et al. 2013). To this end, *SLC26A4* promoter fragments of decreasing size, cloned into luciferase reporter vectors, were transfected into HEK-Blue cells. Stimulation of these cells with IL-4 and IL-13 markedly increased *SLC26A4* promoter activity, that was maximal with a construct containing 1894 bp of the *SLC26A4* 5'-flanking region.

IL-4 and IL-13 regulate gene transcription through activation of the transcription factor signal transducer and activator of transcription protein 6 (STAT6; also known as IL-4 nuclear activated factor, IL-4NAF). STAT6 exerts its activity by binding to the N_4GAS motif (consisting of 5' TTC(N_4)GAA 3' where N is any nucleotide) on the promoter of the target gene (Mikita et al. 1996; Ehret et al. 2001). Having identified two STAT6 DNA binding sequences at positions −3472 to −3463 (motif 1) and −1812 to −1803 (motif 2) on the *SLC26A4* promoter, we examined the involvement of these *cis*-acting elements in the IL-4/IL-13–induced effect on the promoter (Nofziger et al. 2011; Vanoni et al. 2013). We first demonstrated loss of IL-4/IL-13 activity on the *SLC26A4* promoter in STAT6-deficient HEK293 Phoenix cells. Mutation of motif 2 of the *SLC26A4* promoter rendered the promoter insensitive to IL-4/IL-13, while fusion of wild-type motif 2 with an IL-4/IL-13–unresponsive promoter sequence conferred cytokine sensitivity to the unresponsive promoter. Motif 1 was found to have no functional consequence on basal or IL-4–stimulated *SLC26A4* promoter activity.

Taken together, these studies provide strong evidence for the importance of the STAT6 pathway in the transcriptional regulation of the pendrin gene by IL-4 and IL-13.

10.4 Epigenetic Regulation of Pendrin Activity

10.4.1 Background

Pendrin expression varies greatly among cell types and as a function of various physiological and/or pathological states. The mechanisms underlying these differences undoubtedly include a component of epigenetic regulation, including heritable alterations in gene expression that do not alter the DNA sequence (Wu and Morris 2001; Sharma et al. 2010). As opposed to the irreversible nature of genetic events, which introduce changes in the primary DNA sequence, epigenetic modifications are reversible and leave the original DNA sequence intact. The complex pattern of silenced/expressed gene regions that exists within the genome of a single cell type is tightly regulated by epigenetic modifications including DNA methylation, histone modification and non-coding RNAs.

Cytosine methylation involves DNA methyltransferase (DNMT)–mediated transfer of methyl groups from S-adenosyl methionine (SAM) to position 5 of the pyrimidine ring of cytosine, yielding the product 5-methylcytosine (5mC). Cytosines followed by guanines in the linear DNA sequence (cytosine-phosphate-guanine, CpG) are the most common targets for methylation, although other nucleotides can also undergo methylation. 5mC can be further metabolized into 5-hydroxymethylcytosine (5-hmC) via ten-eleven translocation (TET) enzymes, the role of which in epigenetic regulation has been increasingly studied within the recent years (Tahiliani et al. 2009).

Post-translational modification of histones (including but not limited to acetylation, methylation, phosphorylation, ubiquitination and sumoylation) determines chromatin compactness, which in turn governs chromatin accessibility to transcriptional machinery. In general, histone acetylation and deacetylation respectively activate or inactivate gene transcription; histone methylation, in contrast, can regulate gene activity either positively or negatively (Rothbart and Strahl 2014).

Whereas DNA methylation and histone modifications determine transcript expression by regulating activity of ongoing gene transcription, micro-RNAs (miRNAs) (a type of non-coding RNA) dictate the amount of mRNA available for translation after transcription (He and Hannon 2004). In other words, miRNAs are capable of degrading newly made transcripts, so that less (or none) of the total pool is available for translation into functional protein.

All the aforementioned types of epigenetic regulation have been linked to diverse pathological states, especially cancer, and multiple therapeutic drugs targeting various molecular entities driving epigenetic changes have either been approved for clinical use or are under clinical development (Campbell and Tummino 2014).

10.4.2 Epigenetics of Pendrin in Thyroid Carcinoma

Carcinoma of the thyroid is the most common endocrine malignancy. Epigenetic modification plays a major role in thyroid carcinomas of histologic types

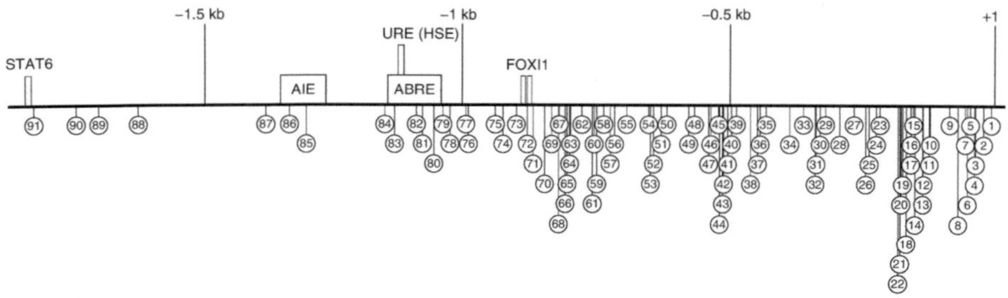

Fig. 10.6 The 2 kb directly upstream of the open reading frame (designated as +1) in the human *SLC26A4* gene (GenBank reference sequence AC08937.1) contains 91 CpG sites denoted by the numbered circles. The consensus STAT6 binding sequence (motif 2), as well as the aldosterone-induced element (*AIE*), acid-base response element (*ABRE*), uroguanylin response element (*URE*), heat shock element (*HSE*) and the two forkhead box transcription factor I1 elements (*FOXI1*) are shown for reference. The methylation status of CpG sites in close proximity to these defined elements may have important pathophysiologic implications

unresponsive to conventional therapeutic intervention, especially thyroid ablation by radioactive I^-, with the follicular lumen as site of action (reviewed in (Catalano et al. 2012)). Pendrin normally drives I^- exit from the thyrocyte into the follicular lumen, where it is organified for subsequent production of the thyroid hormones T3 and T4 (refer to Chaps. 6 and 7 for further information). It is therefore plausible to hypothesize pendrin as one of the culprits that manifests or contributes to these therapeutic insensitivities. Indeed, most studies have shown that the expression of both pendrin mRNA and protein is lower or absent in cancerous thyroid tissue compared to normal thyroid tissue, dependent on the type of thyroid carcinoma type and heterogeneity (Arturi et al. 2001; Bidart et al. 2000b; Porra et al. 2002).

The immediate 5'-UTR of pendrin contains just under 100 CpG sites (Fig. 10.6). To gain insight into the mechanism of decreased pendrin expression in thyroid carcinoma tissues, Xing and collaborators qualitatively assessed the methylation status of a small region of the pendrin promoter (CpG sites 75, 74, 73 and 72 in Fig. 10.6) in numerous thyroid cancer cell line models, as well as human thyroid tumors (Xing et al. 2003). They found that methylation of this region of the pendrin 5'-UTR correlated with decreased pendrin mRNA expression in the majority of tumor tissues, consistent with the gene silencing effect of CpG site methylation.

Changes in pendrin gene expression may not be required, or may not be the only requirement, for thyroid carcinomas to regain susceptibility to radioactive I^-. For example, increased expression and/or activity of the Na^+/I^- symporter (NIS) (mediating the intracellular uptake of I^- in the thyrocyte) can confer sensitivity to radioactive I^-. This was evidenced by Kitazono and collaborators in multiple thyroid cancer cell line models (of follicular and anaplastic origin), in which the expression of NIS was increased following treatment with the histone deacetylase (HDAC) inhibitor, depsipeptide (also known as romidepsin) (Kitazono et al. 2001). Zarnegar and collaborators imposed a similar pharmacological maneuver in thyroid tumor cell lines of various origin (follicular, papillary and Hürthle) with a separate HDAC inhibitor, trichostatin A (Zarnegar et al. 2002). Despite an increase in NIS gene expression, pendrin mRNA expression decreased in this latter study. However, pendrin mRNA was not assessed by Kitazono and collaborators (Kitazono et al. 2001), and the ability to mediate I^- accumulation was not assessed in the study of Zarnegar and collaborators (Zarnegar et al. 2002). Moreover, the isoform specificities of HDAC inhibition by depsipeptide and trichostatin A differ (West and Johnstone 2014). Therefore, the impact of histone modification on pendrin gene expression is difficult to predict, as is the consequence of histone modification in thyroid cancer cells on their ability to accumulate I^-.

In view of the involvement of HSF1 in regulation of the pendrin gene by UGN (see above), it is noteworthy that the HSFs have been recently shown to act as epigenetic regulators in cell cycle-related processes (Das and Bhattacharyya 2014; Vihervaara et al. 2013). Das and collaborators (Das and Bhattacharyya 2014) provided evidence that HSF1-mediated expression of Huntington Yeast Partner K (HYPK; a chaperone protein known to participate in diverse biological processes including cell cycle and cell growth) involves histone modification. By analyzing the genome-wide transcriptional response to stress in human cells, Vihervaara and collaborators (Vihervaara et al. 2013) demonstrated that HSF2 is an epigenetic regulator directing transcription throughout cell cycle progression by modulating chromatin compaction and accessibility. Whether HSF1-mediated modulation of *SLC26A4* activity involves epigenetic regulation is a subject for future investigation.

10.4.3 Epigenetics of Pendrin in Other Organs

Relatively speaking, basal expression levels of pendrin are much less in lung epithelial cells compared to that in intercalated cells of the renal CCD. This observation was confirmed in two continuous cell line models of the kidney (HEK-Blue) and lung (NCI-H292): HEK-Blue cells expressed >5-fold higher levels of pendrin mRNA than in NCI-H292 cells (Nofziger et al. 2011). To gain insight into the mechanisms underlying these differences, we quantitatively assessed the extent of CpG site methylation within a region of the pendrin 5'-UTR (Fig. 10.6) (Lee et al. 2011). While the degree of methylation at the majority of CpG sites was significantly different between the two cell lines, the differences at most sites were minimal to modest, suggesting that other epigenetic mechanisms likely account for the difference in pendrin mRNA expression. The most unexpected finding from this study was the 60 % difference in methylation at two particular CpG sites (sites 91 and 90) located adjacent to a consensus sequence required for IL-4– and IL-13–induced increases in pendrin mRNA production (see Chap. 9 for further information) (Nofziger et al. 2011; Vanoni et al. 2013). Interestingly, NCI-H292 cells are much more IL-4–sensitive than HEK-Blue cells in terms of increased pendrin gene activity (Lee et al. 2011). This sensitivity may in part reflect the lower extent of methylation of CpG sites 91 and 90 in NCI-H292 cells than HEK-Blue cells.

To date, no direct evidence in terms of post-transcriptional epigenetic regulation, *i.e.*, regulation by miRNAs, of pendrin is available.

10.5 Transcriptional Regulation of Other SLC26 Genes

Transcriptional regulation of other SLC26 genes has been reported. The basal activity of the *SLC26A3/DRA* promoter in Caco-2 and LS174T colon carcinoma cells is sustained by nuclear factor HNF-4 (Alrefai et al. 2007). Sodium butyrate stimulates this activity by increasing binding of the zinc finger transcription factors Yin Yang 1 (YY1) and GATA to defined *cis*-elements, whereas interferon gamma (IFN-γ) reduces transcriptional activity (Alrefai et al. 2007). The effect of IFN-γ is mediated through an IFN-γ response element containing a critical STAT1 binding site at nt −933 to −925, mutation of which abrogated the inhibitory activity of IFN-γ (Saksena et al. 2010). An IFN-γ inhibitory response element active in Caco-2 cells was also found in the *SLC26A6* promoter between nt −318 and −300. This is consistent with inflammatory suppression of both intestinal Cl^- reabsorption and HCO_3^- secretion. Lactobacilli exert an anti-inflammatory action on enterocytes. Raheja and collaborators demonstrated that exposure of Caco-2 cells or intact mice to lactobacilli increased *SLC26A3* promoter activity in parallel with increased surface expression of the protein (Raheja et al. 2010). The transcriptional elements involved remain undescribed. Studies using intestinal epithelial Caco-2 cells have demonstrated stimulation of *SLC26A3/DRA* gene transcription by lysophosphatidic acid (LPA) (Singla et al. 2012) and all-*trans*-retinoic acid (ATRA) (Priyamvada et al. 2015) *via* c-FOS- and HNF-1β-dependent pathways, respectively, raising the possibility that these two compounds may act as anti-diarrheal agents.

Transcriptional regulation of the pendrin gene by IFN-γ in respiratory tract has not been reported. However, like the *SLC26A4*/pendrin gene, the outer hair cell gene encoding the mechanotransducer protein SLC26A5/prestin is induced during development by thyroxine. Thyroxine activates *SLC26A5* gene transcription through the TRβ receptor (Winter et al. 2006). *SLC26A3/DRA* mRNA abundance in Caco-2 colon carcinoma cells is also increased by thyroxine by an unknown mechanism (Alrefai et al. 2001).

The SLC26A2/DTDST sulfate/anion exchanger is induced during differentiation of C3H10T1/2 chondrocytes by bone morphogenetic protein-2 (BMP-2). The *SLC26A2* promoter contains a TATA box preceded by a GC-rich region with two specificity protein-1 (SP-1) binding sites and a core binding factor alpha 1 (CFBA1) binding site. Basal transcriptional activity required the region between nt −309 and −275, containing a xenobiotic responsive element (XRE). BMP-2 enhancement of transcriptional activity was associated with increased binding of undefined nuclear proteins to a fragment including this XRE (Kobayashi et al. 1997).

Colonic expression of the *SLC26A2/DTDST* gene is also decreased in colon cancer, in parallel with decreased expression of sialyl 6-sulfo-Lewisx antigen. This transformation-associated downregulation, as well as engineered suppression of transcription, is associated with markedly increased growth rate. However, most cultured colon carcinoma cells exhibit increased SLC26A2 expression after treatment with inhibitors of HDAC, consistent with epigenetic control of gene expression (Yusa et al. 2010). Direct transcriptional regulation of *SLC26A2* has not been reported.

10.6 Potential Therapeutic Implications of Pendrin Gene Modulation

Collectively, the accumulating data on the major role of pendrin in the regulation of fluid and electrolyte balance and the control of blood pressure (see above), including its interaction with several Na^+-dependent transport processes in the distal tubule (Kim et al. 2007; Pech et al. 2010; Leviel et al. 2010; Soleimani et al. 2012; Grimm et al. 2015), raise the possibility that specific inhibitors of pendrin, including new drugs modulating pendrin activity at the transcriptional or epigenetic levels, may have a strong diuretic effect in conditions associated with elevated blood pressure and fluid retention states such as renal failure, heart failure and hepatic disease.

It remains to be determined whether the GN peptides, with their natriuretic/diuretic action, in general, and their pendrin inhibiting/chloriuretic effect, in particular (Fig. 10.5), may themselves become candidate diuretic agents. It is noteworthy, in this regard, that patients with chronic renal failure (Nakazato et al. 1994; Fukae et al. 2000), glomerulonephritis (Kinoshita et al. 1997) and nephrotic syndrome (Kinoshita et al. 1997) display increased plasma levels of GN peptides. In some of these disease states, such as in nephrotic syndrome (Kinoshita et al. 1997; Kikuchi et al. 2005), UGN may be mobilized to act as a compensatory natriuretic factor in the setting of powerful and pathological salt retention stimuli.

In view of the inhibitory effect of UGN on pendrin gene transcription in the kidney, it will be very interesting to explore whether the GN peptides achieve their beneficial effect on airway function and disease at least in part by transcriptional inhibition of pendrin expression and activity in airway epithelium. Such a finding could potentially lead to the development of a novel, GN peptide–based therapeutic intervention in asthma, chronic obstructive lung disease and other airway diseases.

Future research into the molecular mechanisms underlying the transcriptional and epigenetic control of pendrin gene by various modulators may unravel yet additional modes of operation and biological roles for these modulators in various organs, suggesting new prospects for novel therapeutic approaches in multiple disease states.

Conclusion

The renal anion exchanger pendrin, thought by some after its genetic identification in 1997 (Everett et al. 1997) to participate solely in acid–base balance as a bicarbonate secretory pathway, has emerged as a major contributor to renal CCD Cl^- reabsorption, whole body volume regulation and blood pressure control. Pendrin is also essential for epithelial function in inner ear, thyroid and airways, thus regulating endolymph ion balance, thyroid hormone synthesis and ASL composition. Systemic pH, ambient Cl^- concentration, the hormones aldosterone and UGN and the cytokines IL-4 and IL-13 modulate pendrin activity in the kidney and other pendrin-expressing epithelia at the transcriptional level by acting on distinct regulatory elements residing in the *SLC26A4* promoter. Transcriptional regulation of the pendrin gene by UGN creates a novel and unique connection between the intestine and the kidney and an important functional link between two systems. The effect of all of these modulators on pendrin gene activity may play important roles in acid-base balance, extracellular fluid homeostasis, blood pressure control and airway function in health, during development and in disease states.

The precise molecular pathways controlling transcriptional and epigenetic control of pendrin function in pendrin-expressing epithelia remain to be clarified. Unraveling the molecular mechanisms regulating the pendrin gene will provide insight into the role of this important anion exchanger in various biological processes. These studies may provide clues for future drug development and/or treatment strategies targeting aberrations in distal nephron salt handling and blood pressure regulation and, possibly, abnormalities in airway epithelial transport and airway reactivity. Results from these studies will have import for our understanding and potential treatment of hypertension, heart failure, hepatic failure and other fluid retention syndromes, as well as for treatment of asthma, chronic obstructive lung disease and other airway diseases.

Acknowledgments J. Rozenfeld and O. Kladnitsky were supported by donations from Dora and Sidney Gabrel and Martin Kolinsky, respectively, London, UK, and by the L & A Kessler Family Foundation, Berkeley, CA, USA; SL Alper was supported by NIH grants DK43495 and DK34854 (Harvard Digestive Disease Center) and by the USA-Israel Binational Science Foundation; I. Zelikovic was supported by the USA-Israel Binational Science Foundation, by the Rappaport Institute for Research in the Medical Sciences and by the Dr. Y. Rabinovitz Research Fund, Technion – Israel Institute of Technology.

References

Adams KM, Abraham V, Cohen N, Kolls JK, Kreindler JL (2013) Pendrin is a participant in antimicrobial responses in the lung. Am J Respir Crit Care Med 187:A4741

Adams KM, Abraham V, Spielman D, Kolls JK, Rubenstein RC, Conner GE, Cohen NA, Kreindler JL (2014) IL-17A induces pendrin expression and chloride-bicarbonate exchange in human bronchial epithelial cells. PLoS One 9:e103263. doi:10.1371/journal.pone.01032632014

Adler L, Efrati E, Zelikovic I (2008) Molecular mechanisms of epithelial cell-specific expression and regulation of the human anion exchanger (pendrin) gene. Am J Physiol Cell Physiol 294:C1261–C1276

Alper SL, Sharma AK (2013) The *SLC26* gene family of anion transporters and channels. Mol Aspects Med 34:494–515

Alpern RJ et al (2000) Renal acidification mechanisms. In: Brenner BM, Rector FC (eds) Brenner & Rector's the kidney. Saunders, Philadelphia, pp 455–519

Alrefai WA, Tyagi S, Mansour F, Saksena S, Syed I, Ramaswamy K, Dudeja PK (2001) Sulfate and chloride transport in Caco-2 cells: differential regulation by thyroxine and the possible role of DRA gene. Am J Physiol Gastrointest Liver Physiol 280:G603–G613

Alrefai WA, Wen X, Jiang W, Katz JP, Steinbrecher KA, Cohen MB, Williams IR, Dudeja PK, Wu GD (2007) Molecular cloning and promoter analysis of downregulated in adenoma (*DRA*). Am J Physiol Gastrointest Liver Physiol 293:G923–G934

Anckar J, Sistonen L (2007) Heat shock factor 1 as a coordinator of stress and developmental pathways. Adv Exp Med Biol 594:78–88

Anwar S, Riazuddin S, Ahmed ZM, Tasneem S, Ateeq Ul J, Khan SY, Griffith AJ, Friedman TB, Riazuddin S (2009) *SLC26A4* mutation spectrum associated with DFNB4 deafness and Pendred's syndrome in Pakistanis. J Hum Genet 54:266–270

Arturi F, Russo D, Bidart JM, Scarpelli D, Schlumberger M, Filetti S (2001) Expression pattern of the pendrin and sodium/iodide symporter genes in human thyroid carcinoma cell lines and human thyroid tumors. Eur J Endocrinol 145:129–135

Bidart JM, Lacroix L, Evain-Brion D, Caillou B, Lazar V, Frydman R, Bellet D, Filetti S, Schlumberger M (2000a) Expression of Na^+/I^- symporter and Pendred syndrome genes in trophoblast cells. J Clin Endocrinol Metab 85:4367–4372

Bidart JM, Mian C, Lazar V, Russo D, Filetti S, Caillou B, Schlumberger M (2000b) Expression of pendrin and the Pendred syndrome (*PDS*) gene in human thyroid tissues. J Clin Endocrinol Metab 85:2028–2033

Blomqvist SR, Vidarsson H, Fitzgerald S, Johansson BR, Ollerstam A, Brown R, Persson AEG, Bergstrom G, Enerback S (2004) Distal renal tubular acidosis in mice that lack the forkhead transcription factor Foxi1. J Clin Invest 113:1560–1570

Campbell RM, Tummino PJ (2014) Cancer epigenetics drug discovery and development: the challenge of hitting the mark. J Clin Invest 124:64–69

Campbell C, Cucci RA, Prasad S, Green GE, Edeal JB, Galer CE, Karniski LP, Sheffield VC, Smith RJ (2001) Pendred syndrome, DFNB4, and *PDS/SLC26A4* identification of eight novel mutations and possible genotype-phenotype correlations. Hum Mutat 17:403–411

Carlin RW, Sedlacek RL, Quesnell RR, Pierucci-Alves F, Grieger DM, Schultz BD (2006) PVD9902, a porcine vas deferens epithelial cell line that exhibits neurotransmitter-stimulated anion secretion and expresses numerous HCO_3(-) transporters. Am J Physiol Cell Physiol 290:C1560–C1571

Carrithers SL, Ott CE, Hill MJ, Johnson BR, Cai W, Chang JJ, Shah RG, Sun C, Mann EA, Fonteles MC, Forte LR, Jackson BA, Giannella RA, Greenberg RN (2004) Guanylin and uroguanylin induce natriuresis in mice lacking guanylyl cyclase-C receptor. Kidney Int 65:40–53

Catalano MG, Fortunati N, Boccuzzi G (2012) Epigentics modifications and therapeutic prospects in human thyroid cancer. Front Endocrinol 3:1–8

Cetin Y, Kulaksiz H, Redecker P, Bargsten G, Adermann K, Grube D (1995) Bronchiolar nonciliated secretory (Clara) cells: source of guanylin in the mammalian lung. Proc Natl Acad Sci U S A 92:5925–5929

Choi BY, Stewart AK, Madeo AC, Pryor SP, Lenhard S, Kittles R, Eisenman D, Kim HJ, Niparko J, Thomsen J, Arnos KS, Nance WE, King KA, Zalewski CK, Brewer CC, Shawker T, Reynolds JC, Butman JA, Karniski LP, Alper SL, Griffith AJ (2009a) Hypo-functional SLC26A4 variants associated with nonsyndromic hearing loss and enlargement of the vestibular aqueduct: genotype-phenotype correlation or coincidental polymorphisms? Hum Mutat 30:599–608

Choi BY, Stewart AK, Nishimura KK, Cha WJ, Seong MW, Park SS, Kim SW, Chun YS, Chung JW, Park SN, Chang SO, Kim CS, Alper SL, Griffith AJ, Oh SH (2009b) Efficient molecular genetic diagnosis of enlarged vestibular aqueducts in East Asians. Genet Test Mol Biomarkers 13:679–687

Choi BY, Kim HM, Ito T, Lee KY, Li X, Monahan K, Wen Y, Wilson E, Kurima K, Saunders T, Petralia RS, Wangemann P, Friedman TB, Griffith A (2011) Mouse model of enlarged vestibular aqueducts defines temporal requirement of Slc26a4 expression for hearing acquisition. J Clin Invest 121:4516–4525

Comrie MM, Cutler CP, Cramb G (2001) Cloning and expression of guanylin from the European eel (*Anguilla anguilla*). Biochem Biophys Res Commun 281:1078–1085

Dai P, Stewart AK, Chebib F, Hsu A, Rozenfeld J, Huang D, Kang D, Lip V, Fang H, Shao H, Liu X, Shen Y, Yang W, Zelikovic I, Platt OS, Han D, Alper SL, Wu BL (2009) Distinct and novel SLC26A4/Pendrin mutations in Chinese and U.S. patients with non-syndromic hearing loss. Physiol Genomics 38:281–290

Das S, Bhattacharyya NP (2014) Transcription regulation of HYPK by heat shock factor 1. PLoS One 9(1):e85552. doi:10.1371/journal.pone.0085552

Dentice M, Luongo C, Elefante A, Ambrosio R, Salzano S, Zannini M, Nitsch R, Di Lauro R, Rossi G, Fenzi G, Salvatore D (2005) Pendrin is a novel *in vivo* downstream target gene of the TTF-1/Nkx-2.1 homeodomain transcription factor in differentiated thyroid cells. Mol Cell Biol 25:10171–10182

Dossena S, Rodighiero S, Vezzoli V, Nofziger C, Salvioni E, Boccazzi M, Grabmayer E, Botta G, Meyer G, Fugazzola L, Peccoz PB, Paulmichl M (2009) Functional characterization of wild-type and mutated pendrin (SLC26A4), the anion transporter involved in pendred syndrome. J Mol Endocrinol 43:93–103

Dou H, Xu J, Wang Z, Smith AN, Soleimani M, Karet FE, Greinwald JH Jr, Choo D (2004) Co-expression of pendrin, vacuolar H-ATPase alpha4- subunit and carbonic anhydrase II in epithelial cells of the murine endolymphatic sac. J Histochem Cytochem 52:1377–1384

Efrati E, Adler L, Tal O, Zelikovic I (2007) The pendrin gene, *PDS*, is transcriptionally regulated by ambient pH and chloride. J Am Soc Nephrol 8:6A

Ehret GB, Reichenbach P, Schindler U, Horvath CM, Fritz S, Nabholz M, Bucher P (2001) DNA binding specificity of different STAT proteins. Comparison of *in vitro* specificity with natural target sites. J Biol Chem 276:6675–6688

Eladari D, Chambrey R, Peti-Peterdi J (2012) A new look at electrolyte transport in the distal tubule. Annu Rev Physiol 74:325–349

Espiritu DJ, Bernardo AA, Robey RB, Arruda JA (2002) A central role for Pyk2-Src interaction in coupling diverse stimuli to increased epithelial NBC activity. Am J Physiol Renal Physiol 283:F663–F670

Everett LA, Glaser B, Beck JC, Idol JR, Buchs A, Heyman M, Adawi F, Hazani E, Nassir E, Baxevanis AD, Sheffield VC, Green ED (1997) Pendred syndrome is caused by mutations in a putative sulphate transporter gene (*PDS*). Nat Genet 17:411–422

Everett LA, Morsli H, Wu DK, Green ED (1999) Expression pattern of the mouse ortholog of the Pendred's syndrome gene (*Pds*) suggests a key role for pendrin in the inner ear. Proc Natl Acad Sci U S A 96:9727–9732

Fernandez-Lazo Y, Aguilera G, Pham TD, Park AY, Beierwalters WH, Sutliff RL, Verlander JW, Pacack K, Osunkoya AO, Ellis CL, Kim YH, Shipley GL, Wynne BM, Hoover RS, Sen SK, Plotsky PM, Wall SM (2015) Pendrin localizes to the adrenal medulla and modulates catecholamine release. Am J Physiol Endocrinol Metab 309(6):E534–E545. doi:10.1152/ajpendo.00035.2015 [Epub ahead of print]

Fonteles MC, Falcao do Nascimento NR (2011) Guanylin peptide family: history, interaction with ANP, and new pharmacological perspectives. Can J Physiol Pharmacol 89:575–585

Fonteles MC, Greenberg RN, Monteiro HS, Currie MG, Forte LR (1998) Natriuretic and kaliuretic activities of guanylin and uroguanylin in the isolated perfused rat kidney. Am J Physiol Renal Physiol 275:F191–F197

Forte LR (2003) A novel role for uroguanylin in the regulation of sodium balance. J Clin Invest 112:1138–1141

Forte LR (2004) Uroguanylin and guanylin peptides: pharmacology and experimental therapeutics. Pharmacol Ther 104:137–167

Forte LR, Fan X, Hamra FK (1996) Salt and water homeostasis: uroguanylin is a circulating peptide hormone with natriuretic activity. Am J Kidney Dis 28:296–304

Frische S, Kwon TH, Frokiaer J, Madsen KM, Nielsen S (2003) Regulated expression of pendrin in rat kidney in response to chronic NH4Cl or NaHCO3 loading. Am J Physiol Renal Physiol 284:F584–F593

Fruhbeck G (2012) Uroguanylin-a new gut-derived weapon against obesity? Nat Rev Endocrinol 8:5–6

Fukae H, Kinoshita H, Fujimoto S, Nakazato M, Eto T (2000) Plasma concentration of uroguanylin in patients on maintenance dialysis therapy. Nephron 84:206–210

Garnett JP, Hickman E, Burrows R, Hegyi P, Tiszlavicz L, Cuthbert AW, Fong P, Gray MA (2011) Novel role for pendrin in orchestrating bicarbonate secretion in cystic fibrosis transmembrane conductance regulator (CFTR)-expressing airway serous cells. J Biol Chem 286: 41069–41082

Gennari FJ, Maddox DA (2000) Renal regulation of acid-base homeostasis integrated response. In: Seldin DW, Giebisch GH (eds) The kidney: physiology and pathophysiology, 3rd edn. Lippincott Williams & Wilkins, Philadelphia, pp 2015–2053

Gluck SL (2004) Acid sensing in renal epithelial cells. J Clin Invest 114:1696–1699

Gorbunov D, Sturlese M, Nies F, Kluge M, Bellanda M, Battistutta R, Oliver D (2014) Molecular architecture and the structural basis for anion interaction in prestin and SLC26 transporters. Nat Commun 5:3622. doi:10.1038/ncomms4622

Griffith AJ, Wangemann P (2011) Hearing loss associated with enlargement of the vestibular aqueduct: mechanistic insights from clinical phenotypes, genotypes, and mouse models. Hear Res 281:11–17

Grimm PR, Lazo-Fernandez Y, Delpire E, Wall SM, Dorsey SG, Weinman EJ, Coleman R, Wade JB, Welling PA (2015) Integrated compensatory network is activated in the absence of NCC phosphorylation. J Clin Invest 125:2136–2150

Haila S, Höglund P, Scherer SW, Lee JR, Kristo P, Coyle B, Trembath R, Holmberg C, de la Chapelle A, Kere J (1998) Genomic structure of the human congenital chloride diarrhea (*CLD*) gene. Gene 214:87–93

Hamm LL, Nakhoul NL (2008) Renal acidification. In: Brenner BM (ed) Brenner & Rector's the kidney, 8th edn. Saunders, Philadelphia, pp 248–279

He L, Hannon GJ (2004) MicroRNAs: small RNAs with a big role in gene regulation. Nat Rev Genet 5:522–531

Huang J, Shan J, Kim D, Liao J, Evagelidis A, Alper SL, Hanrahan JW (2012) Basolateral chloride loading by the anion exchanger type 2: role in fluid secretion by the human airway epithelial cell line calu-3. J Physiol 590:5299–5316

Hulander M, Kiernan AE, Blomqvist SR, Carlsson P, Samuelsson EJ, Johansson BR, Steel KP, Enerback S (2003) Lack of pendrin expression leads to deafness and expansion of the endolymphatic compartment in inner ears of *Foxi1* null mutant mice. Development 130:2013–2025

Ishida A, Ohta N, Suzuki Y, Kakehata S, Okubo K, Ikeda H, Shiraishi H, Izuhara K (2012) Expression of pendrin and periostin in allergic rhinitis and chronic rhinosinusitis. Allergol Int 61:589–595

Jacques T, Picard N, Miller RL, Riemondy KA, Houillier P, Sohet F, Ramakrishnan SK, Busst CJ, Jayat M, Corniere N, Hassan H, Aronson PS, Hennings JC, Hubner CA, Nelson RD, Chambrey R, Eladari D (2013) Overexpression of pendrin in intercalated cells produces chloride-sensitive hypertension. J Am Soc Nephrol 24:1104–1113

Johnsen T, Larsen C, Friis J, Hougaard-Jensen F (1987) Pendred's syndrome. Acoustic, vestibular and radiological findings in 17 unrelated patients. J Laryngol Otol 101:1187–1192

Kandasamy N, Fugazzola L, Evans M, Chatterjee K, Karet F (2011) Life-threatening alkalosis in Pendred syndrome. Eur J Endocrinol 165:167–170

Kikuchi M, Fujimoto S, Fukae H, Kinoshita H, Kita T, Nakazato M, Eto T (2005) Role of uroguanylin, a peptide with natriuretic activity, in rats with experimental nephrotic syndrome. J Am Soc Nephrol 16:392–397

Kim YH, Pech V, Spencer KB, Beierwaltes WH, Everett LA, Green ED, Shin W, Verlander JW, Sutliff RL, Wall SM (2007) Reduced ENaC protein abundance contributes to the lower blood pressure observed in pendrin-null mice. Am J Physiol Renal Physiol 293:F1314–F1324

Kim YH, Pham TD, Zheng W, Hong S, Baylis C, Pech V, Beierwaltes WH, Farley DB, Braverman LE, Verlander JW, Wall SM (2009) Role of pendrin in iodide balance: going with the flow. Am J Physiol Renal Physiol 297:F1069–F1079

Kim D, Kim J, Burghardt B, Best L, Steward MC (2014) Role of anion exchangers in Cl^- and HCO_3^- secretion by the human airway epithelial cell line Calu-3. Am J Physiol Cell Physiol 307:C208–C219

Kinoshita H, Fujimoto S, Nakazato M, Yokota N, Date Y, Yamaguchi H, Hisanaga S, Eto T (1997) Urine and plasma levels of uroguanylin and its molecular forms in renal diseases. Kidney Int 52:1028–1034

Kita T, Smith CE, Fok KF, Duffin KL, Moore WM, Karabatsos PJ, Kachur JF, Hamra FK, Pidhorodeckyj NV, Forte LR, Currie MG (1994) Characterization of human uroguanylin: a member of the guanylin peptide family. Am J Physiol Ren Fluid Electrolyte Physiol 266:F342–F348

Kita T, Kitamura K, Sakata J, Eto T (1999) Marked increase of guanylin secretion in response to salt loading in the rat small intestine. Am J Physiol Gastrointest Liver Physiol 277:G960–G966

Kitazono M, Robey R, Zhan Z, Sarlis NJ, Skarulis MC, Aikou T, Bates S, Fojo T (2001) Low concentrations of the histone deacetylase inhibitor, depsipeptide (FR901228), increase expression of the Na(+)/I(-) symporter and iodine accumulation in poorly differentiated thyroid carcinoma cells. J Clin Endocrinol Metab 86:3430–3435

Knepper MA (2015) Systems biology of diuretic resistance. J Clin Invest 125:1793–1795

Kobayashi T, Sugimoto T, Saijoh K, Fujii M, Chihara K (1997) Cloning and characterization of the 5'-flanking region of the mouse diastrophic dysplasia sulfate transporter gene. Biochem Biophys Res Commun 238(3):738–743

Kopp P (2000) Pendred's syndrome and genetic defects in thyroid hormone synthesis. Rev Endocr Metab Disord 1:109–121

Krause WJ, Freeman RH, Forte LR (1990) Autoradiographic demonstration of specific binding sites for *E. coli* enterotoxin in various epithelia of the North American opossum. Cell Tissue Res 260:387–394

Kulaksiz H, Schmid A, Honscheid M, Ramaswamy A, Certin Y (2002) Clara cell impact in air-side activation of CFTR in small pulmonary airways. Proc Natl Acad Sci U S A 99:6796–6801

Kuperman DA, Lewis CC, Woodruff PG, Rodriguez MW, Yang YH, Dolganov GM, Fahy JV, Erle DJ (2005) Dissecting asthma using focused transgenic modeling and functional genomics. J Allergy Clin Immunol 116:305–311

Lacroix L, Michiels S, Mian C, Arturi F, Caillou B, Filetti S, Schlumberger M, Bidart JM (2006) *HEX, PAX-8* and *TTF-1* gene expression in human thyroid tissues: a comparative analysis with other genes involved in iodide metabolism. Clin Endocrinol (Oxf) 64:398–404

Lee A, Nofziger C, Dossena S, Vanoni S, Diasio R, Paulmichl M (2011) Methylation of the human pendrin promoter. Cell Physiol Biochem 28:397–406

Lee HJ, Yoo JE, Namkung W, Cho HJ, Kim K, Kang JW, Yoon JH, Choi JY (2015) Thick airway surface liquid volume and weak mucin expression in pendrin-deficient human airway epithelia. Physiol Rep 3(8):e12840. doi:10.14814/phy2.12480

Leviel F, Hubner CA, Houillier P, Morla L, El Moghrabi S, Brideau G, Hatim H, Parker MD, Kurth I, Kougioumtzes A, Sinning A, Pech V, Riemondy KA, Miller RL, Hummler E, Shull GE, Aronson PS, Doucet A, Wall SM, Chambrey R, Eladari D (2010) The Na^+-dependent chloride-bicarbonate exchanger SLC4A8 mediates an electroneutral Na^+ reabsorption process in the renal cortical collecting ducts of mice. J Clin Invest 120:1627–1635

Li W, Lee J, Vikis HG, Lee SH, Liu G, Aurandt J, Shen TL, Fearon ER, Guan JL, Han M, Rao Y, Hong K, Guan KL (2004) Activation of FAK and Src are receptor-proximal events required for netrin signaling. Nat Neurosci 7:1213–1221

Madeo AC, Manichaikul A, Pryor SP, Griffith AJ (2009) Do mutations of the pendred syndrome gene, *SLC26A4*, confer resistance to asthma and hypertension? J Med Genet 46:405–406

McCallum L, Lip S, Padmanabhan S (2015) The hidden hand of chloride in hypertension. Pflugers Arch Eur J Physiol 467:595–603

Mikita T, Campbell D, Wu P, Williamson K, Schindler U (1996) Requirements for interleukin-4-induced gene expression and functional characterization of Stat6. Mol Cell Biol 16:5811–5820

Mohebbi N, Perna A, van der Wijst J, Becker HM, Cappaso G, Wagner CA (2013) Regulation of two renal chloride transporters, AE1 and pendrin, by electrolytes and aldosterone. PLoS One 8(1):e55286. doi:10.1371/journal.pone.0055286

Morimoto RI (2008) Proteotoxic stress and inducible chaperone networks in neurodegenerative disease and aging. Genes Dev 22:1427–1438

Morimoto RI, Santoro MG (1998) Stress-inducible responses and heat shock proteins: new pharmacologic targets for cytoprotection. Nat Biotechnol 16:833–838

Moss NG, Fellner RC, Qian X, Yu SJ, Li Z, Nakazato M, Goy MF (2008) Uroguanylin, an intestinal natriuretic peptide, is delivered to the kidney as an unprocessed propeptide. Endocrinology 149:4486–4498

Mueller T, Dieplinger B (2012) The guanylin peptide family and proposed gastrointestinal-renal natriuretic signaling axis. Kidney Int 82:1253–1255

Muscella A, Verri SM, Urso L, Dimitri C, Botta G, Paulmichl M, Beck-Peccoz P, Fugazzola L, Storelli C (2008) PKC-ε-dependent cytosol-to-membrane translocation of pendrin in rat thyroid PC Cl3 cells. J Cell Physiol 217:103–112

Nakagami Y, Favoreto S Jr, Zhen G, Park SW, Nguyenvu LT, Kuperman DA, Dolganov GM, Huang X, Boushey HA, Avila PC, Erle DJ (2008) The epithelial anion transporter pendrin is induced by allergy and rhinovirus infection, regulates airway surface liquid, and increases airway reactivity and inflammation in an asthma model. J Immunol 181:2203–2210

Nakao I, Kanaji S, Ohta S, Matsushita H, Arima K, Yuyama N, Yamaya M, Nakayama K, Kubo H, Watanabe M, Sagara H, Sugiyama K, Tanaka H, Toda S, Hayashi H, Inoue H, Hoshino T, Shiraki A, Inoue M, Suzuki K, Aizawa H, Okinami S, Nagai H, Hasegawa M, Fukuda T, Green ED, Izuhara K (2008) Identification of pendrin as a common mediator for mucus production in bronchial asthma and chronic obstructive pulmonary disease. J Immunol 180:6262–6269

Nakaya K, Harbidge DG, Wangemann P, Schultz BD, Green ED, Wall SM, Marcus DC (2007) Lack of pendrin HCO_3^- transport elevates vestibular endolymphatic [$Ca2^+$] by inhibition of acid-sensitive TRPV5 and TRPV6 channels. Am J Physiol Renal Physiol 292:F1314–F1321

Nakazato M, Yamaguchi H, Shiomi K, Date Y, Fujimoto S, Kangawa K, Matsuo H, Matsukura S (1994) Identification of 10-kDa proguanylin as a major guanylin molecule in human intestine and plasma and its increase in renal insufficiency. Biochem Biophys Res Commun 205: 1966–1975

Nofziger C, Vezzoli V, Dossena S, Schonherr T, Studnicka J, Nofziger J, Vanoni S, Stephan S, Silva ME, Meyer G, Paulmichl M (2011) STAT6 links IL-4/IL-13 stimulation with pendrin expression in asthma and chronic obstructive pulmonary disease. Clin Pharmacol Ther 90:399–405

Ohbayashi K, Yamaki KI (2000) Both inhalant and intravenous uroguanylin inhibit leukotriene C 4-induced airway changes. Peptides 21:1467–1472

Ohbayashi K, Yamaki K, Suzuki R, Takagi K (1998) Effects of uroguanylin and guanylin against antigen-induced bronchoconstriction and airway microvascular leakage in sensitized guinea pigs. Life Sci 62:1833–1844

Park HJ, Shaukat S, Liu XZ, Hahn SH, Naz S, Ghosh M, Kim HN, Moon SK, Abe S, Tukamoto K, Riazuddin S, Kabra M, Erdenetungalag R, Radnaabazar J, Khan S, Pandya A, Usami SI, Nance WE, Wilcox ER, Riazuddin S, Griffith AJ (2003) Origins and frequencies of *SLC26A4* (*PDS*) mutations in east and south Asians: global implications for the epidemiology of deafness. J Med Genet 40:242–248

Pech V, Kim HY, Weinstein MA, Everett AL, Pham DT, Wall MS (2007) Angiotensin II increases chloride absorption in the cortical collecting duct in mice through a pendrin-dependent mechanism. Am J Physiol Renal Physiol 292:F914–F920

Pech V, Pham TD, Hong S, Weinstein AM, Spencer KB, Duke BJ, Walp E, Kim YH, Sutliff RL, Bao HF, Eaton DC, Wall SM (2010) Pendrin modulates ENaC function by changing luminal HCO_3^-. J Am Soc Nephrol 21:1928–1941

Pedemonte N, Caci E, Sondo E, Caputo A, Rhoden K, Pfeffer U, Di Candia M, Bandettini R, Ravazzolo R, Zegarra-Moran O, Galietta LJ (2007) Thiocyanate transport in resting and IL-4-stimulated human bronchial epithelial cells: role of pendrin and anion channels. J Immunol 178:5144–5153

Pela I, Bigozzi M, Bianchi B (2008) Profound hypokalemia and hypochloremic metabolic alkalosis during thiazide therapy in a child with Pendred syndrome. Clin Nephrol 69:450–453

Petrovic S, Wang Z, Ma L, Soleimani M (2003) Regulation of the apical Cl-/HCO3- exchanger pendrin in rat cortical collecting duct in metabolic acidosis. Am J Physiol Renal Physiol 284:F103–F112

Piala AT, Moon TM, Akella R, He H, Cobb MH, Goldsmith EJ (2014) Chloride sensing by WNK1 involves inhibition of autophosphorylation. Sci Signal 7(324):ra41. doi:10.1126/scisignal.2005050

Porra V, Bernier-Valentin F, Trouttet-Masson S, Berger-Dutrieux N, Peix JL, Perrin A, Selmi-Ruby S, Rousset B (2002) Characterization and semiquantitative analyses of pendrin expressed in normal and tumoral human thyroid tissues. J Clin Endocrinol Metab 87:1700–1707

Potthast R, Ehler E, Scheving LA, Sindic A, Schlatter E, Kuhn M (2001) High salt intake increases uroguanylin expression in mouse kidney. Endocrinology 142:3087–3097

Priyamvada S, Anbazhagan AN, Gujral T, Borthakur A, Saksena S, Gil RK, Alrefai WA, Dudjena PK (2015) All-trans-retinoic acid increases SLC26A3 DRA (down-regulated in adenoma) expression in intestinal epithelial cells via HNF-1β. J Biol Chem 290:15066–15077

Qian X, Moss NG, Fellner RC, Goy MF (2008) Circulating prouroguanylin is processed to its active natriuretic form exclusively within the renal tubules. Endocrinology 149:4499–4509

Qian X, Moss NC, Fellner RC, Taylor-Blake B, Goy MF (2011) The rat kidney contains high levels of prouroguanylin (the uroguanylin precursor) but does not express GC-C (the enteric uroguanylin receptor). Am J Physiol Renal Physiol 300:F561–F573

Quentin F, Chambrey R, Trinh-Trang-Tan MM, Fysekidis M, Cambillau M, Paillard M, Aronson PS, Eladari D (2004) The Cl^-/HCO_3- exchanger pendrin in the rat kidney is regulated in response to chronic alterations in chloride balance. Am J Physiol Renal Physiol 287: F1179–F1188

Raheja G, Singh V, Ma K, Boumendjel R, Borthakur A, Gill RK, Saksena S, Alrefai WA, Ramaswamy K, Dudeja PK (2010) *Lactobacillus acidophilus* stimulates the expression of *SLC26A3 via* a transcriptional mechanism. Am J Physiol Gastrointest Liver Physiol 298:G395–G401

Reardon W, Trembath RC (1996) Pendred syndrome. J Med Genet 33:1037–1040

Reimold FR, Heneghan JF, Stewart AK, Zelikovic I, Vandorpe DH, Shmukler BE, Alper SL (2011) Pendrin function and regulation in *Xenopus* oocytes. Cell Physiol Biochem 28:435–440

Rillema JA, Hill MA (2003) Prolactin regulation of the pendrin-iodide transporter in the mammary gland. Am J Physiol Endocrinol Metab 284:E25–E28

Rothbart SB, Strahl BD (2014) Interpreting the language of histone and DNA modifications. Biochim Biophys Acta 1839:627–643

Royaux IE, Suzuki K, Mori A, Katoh R, Everett LA, Kohn LD, Green ED (2000) Pendrin, the protein encoded by the Pendred syndrome gene (*PDS*), is an apical porter of iodide in the thyroid and is regulated by thyroglobulin in FRTL-5 cells. Endocrinology 141:839–845

Royaux IE, Wall SM, Karniski LP, Everett LA, Suzuki K, Knepper MA, Green ED (2001) Pendrin, encoded by the Pendred syndrome gene, resides in the apical region of renal intercalated cells and mediates bicarbonate secretion. Proc Natl Acad Sci U S A 98:4221–4226

Rozenfeld J, Efrati E, Adler L, Tal O, Carrithers SL, Alper SL, Zelikovic I (2011) Transcriptional regulation of the pendrin gene. Cell Physiol Biochem 28:385–396

Rozenfeld J, Tal O, Kladnitsky O, Adler L, Efrati E, Carrithers SL, Alper SL, Zelikovic I (2012) The pendrin anion exchanger gene is transcriptionally regulated by uroguanylin: a novel enterorenal link. Am J Physiol Renal Physiol 302:F614–F624

Rozenfeld J, Tal O, Kladnitsky O, Adler L, Efrati E, Carrithers SL, Alper SL, Zelikovic I (2013) Pendrin, a novel transcriptional target of the uroguanylin system. Cell Physiol Biochem 32(Suppl 1):221–237

Saksena S, Singla A, Goyal S, Katyal S, Bansal N, Gill RK, Alrefai WA, Ramaswamy K, Dudeja PK (2010) Mechanisms of transcriptional modulation of the human anion exchanger *SLC26A3* gene expression by IFN-{gamma}. Am J Physiol Gastrointest Liver Physiol 298:G159–G166

Scanlon KM, Gau Y, Zhu J, Skerry C, Wall SM, Soleimani M, Carbonetti NH (2014) Epithelial anion transporter pendrin contributes to inflammatory lung pathology in mouse models of *Bordetella pertussis* infection. Infect Immun 82:4212–4221

Schwartz GJ (2001) Plasticity of intercalated cell polarity: effect of metabolic acidosis. Nephron 87:304–313

Scott DA, Wang R, Kreman TM, Sheffield VC, Karniski LP (1999) The Pendred syndrome gene encodes a chloride-iodide transport protein. Nat Genet 21:440–443

Shan J, Liao J, Robert R, Palmer ML, Fahrenkrug SC, O'Grady SM, Hanrahan JW (2012) Bicarbonate-dependent chloride transport drives fluid secretion by the human airway epithelial cell line Calu-3. J Physiol 590:5273–5297

Sharma S, Kelly TK, Jones PA (2010) Epigenetics in cancer. Carcinogenesis 31:27–36

Shcheynikov N, Yang D, Wang Y, Zeng W, Karniski LP, So I, Wall SM, Muallem S (2008) The Slc26a4 transporter functions as an electroneutral $Cl^-/I^-/HCO_3^-$ exchanger: role of Slc26a4 and Slc26a6 in I^- and HCO_3^- secretion and in regulation of CFTR in the parotid duct. J Physiol 586:3813–3824

Sheffield VC, Kraiem Z, Beck JC, Nishimura D, Stone EM, Salameh M, Sadeh O, Glaser B (1996) Pendred syndrome maps to chromosome 7q21-34 and is caused by an intrinsic defect in thyroid iodine organification. Nat Genet 12:424–426

Shelden MC, Howitt SM, Price GD (2010) Membrane topology of the cyanobacterial bicarbonate transporter, BicA, a member of the SulP (SLC26A) family. Mol Membr Biol 27:12–23
Sindic A, Schlatter E (2006) Cellular effects of guanylin and uroguanylin. J Am Soc Nephrol 17:607–616
Sindic A, Schlatter E (2007) Renal electrolyte effects of guanylin and uroguanylin. Curr Opin Nephrol Hypertens 16:10–15
Sindic A, Velic A, Basoglu C, Hirsch JR, Edemir B, Kuhn M, Schlatter E (2005) Uroguanylin and guanylin regulate transport of mouse cortical collecting duct independent of guanylate cyclase C. Kidney Int 68:1008–1017
Singla A, Kumar A, Priyamvada S, Tahniyath M, Saksena S, Gill RK, Alrefai WA, Dudeja K (2012) LPA stimulates intestinal DRA gene transcription via LPA2 receptor, PI3K/AKT, and c-Fos-dependent pathway. Am J Physiol Gastrointest Liver Physiol 302:G618–G627
Soleimani M, Xu J (2006) SLC26 chloride/base exchangers in the kidney in health and disease. Sem Nephrol 26:375–385
Soleimani M, Barone S, Xu J, Shull GF, Siddiqui F, Zahedi K, Amlal H (2012) Double knockout of pendrin and Na-Cl cotransporter (NCC) causes severe salt wasting, volume depletion, and renal failure. Proc Natl Acad Sci U S A 109:13368–13373
Suzuki K, Royaux IE, Everett LA, Mori-Aoki A, Suzuki S, Nakamura K, Sakai T, Katoh R, Toda S, Green ED, Kohn LD (2002) Expression of PDS/Pds, the Pendred syndrome gene, in endometrium. J Clin Endocrinol Metab 87:938–941
Tahiliani M, Koh KP, Shen Y, Pastor WA, Bandukwala H, Brudno Y, Agarwal S, Iyer LM, Liu DR, Aravind L, Rao A (2009) Conversion of 5-methylcytosine to 5-hydroxymethylcytosine in mammalian DNA by MLL partner TET1. Science 324:930–935
Vallet M, Picard N, Loffing-Cueni D, Fysekidis M, Bloch-Faure M, Deschênes G, Breton S, Meneton P, Loffing J, Aronson PS, Chambrey R, Eladari D (2006) Pendrin regulation in mouse kidney primarily is chloride-dependent. J Am Soc Nephrol 17:2153–2163
Vanoni S, Nofziger C, Dossena S, Soyal SM, Patsch W, Plevani P, Duschl A, Paulmichl M (2013) The human pendrin promoter contains two N4GAS motifs with different functional relevance. Cell Physiol Biochem 32:238–248
Verlander JW, Hassell KA, Royaux IE, Glapion DM, Wang ME, Everett LA, Green ED, Wall SM (2003) Deoxycorticosterone upregulates PDS (Slc26a4) in mouse kidney: role of pendrin in mineralocorticoid-induced hypertension. Hypertension 42:356–362
Verlander JW, Kim YH, Shin W, Pham TD, Hassell KA, Beierwaltes WH, Green ED, Everett L, Matthews SW, Wall SM (2006) Dietary Cl- restriction upregulates pendrin expression within the apical plasma membrane of type B intercalated cells. Am J Physiol Renal Physiol 291:F833–F839
Verlander JW, Hong S, Pech V, Bailey JL, Agazatian D, Matthews SW, Coffman TM, Le T, Inagami T, Whitehill FM, Weiner ID, Farley DB, Kim YH, Wall SM (2011) Angiotensin II acts through the angiotensin 1a receptor to upregulate pendrin. Am J Physiol Renal Physiol 301: F1314–F1325
Vihervaara A, Sergelius C, Vasara J, Blom MA, Elsing AN, Roos-Mattjus P, Sistonen L (2013) Transcriptional response to stress in the dynamic chromatin environment of cycling and mitotic cells. Proc Natl Acad Sci U S A 110:E3388–E3397
Wagner CA, Finberg KE, Stehberger PA, Lifton RP, Giebisch GH, Aronson PS, Geibel JP (2002) Regulation of the expression of the Cl^-/anion exchanger pendrin in mouse kidney by acid-base status. Kidney Int 62:2109–2117
Wall SM, Lazo-Fernandez Y (2015) The role of pendrin in renal physiology. Annu Rev Physiol 77:363–378
Wall SM, Hassell KA, Royaux IE, Green ED, Chang JY, Shipley GL, Verlander JW (2003) Localization of pendrin in mouse kidney. Am J Physiol Renal Physiol 284(1):F229–F241
Wangemann P (2011) The role of pendrin in the development of the murine inner ear. Cell Physiol Biochem 28:527–534
West AC, Johnstone RW (2014) New and emerging HDAC inhibitors for cancer treatment. J Clin Invest 124:30–39

Winter H, Braig C, Zimmermann U, Geisler HS, Fränzer JT, Weber T, Ley M, Engel J, Knirsch M, Bauer K, Christ S, Walsh EJ, McGee J, Köpschall I, Rohbock K, Knipper M (2006) Thyroid hormone receptors TRalpha1 and TRbeta differentially regulate gene expression of *Kcnq4* and prestin during final differentiation of outer hair cells. J Cell Sci 119:2975–2984

Wu C, Morris JR (2001) Genes, genetics, and epigenetics: a correspondence. Science 293: 1103–1105

Xing M, Tokumaru Y, Wu G, Westra WB, Ladenson PW, Sidransky D (2003) Hypermethylation of the Pendred syndrome gene *SLC26A4* is an early event in thyroid tumorigenesis. Cancer Res 63:2312–2315

Yang T, Vidarsson H, Rodrigo-Blomqvist S, Rosengren SS, Enerback S, Smith RJ (2007) Transcriptional control of *SLC26A4* is involved in Pendred syndrome and nonsyndromic enlargement of vestibular aqueduct (DFNB4). Am J Hum Genet 80:1055–1063

Yoshida A, Taniguchi S, Hisatome I, Royaux IE, Green ED, Kohn LD, Suzuki K (2002) Pendrin is an iodide-specific apical porter responsible for iodide efflux from thyroid cells. J Clin Endocrinol Metab 87:3356–3361

Yoshida A, Hisatome I, Taniguchi S, Sasaki N, Yamamoto Y, Miake J, Fukui H, Shimizu H, Okamura T, Okura T, Igawa O, Shigemasa C, Green ED, Kohn LD, Suzuki K (2004) Mechanism of iodide/chloride exchange by pendrin. Endocrinology 145:4301–4308

Yuge S, Inoue K, Hyodo S, Takei Y (2003) A novel guanylin family (guanylin, uroguanylin, and renoguanylin) in eels: possible osmoregulatory hormones in intestine and kidney. J Biol Chem 278:22726–22733

Yusa A, Miyazaki K, Kimura N, Izawa M, Kannagi R (2010) Epigenetic silencing of the sulfate transporter gene *DTDST* induces sialyl Lewisx expression and accelerates proliferation of colon cancer cells. Cancer Res 70:4064–4073

Zarnegar R, Brunaud L, Kanauchi H, Wong M, Fung M, Ginzinger D, Duh QY, Clark OH (2002) Increasing the effectiveness of radioactive iodine therapy in the treatment of thyroid cancer using Trichostatin A, a histone deacetylase inhibitor. Surgery 132:984–990; discussion 990

Zhang ZH, Jow F, Numann R, Hinson J (1998) The airway epithelium: a novel site of action by guanylin. Biochem Biophys Res Commun 244:50–56

The Pendrin Polypeptide

11

Silvia Dossena, Emanuele Bernardinelli, Alok K. Sharma, Seth L. Alper, and Markus Paulmichl

Abstract

Pendrin (SLC26A4) is an electroneutral anion exchanger in which function-altering mutations cause inherited forms of non-syndromic (DFNB4) and syndromic deafness (Pendred syndrome) characterized by inner ear malformations. Elevated pendrin expression has also been identified as a risk modifier for inflammatory and infectious lung diseases. In addition, pharmacological inhibition of pendrin activity in the kidney is under development as a strategy for treatment of hypertension and diuretic-resistant fluid overload conditions. Although pendrin is an attractive therapeutic target for pharmaceutical development, the molecular structure and pharmacological inhibition profile of pendrin have remained elusive. Here, we present a new model of human pendrin structure based on the crystal structures of the transmembrane domain of the SLC26Dg fumarate transporter of *Deinococcus geothermalis* and of the cytosolic (carboxy terminal) STAS domain of rat prestin. The model proposes 14 membrane-spanning α-helices, with carboxy-terminal and amino-terminal regions located within the intracellular space. In addition, we summarize current knowledge on small-molecule modifiers of pendrin activity and on *SLC26A4* gene sequence alterations encoding protein variants with modified functional and molecular features.

S. Dossena (✉) • E. Bernardinelli • M. Paulmichl
Institute of Pharmacology and Toxicology, Paracelsus Medical University, Salzburg, Austria
e-mail: silvia.dossena@pmu.ac.at

A.K. Sharma • S.L. Alper (✉)
Division of Nephrology and Vascular Biology Research Center, Departments of Medicine, Beth Israel Deaconess Medical Center and Harvard Medical School, Boston, MA 02215, USA
e-mail: salper@bidmc.harvard.edu

S. Dossena, M. Paulmichl (eds.), *The Role of Pendrin in Health and Disease*,
DOI 10.1007/978-3-319-43287-8_11

A precise understanding of the structure and function of pendrin and the molecular events leading to alterations in its function will facilitate and focus the quest for future pendrin-targeted therapeutic approaches.

11.1 Models for Human Pendrin Structure

11.1.1 Background

Pendrin/SLC26A4 is an electroneutral anion exchanger and member of the solute carrier 26/sulfate permease (*SLC26/SulP*) gene family conserved throughout phylogeny (Sherman et al. 2005; Felce and Saier 2004; Dorwart et al. 2008b; Alper and Sharma 2013). The 11 human *SLC26* genes encode 10 proteins and their isoforms, each expressed in a tissue-specific manner. SLC26/SulP protein structures consist of an N-terminal transmembrane domain (TMD) of 14 transmembrane-spanning helices (Gorbunov et al. 2014) linked to a C-terminal cytoplasmic Sulfate Transporter anti-Sigma factor Antagonist (STAS) domain (Fig. 11.1).

The first mammalian SLC26 gene, *SLC26A2/DTDST*, was discovered by positional cloning (Hastbacka et al. 1994). The primary sequence was noted to resemble that of a previously described sulfate transporter of *Neurospora crassa*. SLC26 proteins transport halides (Cl^-, I^-, Br-), thiocyanate (SCN^-), monovalent oxyanions (OH^-, HCO_3^-, NO_3^-, formate, glyoxylate), and divalent oxyanions (SO_4^{2-}, oxalate), with narrow or broad anion selectivities and mechanisms (electroneutral or electrogenic anion exchange, anion conductance) characteristic of each gene product (Alper and Sharma 2013). Mutations in at least four human SLC26 genes cause early-onset Mendelian diseases of chondro-dysplasia (*SLC26A2/DTDST*), chloride-losing diarrhea (*SLC26A3/DRA*), deafness with enlargement of the vestibular aqueduct and low-penetrance goiter (*SLC26A4*/pendrin), and a rare form of impaired male fertility (*SLC26A8*). *SLC26A9* has been identified as a strong risk modifier for cystic fibrosis lung disease. *SLC26A4*/pendrin has been investigated as a co-mediator or risk modifier of systemic fluid balance, blood pressure, and asthma (Alper and Sharma 2013).

The phylogenetically ancient *SLC26/SulP* gene family (Felce and Saier 2004; Sharma et al. 2011a) includes major sulfate transporters in yeast (Shibagaki and Grossman 2010) and plants (Shibagaki and Grossman 2004; Gigolashvili and Kopriva 2014) and organic oxyanion transporters in bacteria (Babu et al. 2010; Geertsma et al. 2015; Karinou et al. 2013; Zolotarev et al. 2008). The transmembrane domains of *SLC26/SulP* genes have been shown to transport a wide spectrum of anions, including bicarbonate, halides, monocarboxylates, and dicarboxylates.

All SLC26 and SulP polypeptides have, following a short N-terminal hydrophilic tail and a complex polytopic transmembrane domain, a C-terminal cytoplasmic hydrophilic region of 125–250 amino acids (aa) including a conserved Sulfate Transporter and Anti-Sigma factor antagonist (STAS) domain. The STAS domains of SLC26/SulP transporters are homologous to bacterial anti-sigma factor

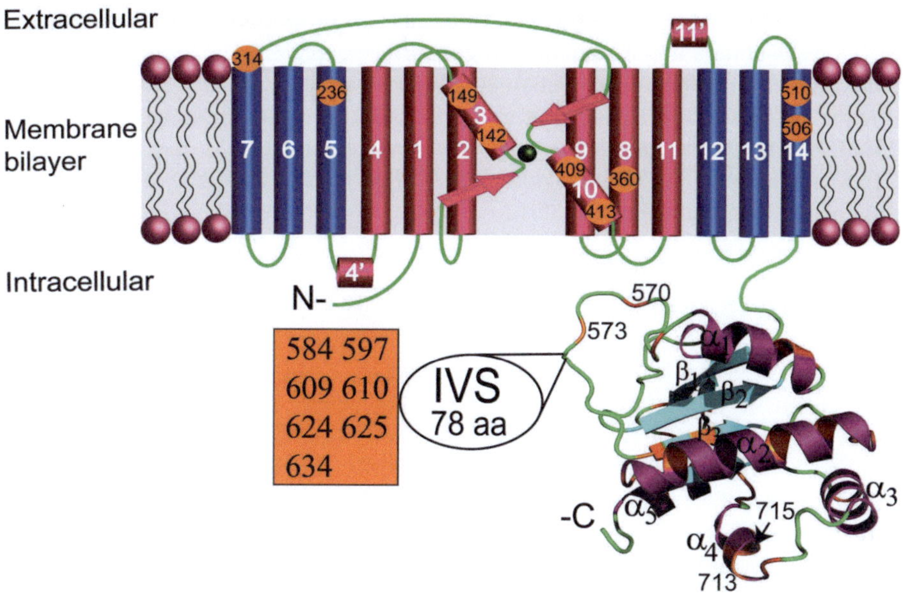

Fig. 11.1 Schematic of the topographical transmembrane domain (TMD) structure of human pendrin (hPDS/SLC26A4) anion transporter, modeled on the crystal structure of fumarate transporter *Deinococcus geothermalis* (SLC26Dg) in the inward-facing conformation (Geertsma et al. 2015). The hPDS TMD is shown with 14 membrane-spanning segments (mostly α-helices as columns) embedded within the lipid bilayer, connecting to the C-terminal cytoplasmic STAS domain modeled on the crystal and NMR structures of rPres/SLC26A5 (Pasqualetto et al. 2010). Note that the loop regions between TMD helices α3 and α4, and between α6 and α7 are relatively longer in hPDS than in SLC26Dg. The 14 transmembrane spans of the TMD are arranged as two inverted repeat units, each consisting of 7 transmembrane spans (1–7 and 8–14). Inner helices 1–4 and 8–11 pack to constitute the *core domain* (*magenta*) of hPDS. The core domain is surrounded on one side by the *gate domain* consisting of outer helices 5–7 and 12–14 (*marine*). The core domain includes the centrally located substrate translocation pathway (containing a hypothetical-bound Cl^- ion in a possible binding site). The C-terminal cytoplasmic STAS domain (lower right, backbone structure model generated in PyMOL with helices in purple, strands in cyan, and loops in green) encompasses hPDS aa 515–734, but excludes the disordered intervening sequence (IVS) of aa 574–652 between helix α1 and strand β3 (Sharma et al. 2014). Backbone atom superposition of hPDS TMD helices onto those of SLC26Dg yields an overall root mean square deviation (RMSD) of 4.32 Å. The numbered orange ovals denote TMD sites of disease-associated hPDS mutations (*see text for details*). The orange box lists sites of disease-associated mutations within the hPDS STAS IVS. The numbered residues within the STAS domain structural model are sites of four disease-associated hPDS mutations aligning with SLC26-related SulP protein Rv1739c STAS residues chemically perturbed upon GDP binding (Sharma et al. 2011b) and exhibiting high root mean square fluctuation (RMSF) during MD simulation (Sharma et al. 2014)

antagonists (ASAs) such as SpoIIAA of *B. subtilis* (Aravind and Koonin 2000; Sharma et al. 2011a). Anti-sigma factor antagonists counter the activity of repressors (anti-sigma factors) of sigma (transcription) factors in the bacterial sporulation stress response pathway (Masuda et al. 2004). STAS domains are essential for plasmalemmal targeting, contribute to anion transport function, and serve as protein-protein interaction modules. Mutations in STAS domains of mammalian SLC26 transporters impair function and/or surface expression of these transporters (Chernova et al. 2003; Dorwart et al. 2008b; Dorwart et al. 2008a), and the same is true for STAS domains of bacterial transporters such as SLC26Dg (Geertsma et al. 2015). As true for bacterial anti-sigma factor antagonists (ASAs) such as SpoIIAA, STAS domains of SLC26 transporters have also been hypothesized to serve as protein-protein interaction modules.

11.1.2 Structure of STAS Domains

SpoIIAA-like STAS domain structures solved by X-ray crystallography include those from dicarboxylate and bicarbonate transporter (Karinou et al. 2013; Srinivasan et al. 2016) DauA(YchM) of *E. coli* (Babu et al. 2010) and from rat SLC26A5/prestin (Pasqualetto et al. 2010). NMR solution structures have been solved for the STAS domains of rat prestin (Pasqualetto et al. 2010) and of putative sulfate transporter Rv1739c from *M. tuberculosis* (Sharma et al. 2011b). Mammalian STAS domains differ from anti-sigma factor antagonists in the presence only in the mammalian proteins of the nominally unstructured intervening sequence (IVS) between helix α1 and β3 [nomenclature from (Pasqualetto et al. 2010)]. No function has yet been reported for the IVS, and its deletion was required for production of the first STAS domain crystals diffracting to high resolution (Pasqualetto et al. 2010). Mammalian and bacterial STAS domain solution structures reported to date have been monomeric and adopt a common structural fold resembling *B. subtilis* SpoIIAA ASA structure (Kovacs et al. 1998). SulP STAS domains are nucleotide-binding proteins (Sharma et al. 2011a; Srinivasan et al. 2016), as previously demonstrated for SpoIIAA and ASAs (Sharma et al. 2011a).

11.1.2.1 STAS Domain-Binding Proteins

SLC26/SulP anion transporter STAS domains interact in pulldown assays with multiple cellular proteins. The STAS domain of bacterial oxyanion transporter DauA (YchM) interacts with acyl carrier protein (Babu et al. 2010), suggesting their joint participation in fatty acid biosynthetic pathways. The C-terminal PDZ recognition motifs of STAS domains of SLC26A3 and SLC26A6 bind the PDZ proteins NHERF1, NHERF2, and PDZK1, with cytoskeletal signaling and scaffolding functions linking STAS domains indirectly to the Cl^- channel, cystic fibrosis transmembrane regulator (CFTR). The STAS domains of anion exchangers SLC26A3, SLC26A4 (Garnett et al. 2011), and SLC26A6, of anion channel/exchanger SLC26A9, and of SLC26A8 (of still unknown function) can also bind directly to the regulatory R domain of CFTR (Dorwart et al. 2008b; Alper and Sharma 2013). The STAS domain of SLC26A6 has also been shown to bind to SLC13A2/NaDC1, as part of a succinate transport metabolon (Ohana et al. 2013), as well as to the Gq-IP_3-activated regulatory hub protein IRBIT (Hong et al. 2013). The STAS domain of SLC26A5/prestin, the Cl^--dependent mechanotransducer of the cochlear outer hair cell, has been reported to bind to MAP-1S, VAP33 (Sharma et al. 2011a), and to calmodulin (Keller et al. 2014). The STAS domain of the high-affinity SulP sulfate transporter SULTR1;2 of the roots of the plant *A. thaliana* binds to the enzyme cysteine synthase (O-acetylserine (thiol) lyase, OASTL). This interaction reduces sulfate transport activity without reduction in transporter surface expression, while increasing OASTL activity (Shibagaki and Grossman 2010), and may contribute to plant sulfate sensing (Zheng et al. 2014). The STAS domain of the sperm anion transporter, SLC26A8, binds to the GTPase activating protein, MgcRacGap (Toure et al. 2001). Preliminary data suggest that the widely expressed GTPase activator and scaffold protein, IQGAP-1, binds to the STAS domain of hPDS/SLC26A4 and reduces its anion exchange activity (Xu et al. 2015; Karinou et al. 2013). Disease-associated mutations in STAS domains of multiple mammalian SLC26 transporter proteins disrupt trafficking and stable accumulation of transporters at the plasma membrane (Sharma et al. 2011a).

11.1.3 SLC26Dg Crystal Structure

The first reported structure of a full-length SLC26/SulP polypeptide, including both TMD and cytoplasmic STAS domain, was from SLC26 of *Deinoccocus geothermalis* (SLC26Dg) (Geertsma et al. 2015). The SLC26Dg TMD shares 46 % and 57 % amino acid sequence similarity with the corresponding regions of human prestin and of *E. coli* DauA. Initial crystallization attempts achieved a structural resolution of only 7.2 Å. Co-crystallization with an immunospecific anti-SLC26Dg nanobody allowed structural resolution of 3.2 Å, without alteration of the TMD orientation of SLC26Dg.

The SLC26Dg structure reveals a UraA fold similar to that previously predicted and used as a modeling template for rat and chicken prestin/SLC26A5 (Gorbunov et al. 2014) and for human AE1/SLC4A1 (Barneaud-Rocca et al. 2013). Indeed, the X-ray crystal structure of human AE1 ultimately confirmed its UraA fold (Arakawa et al. 2015). The UraA fold shared by SLC26Dg and by AE1/SLC4A1 consists of 14 TMD α-helices of variable length arranged as two inverted repeats, each containing 7 α-helices. Pseudosymmetry-related helices 1–4 and 8–11 form a "core domain," whereas helices 5–7 and 12–14 from an elongated "gate domain" that shields the core domain. Each inverted repeat includes one "broken" transmembrane span (spans 3 and 10) comprising a short α helix joined by flexible linker to a short β-sheet oriented at divergent angles within the bilayer, and contributing to a proposed anion substrate-binding site (no ligand was detected in the SLC26Dg crystals). The SLC26Dg crystal structure revealed the polypeptide in its inward-facing conformation, exposing the substrate anion-binding site to the cytosol, and occluding it from the extracellular medium.

Purified SLC26Dg was monomeric in dodecyl maltoside but reconstituted into proteoliposomes as a dimer, consistent with previous evidence of stable dimers in solution, based on pulsed electron–electron double resonance spectroscopy and small-angle neutron scattering experiments (Compton et al. 2014; Compton et al. 2011). TMD monomeric dimensions of 45 Å × 60 Å × 40 Å suggest minimal protrusion of the polypeptide beyond either surface of the lipid bilayer (Geertsma et al. 2015). Each monomer is believed to harbor an independent anion translocation pathway. However, one monomer can influence its dimeric partner's conformation as also suggested by prestin, for which co-expression of two mutant polypeptides of distinct voltage-dependences yielded a unique electrical signature, consistent with functional inter-protomer interaction within the oligomer (Detro-Dassen et al. 2008). Regulated modulation of SLC26 oligomeric state has been postulated, but not yet reported.

11.1.4 Modeling of Human Pendrin (hPDS) Structure

The availability of the SLC26Dg crystal structure has allowed us to model on this template a three-dimensional structure of human pendrin (hPDS; Fig. 11.1). Most mammalian and plant SLC26 polypeptides, such as the 780 aa hPDS and 744 aa rat prestin (rPres), are longer than bacterial SulP proteins (such as the 499 aa SLC26Dg), with longer N-terminal cytoplasmic domains, longer STAS domains within longer C-terminal cytoplasmic domains, and longer loops connecting transmembrane spans. The latter include the ecto-loop between TM3-4 and the endo-loop between TM6-7, each of length ~40 aa *vs.* only 8 aa in SLC26Dg. The structures of these long loops remain unknown. The predicted net surface charges of hPDS (−4) and rPres holoproteins (−11) with correspondingly acidic predicted isoelectric points (pI) contrast with the net surface charge of +2 and pI of 8.6 for SLC26Dg holoprotein. Similarly, the 79 aa hPDS cytoplasmic N-terminal domain, with 2–3 predicted α-helices each of 10 aa length, contrasts with the putatively unstructured N-terminal SLC26Dg cytosolic domain of 13 aa. The functional role of the hPDS N-terminal cytosolic domain is not known.

We have modeled the STAS domain of hPDS (excluding its IVS region) on the template of the rPres STAS domain crystal structure (Fig. 11.1) (Pasqualetto et al. 2010). The modeled hPDS STAS domain is remarkable for several differences from the SLC26Dg STAS domain, including its much longer putatively unstructured IVS region. The net surface charge of −12 in hPDS STAS contrasts with the surface charge of −2 in the SLC26Dg STAS domain. Unlike the well-structured STAS C-terminal helical regions in Rv1739c (Sharma et al. 2011b) and rPres (Pasqualetto et al. 2010), and unlike the modeled hPDS structure (Fig 11.1), the C-terminal region of SLC26Dg STAS was undetected in its crystal structure. This may reflect its likely artifactual interaction with co-crystallized nanobody, resulting in ectopic location of the SLC26Dg STAS domain in the crystal, in a position that would place it within the lipid bilayer of a cell (Geertsma et al. 2015). The length of SLC26Dg STAS helix α1 (Geertsma et al. 2015) is only half of those in STAS domains of DauA (Karinou et al. 2013) and Rv1739c (Sharma et al. 2011b).

A notable difference between SLC26Dg and hPDS is in cysteine (Cys) content. SLC26Dg has a single Cys residue in the TMD located in the ecto-loop between TM9 and TM10. In contrast, hPDS has 12 Cys residues distributed across its entire sequence, with four each in the N-terminal cytoplasmic region, the TMD, and the

C-terminal STAS domain. Other mammalian SLC26 polypeptides also have multiple Cys residues, which may help govern protein oligomerization state, contribute to regulation of protein half-life, and govern ligand binding.

11.1.5 Proposed Substrate-Binding Site in hPDS Structure

A fumarate transport substrate-binding site within the TMD of SLC26Dg may be constituted by two pseudosymmetry-related glutamates (E38 in α1 and E241 in α8) and by one glutamine (Q287 in α10). However, the SLC26Dg double mutant p.E38A/E241A exhibited increased fumarate-uptake activity that was further enhanced by imposition of an inwardly directed pH gradient. Thus, the carboxylate packing of E38 and E241 around the substrate-binding cavity might function as a brake on substrate binding or translocation. The lack of conservation of these two glutamate residues in hPDS correlates with, and may contribute to, the absence in hPDS of detectable H^+-anion symport of the type believed to represent the transport mechanism of SLC26Dg.

The residues involved in anion binding in rPres were selected for investigation based upon the uracil-binding site detected in the UraA crystal structure (Lu et al. 2011). The rPres p.S398C mutation reduced oxalate sensitivity of non-linear capacitance (NLC), while the corresponding p.S404C mutation of chicken prestin abolished its sulfate transport. rPres mutation p.R398S abrogated salicylate inhibition of NLC, and mutants p.R399T and p.R399C also decreased salicylate block of NLC, while corresponding chicken prestin mutant p.R405C abolished anion transport. Based on these data, the anion-binding site in rPres was proposed to involve S398 and R399 of TMD helix α10 (Gorbunov et al. 2014), a motif conserved in other mammalian orthologs.

Homology modeling on the SLC26Dg template allows proposal of a substrate-binding site in hPDS (Table 11.1 and Fig. 11.2) comprising residues Q101 (α1), I363 (α8), and R409 (α10). Interestingly, R409 is the site of a disease-associated mutation, as is P142 in helix α3, located close to the proposed binding site (Fig. 11.1). The planar 5-membered ring of P142 orients along the axis of the proposed anion-transport pathway. The R409 guanidine is slightly inclined toward the binding site, so as to electrostatically favor anion approach and/or binding. Additional disease-associated residues G149 (α3) and Q413 (α10) are modeled at the respective extracellular and intracellular edges of the binding cavity (Fig. 11.1). Disease mutation-associated residue A360 (α8) is also found in proximity to the binding site, near two β-strands. Other sites of disease-associated mutation that are involved in loss of anion exchange activity, including L236 (α5), S314 (α7), L506 (α14), and V510 (α14), are within the gate domain at locations farther from the binding site. Mapping of the above hPDS disease-associated residues onto the hPDS structural model supports the idea that packing of core domain helices α3 and α10 is crucial to integrity and conformational flexibility of the substrate translocation pathway (Fig. 11.1). These predictions further support the importance of co-ordinated movement of gate and core domains in anion translocation (Fig. 11.3). Helices

Table 11.1 Amino acid (aa) residues defining transmembrane-spanning segments (TMD) of the 780 aa human pendrin polypeptide (hPDS/SLC26A4) according to the tertiary structural model in Fig. 11.2, as modeled on the template of the SLC26Dg X-ray crystal structure (Geertsma et al. 2015)

TMD (α-helix)	Amino acid range
1	88–108
2	113–129
3	141–147 (149)
4	188–206
5	217–240
6	244–260
7	303–316
8	346–374
9	381–397
10	407–416
11	421–441
12	447–461
13	479–491
14	494–511

Note that TMD 3 α-helix encompassed aa 141–147 in 70 % of models, and 141–149 in 30 % of models

α1 and α10 are electroneutral in both SLC26Dg and human pendrin. In contrast, helix α8 carries net charge of −3 in SLC26Dg and +1 in hPDS, possibly reflecting differences in substrate and transport mechanism.

The inward-facing structure of SLC26Dg resembles the uracil-bound inward-facing structure of UraA. Substrate-free UraA was proposed to adopt an outward-facing conformation upon deprotonation of UraA residues E241 and E290. Co-ordinated movements of core and gate domains controlling transition between inward and outward facing conformations were proposed to reflect or to be controlled by the protonation states of these glutamate residues (likely transitions in hPDS; Fig. 11.3). The outward-facing conformation of hPDS might be best modeled on the recently reported UraA-like structure of human erythrocyte Band 3/SLC4A1, locked in its outward-facing conformation by covalently bound inhibitor, H_2DIDS (Arakawa et al. 2015).

Molecular modeling and molecular dynamics (MD) simulations will be useful approaches toward initial understanding of the outward-facing conformation of hPDS, further enhanced in value by the recent development and validation of the first inhibitors of hPDS with promising potency and specificity (Haggie et al. 2016). However, μsec timescale MD simulations will likely reveal more about anion binding and unbinding than about the inward-outward conformational change of hPDS. Modeling of the steady-state inward and outward-facing conformations will aid future MD simulation of the hPDS TMD conformational transition cycle, perhaps using iterative MD simulations with the post-hoc string method to efficiently extract the most relevant conformation sets generated by bias-exchange umbrella

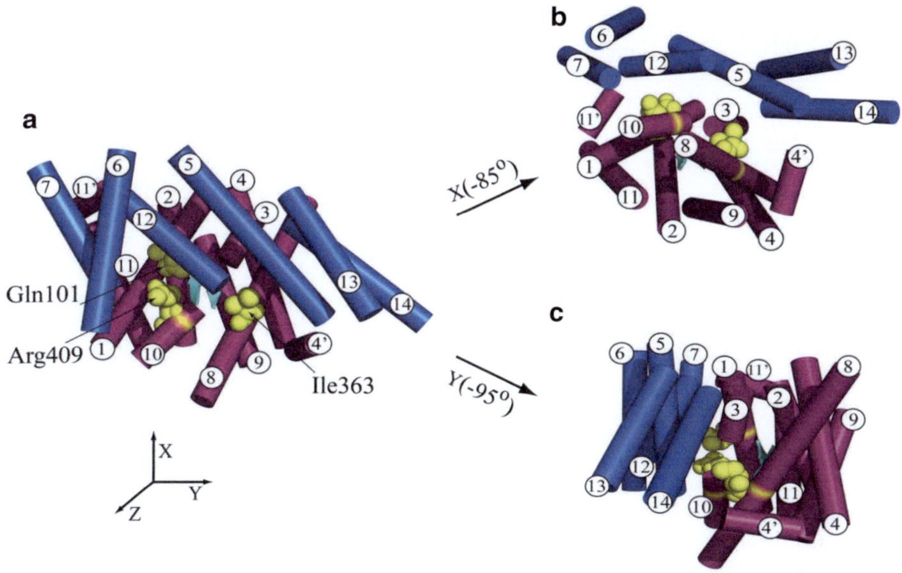

Fig. 11.2 Three-dimensional homology model of hPDS/SLC26A4 TMD α-helices (as cylinders) with core domain helices in magenta and gate domain helices in marine. Residues Q101 (α1), I363 (α8), and R409 (α10) portrayed in CPK (*yellow*) partially demarcate the proposed anion transport substrate-binding site (Geertsma et al. 2015). Loop regions connecting the transmembrane spans are removed for clarity. The two anti-parallel β-strands of TMD spans 3 and 10 are in cyan. (**a**) TMD structure viewed from within the plane of the lipid bilayer. (**b**) TMD structure viewed from the cytoplasm. (**c**) TMD structure rotated within the plane of the bilayer, showing empty stereo space between the core and gate domains constituting the proposed substrate-binding site (*adjacent to yellow CPK spheres*). The hPDS/SLC26A4 TMD structure was modeled using as template the inward-facing conformation of SLC26Dg (Geertsma et al. 2015). Primary sequences of SLC26Dg and hPDS/SLC26A4 were aligned using Clustal Omega, then manually adjusted to redefine secondary structural regions as predicted by PSIPRED and ModWeb. MODELLER v9.14 was used to generate hPDS/SLC26A4 TMD structural models. The best stereochemical quality structure of lowest objective function was then energy-minimized. A representative model was visualized and rendered in PyMOL v1.5.0.4 (Molecular Graphics System, Schrödinger, LLC)

sampling, in parallel with the string method with swarms of trajectories and the post-hoc string method (Moradi et al. 2015). Recent advances in single-molecule fluorescence resonance energy transfer imaging (Akyuz et al. 2015; Wang et al. 2014) and in solution nuclear magnetic resonance spectroscopy of polytopic transmembrane receptors (Isogai et al. 2016) suggest that the ability to study conformational transitions of pendrin and other mammalian SLC26 polypeptides in real time at molecular resolution may not be far off.

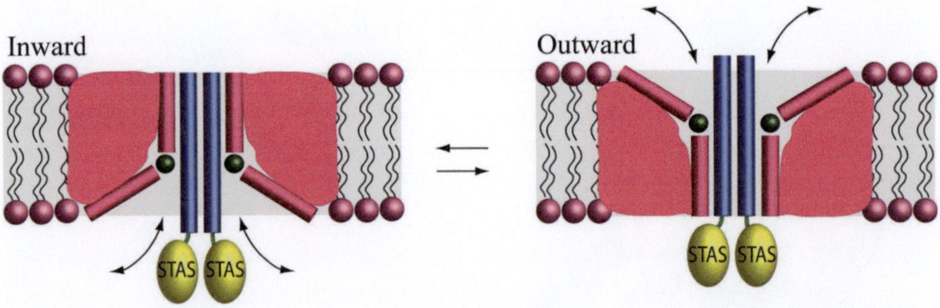

Fig. 11.3 Schematic of the alternating-access mechanism of anion transport postulated for homodimers of hPDS/SLC26A4 and other SLC26 transporters. The core domain (*magenta*) and gate domain (*marine*) regions of the TMD are embedded within the lipid bilayer. The cytosolic STAS domain is represented as a yellow sphere connected to the terminal helix of the gate domain. The bound anion substrate (*green*) can be released to the cytoplasm from the inward-facing conformation (*left*), as modeled on the DgSLC26 structure (Geertsma et al. 2015), or to the extracellular space from the outward-facing conformation of unknown structure (*right*). Conformational transitions between inward- and outward-facing states may traverse a *quasi*-stable, occluded-state intermediate conformation, in which the bound substrate is accessible neither to outer solution nor to cytoplasm (not shown) (Modified from Geertsma et al. 2015)

11.2 Functional and Molecular Properties of Wild-Type Pendrin and Its Variants

11.2.1 Functional Tests for Assessing Pendrin Function

A positional cloning strategy enabled identification of *SLC26A4/PDS* as the gene mutated in Pendred syndrome. The amino acid sequence similarity of the *PDS* gene product, pendrin, to previously identified sulfate transporters led to the initial prediction that pendrin could transport sulfate (Everett et al. 1997). Shortly thereafter, however, Scott et al. showed that pendrin overexpression in *Xenopus laevis* oocytes or insect Sf9 cells, rather than transporting sulfate increased the transport of iodide and chloride in both heterologous cell expression systems (Scott et al. 1999). The inability of pendrin to transport sulfate was later confirmed in mammalian cells (Bogazzi et al. 2000). Accordingly, Kraiem et al. found normal transport of sulfate in thyrocytes obtained from Pendred syndrome patients (Kraiem et al. 1999). *Cis*-inhibition and *trans*-stimulation uptake studies in *Xenopus laevis* oocytes indicated that pendrin exhibited properties of an exchanger of chloride and monovalent anions, rather than of a channel (Scott and Karniski 2000). The electroneutrality of pendrin-mediated ion exchange (the absence of ion currents accompanying ion transport) was later confirmed by others (Dossena et al. 2005; Reimold et al. 2011; Shcheynikov et al. 2008). Intracellular pH measurements in HEK-293 cells stably expressing pendrin demonstrated elevation of both chloride/bicarbonate exchange and chloride/hydroxyl exchange (Soleimani et al. 2001), the former representing the probable physiological transport mode of endogenously expressed pendrin in the inner ear (Chap. 2), kidney (Chap. 8), and airways (Chap. 9). In the thyroid, where the transport of iodide is physiologically relevant, pendrin is thought to act as an iodide/chloride exchanger with a 1:1 stoichiometry (Dossena et al. 2006a) and preference for iodide over other anions (Dossena et al. 2006b; Shcheynikov et al. 2008). In a bicameral system of iodide-loaded polarized cells, pendrin contributes to the vectorial transport of radiolabeled iodide from the basolateral to the apical compartment (Gillam et al. 2004), consistent with pendrin's proposed role in transporting circulating iodide across the apical membrane of the thyrocyte into the follicular lumen (Chap. 7). The ability of pendrin to transport iodide was confirmed by several studies using yellow fluorescent protein (YFP) variants with improved sensitivity for this halide (Dossena et al. 2006a; Pedemonte et al. 2007).

11.2.2 Pendrin Variants with Complete or Partial Loss-of-Function, with No Change in Function, or with Gain-of-Function

Functional studies in heterologous expression systems have been important not only for our growing understanding of the mechanism of action of wild-type pendrin, but also to determine the anion transport properties of pendrin variants identified in patients affected by non-syndromic deafness or Pendred syndrome.

The association of deafness and goiter is not a sufficient diagnostic criterion for Pendred syndrome. The identification of pseudo-Pendred syndromes (that is, the association of deafness and goiter in patients with no pendrin mutations) (Fugazzola et al. 2002; Kara et al. 2010; Davis et al. 2006; Pfarr et al. 2006; Kopp et al. 1999) indicates that the combined presence of deafness and goiter in a single individual or family may arise from distinct genetic and environmental factors. Malformations of the inner ear, such as an enlarged vestibular aqueduct (EVA), are considered a hallmark of Pendred syndrome and non-syndromic deafness caused by mutation of the pendrin gene (Reardon et al. 2000). However, EVA can also be found in individuals in which both alleles of the pendrin gene are wild type (Pryor et al. 2005). Consequently, unequivocal assessment of a link between deafness and pendrin dysfunction requires, in addition to clinical and radiological investigation, the sequencing of the pendrin gene (Fugazzola et al. 2000). However, the detection of a sequence alteration in the pendrin gene does not always imply functional impairment of the encoded pendrin polypeptide. As a general rule, the incidence of pathogenic sequence alterations in any gene associated with recessive disease is relatively low in the healthy population and higher in cohorts of affected individuals. As in many other genes, some pendrin sequence variants are present in healthy and affected individuals at the same frequency, suggesting low pathogenic potential (Pera et al. 2008). Missense sequence variants leading to substitution of highly conserved amino acids do not invariably affect pendrin functionality, as exemplified by p.L597S (Pera et al. 2008) and p.F354S (Dossena et al. 2011a). Moreover, certain pendrin protein variants associated with wild-type anion transport function can be found hearing-impaired individuals at the same frequency as in the general population, as for p.R776C (Pera et al. 2008). These considerations mandate that any pendrin sequence variant found in a hearing-impaired individual (outside the context of a large family in which genetic linkage is possible) be subjected to functional testing of the corresponding pendrin polypeptide in order to unequivocally link the deafness to pendrin dysfunction.

The Human Gene Mutation Database (http://www.hgmd.cf.ac.uk/ac/index.php; accessed on the 6 June 2016) (Stenson et al. 2014) lists nearly 500 sequence variants detected to date within the pendrin gene, of which approximately 67 % are missense/nonsense mutations, 13 % affect splicing, and 16 % are small indels. Less common sequence alterations are found in gene regulatory regions, or are large deletions or mutations of more complex structure. The spectrum of pendrin mutations differs among ethnic populations (Dai et al. 2009), often reflecting founder effects (Rebeh et al. 2010). For instance, whereas amino acid substitutions p.L236P and p.T416p (Van Hauwe et al. 1998) are frequently found in the Caucasian population, p.H723R is the most common mutation in East Asians (Yoon et al. 2008).

The impact on pendrin anion transport function or subcellular trafficking has been reported for only ~17 % of documented pendrin sequence variants (Table 11.2). Most of these lead to reduction or loss of ion transport function and have been linked to Pendred syndrome or non-syndromic deafness (Table 11.2 and Fig. 11.4). Deleterious amino acid substitutions may involve any of the domains of the pendrin polypeptide, suggesting that no structural subdomains of pendrin are dispensable for physiological function. Moreover, a mutation's functional consequence cannot easily be predicted from its location within the polypeptide. However, nonsense mutations leading to pendrin polypeptide truncation, and missense mutations introducing or substituting within the TMD the secondary structure disruptor, proline, invariably lead to complete abrogation of anion transport function (Dossena et al. 2011c) (Table 11.2 and Fig. 11.4) and can be considered likely pathogenic.

Amino acid substitutions leaving wild-type pendrin function undiminished represent ~9 % of functionally characterized variants (Table 11.2 and Fig. 11.4). These functional variants, unlikely to be pathogenic, should lead to consideration of other genetic and/or environmental causes for deafness and EVA, especially when found in *trans* with the wild-type allele (de Moraes et al. 2016). An important *caveat*, however, is that missense mutations tested in heterologous expression systems do not entirely replicate conditions in the native setting. The unavailability of well-differentiated cochlear and vestibular epithelial cell lines leaves open to question the relevance to cochlear-vestibular function of pendrin transport data obtained from *Xenopus* oocytes or immortalized mammalian cell lines. Thus, functional evaluation of mutant pendrin polypeptides in the laboratory must be interpreted in the context of clinical and genetic information.

The amino acid substitutions p.F354S, p.C565Y, p.L597S, and p.R776C are reported either to not affect pendrin anion function or to decrease it only minimally or moderately (Table 11.2 and Fig. 11.4). Lack of agreement among reported results may arise from the technique used to evaluate pendrin ion transport function (isotopic flux, measurement of intracellular pH, measurement of intracellular halide concentration), the expression system used (*Xenopus* oocytes or various immortalized transiently or stably transfected mammalian cell lines), or the specific anion

Table 11.2 Subcellular localization and function of pendrin variants

Nucleotide change	Amino acid change	Localization	Function	Activity compared to WT	Rescue	References
c.C84A c.A82C	p.S28R	PM intracellular	Loss of chloride uptake Loss of Cl^-/I^- exchange activity Loss of Cl^-/HCO_3^- exchange activity	–	Yes[f]	Dossena et al. (2006b) Dossena et al. (2006a) Yoon et al. (2008)
c.G85C	p.E29Q		Reduction of Cl^-/I^- exchange activity	–		Pera et al. (2008)
c.G87C	p.E29D	PM/intracellular	Reduction of Cl^-/I^- exchange activity	–		Cirello et al. (2012)
c.C209T	p.P70L		Loss of Cl^-/I^- exchange activity	–		Dossena et al. (2011a)
c.G259T	p.D87Y		Reduction of formate transport	–		Yuan et al. (2012)
c.G262A	p.V88I		Increased Cl^-/I^- exchange activity Cl^-/OH^- exchange activity not reduced Increased iodide efflux	+		Pera et al. (2008) Dossena et al. (2011a)
c.T279A	p.S93R		Reduction of formate transport	–		Yuan et al. (2012)
c.279delT	p.FS93>96X[a]+ C416-1G_A[b]	Intracellular	Intracellular iodide retention (primary thyrocytes culture)	–		Palos et al. (2008)
c.G304A	p.G102R	ER	Loss of iodide efflux	–		Taylor et al. (2002)
c.C349T	p.L117F	PM	Normal iodide efflux	=		Taylor et al. (2002)
c.C367T	p.P123S	Intracellular	Loss of Cl^-/I^- exchange activity	–	Yes[g]	Ishihara et al. (2010)
c.G412T	p.V138F	ER	Loss of iodide efflux	–		Taylor et al. (2002)

(continued)

Table 11.2 (continued)

Nucleotide change	Amino acid change	Localization	Function	Activity compared to WT	Rescue	References
c.G412C	p.V138L	ER	Strong reduction of Cl^-/Cl^-, Cl^-/I^-, Cl^-/HCO_3^- exchange activity	–		Choi et al. (2009b)
c.C419A	p.P140H		Loss of Cl^-/I^- exchange activity	–		Pera et al. (2008)
c.C425G	p.P142R	Intracellular	Loss of Cl^-/HCO_3^- exchange activity	–	+/–[f]	Yoon et al. (2008)
c.C425T	p.P142L	PM	Reduction of Cl^-/I^- exchange activity	–		de Moraes et al. (2016)
c.A439G	p.M147V	Intracellular Intracellular	Loss of Cl^-/HCO_3^- exchange activity Loss of Cl^-/I^- exchange activity	–	Yes[f] Yes[g]	Yoon et al. (2008) Ishihara et al. (2010)
c.T441C	p.M147T	Intracellular				Rebeh et al. (2010)
c.G445A	p.G149R	PM/Intracellular	Reduction of Cl^-/I^- exchange activity	–		de Moraes et al. (2016)
c.G497A	p.S166N	PM	Normal Cl^-/HCO_3^- exchange activity	=		Yoon et al. (2008)
c.G554C	p.R185T	Intracellular	Severe reduction of Cl^-/I^- exchange activity	–		Cirello et al. (2012)
c.C578T	p.T193I	Intracellular	Loss of Cl^-/I^- exchange activity	–		de Moraes et al. (2016)
c.G626T	p.G209V	PM	Severe reduction of Iodide efflux	–		Taylor et al. (2002)
c.G665T	p.G222V		Reduction of formate transport	–		Yuan et al. (2012)
c.T707C	p.L236P	ER ER Intracellular	Loss of iodide efflux Loss of chloride and iodide uptake Loss of Cl^-/HCO_3^- exchange activity Loss of Cl^-/I^- and Cl^-/HCO_3^- exchange activity	–	No[f, h]	Taylor et al. (2002) Scott et al. (2000) Rotman-Pikielny et al. (2002) Yoon et al. (2008) Choi et al. (2009a)

c.T716A	p.V239D	ER	Severe reduction of Cl^-/I^- exchange activity Loss of Cl^-/OH^- exchange activity	–		Walsh et al. (2006) Dossena et al. (2011b)
c.T749C	p.V250A		Normal Cl^-/HCO_3^- and Cl^-/I^- exchange activity	=		Dai et al. (2009)
c.G796A	p.D266N		Normal Cl^-/HCO_3^- and Cl^-/I^- exchange activity	=		Dai et al. (2009)
c.G845A	p.C282Y	PM	Reduction of Cl^-/I^- exchange activity	–		de Moraes et al. (2016)
c.890delC	p.FS297>302X		Loss of iodide efflux	–		Gillam et al. (2005)
c.C902T	p.P301L		Loss of Cl^-/I^- exchange activity	–		Dossena et al. (2011a)
c.G907C	p.E303Q	PM	Loss of Cl^-/I^- and Cl^-/HCO_3^- exchange activity	–		Dai et al. (2009)
c.916insG	p.FS306>309X	Intracellular	Loss of iodide efflux	–		Gillam et al. (2004)
c.C941T	p.S314L		Reduction of formate transport	–		Yuan et al. (2012)
c.G1001T	p.G334V 335X		Loss of Cl^-/I^- exchange activity Loss of Cl^-/OH^- exchange activity	–		Dossena et al. (2011b)
c.T1003C	p.F335L	PM	Reduction of Cl^-/I^- and Cl^-/HCO_3^- exchange activity	–		Choi et al. (2009a)

(continued)

Table 11.2 (continued)

Nucleotide change	Amino acid change	Localization	Function	Activity compared to WT	Rescue	References
c.T1061C	p.F354S		Mild reduction of Cl^-/HCO_3^- exchange activity Cl^-/I^- exchange activity not reduced Cl^-/OH^- exchange activity not reduced Iodide efflux not reduced	=?		Dai et al. (2009) Dossena et al. (2011a)
c.C1079T	p.A360V		Reduction of formate transport	–		Yuan et al. (2012)
c.A1105G	p.K369E	PM	Normal Cl^-/I^- exchange	=		Ishihara et al. (2010)
c.C1115T	p.A372V	Intracellular	Loss of Cl^-/I^- exchange activity	–	No[g]	Ishihara et al. (2010)
c.A1151G	p.E384G	ER Intracellular	Loss of chloride and iodide uptake Loss of Cl^-/HCO_3^- exchange activity Loss of Cl^-/I^- and Cl^-/HCO_3^- exchange activity	–	No[f]	Scott et al. (2000b) Rotman-Pikielny et al. (2002) Yoon et al. (2008) Choi et al. (2009a)
c.A1174T	p.N392Y	Intracellular	Loss of Cl^-/I^- exchange activity	–	No[g]	Ishihara et al. (2010)
c.G1204A	p.V402M	Intracellular	Loss of Cl^-/HCO_3^- and Cl^-/I^- exchange activity	–		Choi et al. (2009a)
c.C1225T	p.R409C		Reduction of formate transport	–		Yuan et al. (2012)
c.G1226A	p.R409H		Loss of iodide efflux	–		Gillam et al. (2005)
c.G1226A/ c.G262A	p.R409H/ p.V88I[c]		Reduction of Cl^-/I^- exchange activity	–		Pera et al. (2008)
c.C1229T	p.T410M	ER	Loss of iodide efflux	–		Taylor et al. (2002)
c.A1238C	p.Q413P		Loss of Cl^-/I^- exchange activity	–		Pera et al. (2008)

c.A1238G	p.Q413R	Intracellular	Reduction of Cl^-/I^- exchange activity	–		de Moraes et al. (2016)
c.A1246C	p.T416P	ER Intracellular	Loss of chloride and iodide uptake Loss of Cl^-/HCO_3^- exchange activity Loss of Cl^-/HCO_3^- and Cl^-/I^- exchange activity	–	+/–[f]	Scott et al. (2000) Rotman-Pikielny et al. (2002) Yoon et al. (2008) Choi et al. (2009a)
c.G1271A	p.G424D		Reduction of Cl^-/I^- exchange activity	–		Pera et al. (2008)
c.T1334G	p.L445W	Intracellular Intracellular ER	Loss of Cl^-/I^- exchange activity	–		Choi et al. (2009a) Rebeh et al. (2010) de Moraes et al. (2016)
c.A1337G	p.Q446R	ER	Loss of iodide efflux	–		Taylor et al. (2002)
c.T1440A	p.V480D		Reduction of chloride and iodide uptake	–		Scott et al. (2000)
c.C1454G	p.T485R		Reduction of Cl^-/I^- exchange activity	–		Pera et al. (2008)
c.1458_1459insT	p.I487YFSX39 (526X)[d]	ER	Loss of Cl^-/I^- exchange activity Loss of Cl^-/OH^- exchange activity	–		Brownstein et al. (2008) Dossena et al. (2011b)
c.A1468C	p.I490L		Mild reduction of chloride and iodide uptake	–		Scott et al. (2000)
c.G1489A	p.G497S	Intracellular	Strong reduction of chloride and iodide uptake Loss of Cl^-/HCO_3^- exchange activity	–	+/–[f]	Scott et al. (2000) Yoon et al. (2008)

(continued)

Table 11.2 (continued)

Nucleotide change	Amino acid change	Localization	Function	Activity compared to WT	Rescue	References
c.A1468C c.G1489A	p.I490L p.G497S[e]		Strong reduction of chloride and iodide uptake	–		Scott et al. (2000)
c.T1517G	p.L506R		Reduction of formate transport	–		Yuan et al. (2012)
c.T1529A	p.V510D	Intracellular	Severe reduction of Cl^-/HCO_3^- exchange activity	–		Jang et al. (2014)
c.C1541A	p.Q514K		Loss of Cl^-/I^- exchange activity	–		Pera et al. (2008)
c.1561_1571insCTTGGAATGGC	p.FS523>548X		Loss of Cl^-/I^- exchange activity	–		Fugazzola et al. (2007)
c.T1588C	p.Y530H	Intracellular				Choi et al. (2009a)
c.A1589C	p.Y530S	Intracellular				Choi et al. (2009a)
c.C1625G	p.P542R	PM	Cl^-/Cl^-, Cl^-/I^-, Cl^-/HCO_3^- exchange activity not significantly reduced	=		Choi et al. (2009b)
c.A1667G	p.Y556C	Partially PM	Loss of iodide efflux	–		Taylor et al. (2002)
c.G1694A	p.C565Y	PM PM	Reduction of Cl^-/I^- and Cl^-/HCO_3^- exchange activity Normal Cl^-/I^- exchange activity	=?		Choi et al. (2009a) Ishihara et al. (2010)
c.T1790C	p.L597S	PM	Normal Cl^-/I^- exchange activity Reduction of Cl^-/I^- and Cl^-/HCO_3^- exchange activity	=?		Pera et al. (2008) Choi et al. (2009a)
c.T1826C	p.V609G		Reduction of Cl^-/I^- exchange activity Reduction of iodide efflux	–		Dossena et al. (2011a)
c.G1873T	p.E625X	Intracellular	Loss of Cl^-/HCO_3^- exchange activity	–	No[f]	Yoon et al. (2008)
c.T1958C	p.V653A		Reduction of chloride and iodide uptake	–		Scott et al. (2000)

c.G1970A	p.S657N	Intracellular	Loss of Cl^-/I^- exchange activity	–	Yes[g]	Ishihara et al. (2010)
c.G1985A	p.C662Y		Reduction of formate transport	–		Yuan et al. (2012)
c.C1991T	p.A664V		Reduction of formate transport	–		Yuan et al. (2012)
c.C1997T	p.S666F	Intracellular	Loss of Cl^-/I^- exchange activity	–	No[g]	Ishihara et al. (2010)
c.T2000C	p.F667C		Loss of Cl^-/I^- exchange activity	–		Dossena et al. (2011a)
c.G2015A	p.G672E	Partially PM	Loss of iodide efflux	–		Taylor et al. (2002)
c.T2027A	p.L676Q	Intracellular Intracellular	Loss of iodide efflux Loss of Cl^-/HCO_3^- exchange activity	–	±[f]	Gillam et al. (2004) Yoon et al. (2008)
c.G2059T	p.D687Y		Reduction of Cl^-/I^- exchange activity	–		Dossena et al. (2011a)
c.A2090C	p.D697A	Intracellular	Reduction of Cl^-/I^- exchange activity	–		Dai et al. (2009)
c.G2145T	p.K715N	Intracellular	Reduction of Cl^-/I^- exchange activity	–		Dai et al. (2009)
c.C2162T	p.T721M	Intracellular	Loss of Cl^-/I^- exchange activity	–	No[g]	Ishihara et al. (2010)
c.A2168G	p.H723R	Intracellular Intracellular	Loss of Cl^-/HCO_3^- exchange activity Loss of Cl^-/I^- exchange activity	–	Yes[f,h] Yes[g] Yes	Yoon et al. (2008) Ishihara et al. (2010) Jung et al. (2016)
c.A2171G	p.D724G		Loss of Cl^-/I^- exchange activity	–		Pera et al. (2008)
c.G2211C	p.E737D	Intracellular	Reduction of Cl^-/I^- and Cl^-/HCO_3^- exchange activity	–		Dai et al. (2009)
c.G2218A	p.G740S		Increased Cl^-/I^- exchange activity Increased Cl^-/OH^- exchange activity Increased iodide efflux	+		Dossena et al. (2011a)

(continued)

Table 11.2 (continued)

Nucleotide change	Amino acid change	Localization	Function	Activity compared to WT	Rescue	References
c.T2324C	p.M775T	PM	Reduction of Cl^-/I^- and Cl^-/HCO_3^- exchange activity	–		Choi et al. (2009a)
c.C2326T	p.R776C	PM PM	Normal iodide efflux Reduction of Cl^-/I^- and Cl^-/HCO_3^- exchange activity Normal Cl^-/I^- exchange activity	=?		Pfarr et al. (2006) Choi et al. (2009a) de Moraes et al. (2016)

Updated from Dossena et al. (2011c)

Anion transport activity of those pendrin variants tested is compared to that of wild-type pendrin. Also indicated, if tested, is degree of functional rescue observed. The position of each amino acid substitution on the model of pendrin structure presented in Fig. 11.1 is indicated in Fig. 11.4

+ gain of function, – reduction or loss of function, = function unaffected, =*?* lack of agreement among reported results: the allelic variant may be hypofunctional, *del* deletion, *ins* insertion, *PM* plasma membrane, *ER* endoplasmic reticulum

[a]Originally reported as: Ser93ArgfsX3

[b]Acceptor splice site mutation

[c]The two mutations are in *cis*

[d]Originally reported as: Ile487TyrfsX39

[e]The two mutations are in *trans* (the two mutant proteins were co-expressed)

[f]Low temperature

[g]10 mM salicylate

[h]Na butyrate

exchange activity evaluated (chloride/iodide, chloride/bicarbonate, or chloride/hydroxyl exchange). The chloride/bicarbonate exchange activity of pendrin seems to be more sensitive to sequence variation than the chloride/iodide exchange activity. However, such differences between chloride/iodide and chloride/bicarbonate exchange efficiency were considered "moderate," such that "the magnitude and consistency of the hypomorphic variants effect upon exchange of different anions seemed insufficient to infer a causal association with non-syndromic EVA (Choi et al. 2009a)". Thus, the amino acid substitutions p.F354S, p.C565Y, p.L597S, and p.R776C lead to hypomorphic pendrin variants themselves insufficiently pathogenic to cause clinical disease, but potentially pathogenic in conjunction with a strong dysfunctional pendrin variant encoded by the *trans* allele (Choi et al. 2009a).

Two amino acid substitutions (p.V88I and p.G740S) leading to a moderate gain of transport function (Dossena et al. 2011a) (Table 11.2 and Fig. 11.4) were found in a cohort of Spanish deaf patients and in a cohort of unrelated healthy individuals (controls) (Pera et al. 2008). Of note, patients carrying the p.G740S variant were deaf but lacked enlargement of the vestibular aqueduct (EVA) or other bony malformation, whereas the patient harboring the p.V88I variant presented with EVA and was found to be compound heterozygous for the deleterious mutations p.R409H and p.E29Q. Although the molecular feature leading to gain of function is obscure, the amino acid substitutions p.G740S and p.V88I cannot be considered as the determinant of deafness, unless the corresponding protein variant is inactivated by an additional deleterious amino acid substitution, as observed for the pendrin variant p.V88I;p.R409H. The possible contribution of hyperfunctional pendrin variants to hypertension (Chap. 8) or pulmonary dysfunction (Chap. 9) is currently unknown.

11.2.3 Subcellular Localization and Expression Levels of Pendrin Variants

In an early report aimed at identifying the molecular features of pathogenic pendrin variants, Taylor and collaborators determined the consequences of nine disease-causing amino acid substitutions on subcellular localization and transport function. Only two mutants localized to the plasma membrane, whereas five mutants were completely retained in the endoplasmic reticulum and two mutants were partially retained (Taylor et al. 2002). All these mislocalized mutant polypeptides exhibited loss of cellular iodide efflux, thereby demonstrating that inability to reach and/or accumulate at the plasma membrane could be considered a major determinant of loss of pendrin function. These findings were corroborated shortly afterward by Rotman-Pikienly and collaborators (Rotman-Pikielny et al. 2002), who showed that three of the more commonly found Pendred syndrome mutants, *i.e.,* p.L236P, p.T416P, and p.E384G, are retained in the endoplasmic reticulum.

More recently, molecular studies have routinely accompanied functional tests in testing the pathogenic potential of pendrin amino acid substitutions found in deaf patients. Fully functional pendrin variants always correctly target to the plasma membrane, and partial or total retention within subcellular compartments has always associated with

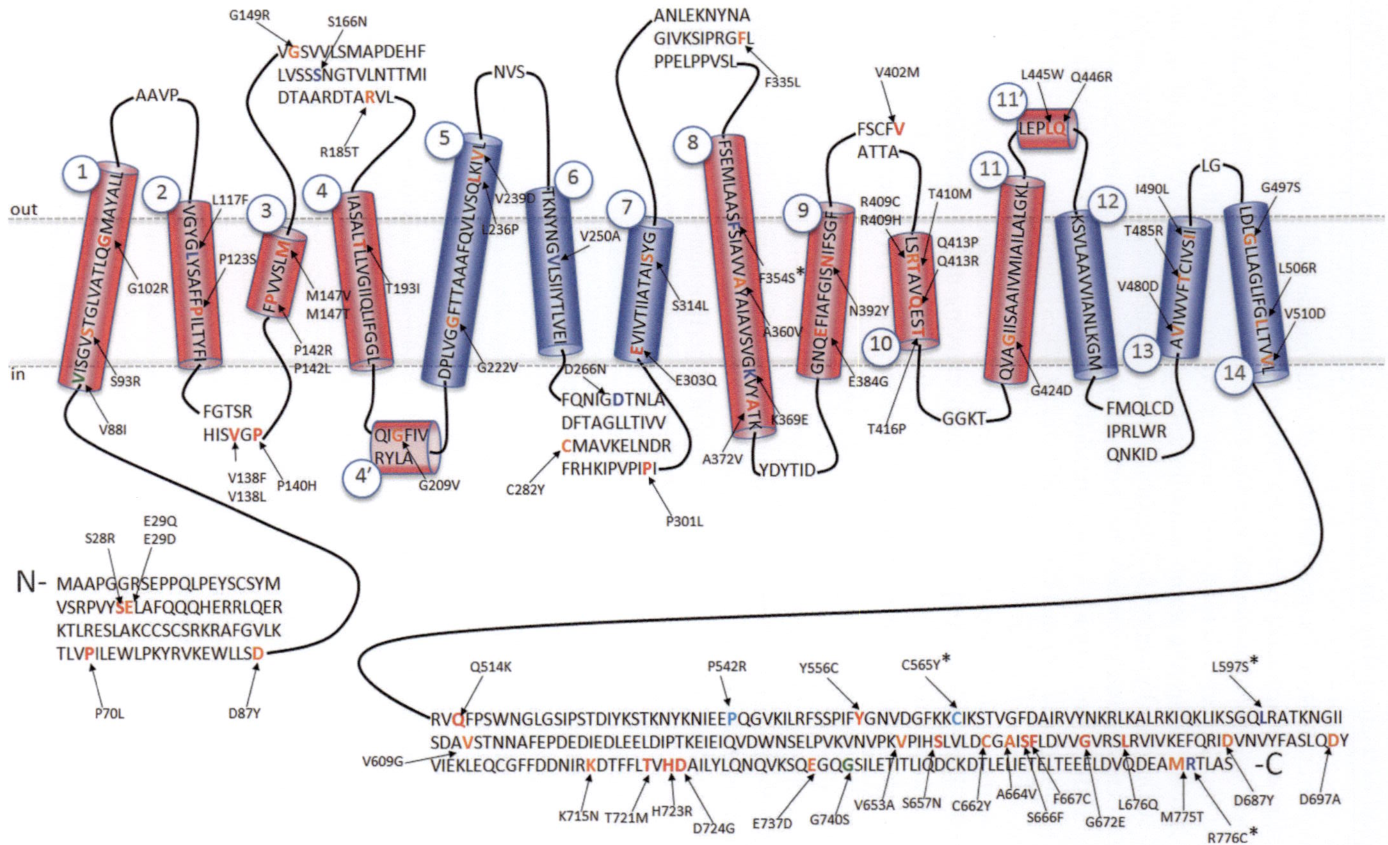

Fig. 11.4 Amino acid sequence and topology of human pendrin. The 14 membrane-spanning α helices, represented as columns, and the N and the C-terminal cytoplasmic domains are depicted based on structural information as in Fig. 11.1 and Table 11.1. Missense variants whose impact on anion transport function has been tested are indicated by arrows. Amino acid font color indicates that substitution of the corresponding amino acid leads to loss of function (*red*), reduced function (*orange*), no change in function (*blue*), or gain of function (*green*). An asterisk (*) denotes those variants with lack of agreement among reported results, and that could be hypofunctional

reduction or loss of anion transport function. However, wild-type targeting to the plasma membrane does not guarantee normal transport function, since some protein variants exhibit impaired function at the cell surface (Table 11.2). We observed in a cohort of deaf Brazilian patients that pendrin variants co-localizing with the plasma membrane (such as p.P142L, p.G149R, p.C282Y) maintain some degree of residual transport activity, whereas predominant retention in the endoplasmic reticulum accompanied by minimal co-localization with plasma membrane (as for p.T193I and p.L445W) correlated well with complete loss of anion transport function (de Moraes et al. 2016).

The wide range of functional dysregulation (Yoon et al. 2008) exhibited by the many pendrin variants reported to date emphasizes the utility of defining molecular features shared among all or most pathogenic pendrin protein variants in planning therapeutic approaches aimed at restoration of anion transport function. We recently showed that total cell expression levels of functionally deficient pendrin protein variants, regardless of their subcellular localization, are significantly lower than those of wild-type pendrin and of fully functional pendrin variant p.R776C. Thus, reduction of variant expression usually correlates with reduction of ion transport function (de Moraes et al. 2016). The mechanism leading to reduced abundance of dysfunctional pendrin variants is currently under investigation (see Sect. 11.3.2).

11.3 Potential Pendrin-Targeted Pharmacological Interventions

As detailed elsewhere in this book, both decreases and increases in pendrin function have been linked to distinct pathological conditions, namely Pendred syndrome and non-syndromic deafness in the case of pendrin loss-of-function (Parts I and II), and inflammatory and infectious lung diseases in the case of pendrin gain-of-function (Chap. 9). Furthermore, the established role of pendrin in regulating blood pressure (Wall 2015) recommends pendrin inhibition as a worthwhile strategy for the treatment of fluid overload states unresponsive to diuretics in current use (Soleimani 2012). These considerations suggest that pharmacological modulators of pendrin activity may have wide clinical utility.

11.3.1 Screening of Pendrin Ligands

For many years, pendrin seemed an "undraggable" target. Since the initial studies aimed at its functional characterization, pendrin has repeatedly exhibited low sensitivity to established inhibitors of ion transport, such as the anion exchange blocker 4,4′-diisothiocyano-2,2′-stilbene-disulfonic acid (DIDS), furosemide, and probenecid (Scott et al. 1999) (Table 11.3). The anti-inflammatory drugs niflumic acid and tenidap (Dossena et al. 2006b; Reimold et al. 2011) remained for a while the only substances capable of a significant inhibitory effect on pendrin activity (Table 11.3), and a first published screening of a large library of compounds contributed no novel pendrin inhibitors (Pedemonte et al. 2007).

Table 11.3 Effect of known inhibitors of ion transport on pendrin

Test compound	Conventional target	Effect on pendrin
Acetazolamide	Carbonic anhydrase (Belsky 1953)	No effect (Pedemonte et al. 2007; Bernardinelli et al. 2016)
Amiloride	ENaC (McDonald et al. 1994; Tamargo et al. 2014a)	No effect (Pedemonte et al. 2007; Bernardinelli et al. 2016)
Bumetanide	NKCC2 Tamargo et al. (2014a)	No effect (Pedemonte et al. 2007; Bernardinelli et al. 2016)
Chlorothiazide	NCC (Monroy et al. 2000; Tamargo et al. 2014b)	No effect (Pedemonte et al. 2007; Bernardinelli et al. 2016)
Eplerenone	Aldosterone antagonist (Tamargo et al. 2014a; Frishman and Stier 2004)	No effect (10 μM) (Bernardinelli et al. 2016)
Furosemide	NKCC2 (Tamargo et al. 2014a; Garay et al. 1998)	No effect (Pedemonte et al. 2007; Bernardinelli et al. 2016) 59 % inhibition (1 mM) (Scott et al. 1999) 39 % inhibition (1 mM) (Bernardinelli et al. 2016)
Hydrochlorothiazide	NCC (Monroy et al. 2000; Tamargo et al. 2014b)	No effect (Pedemonte et al. 2007; Bernardinelli et al. 2016)
Hydroflumethiazide	NKCC2 (Tamargo et al. 2014a)	No effect (Pedemonte et al. 2007; Bernardinelli et al. 2016)
Indapamide	NCC (?) (Tamargo et al. 2014b)	No effect (Pedemonte et al. 2007; Bernardinelli et al. 2016)
Methazolamide	Carbonic anhydrase (Lindskog 1997)	No effect (Pedemonte et al. 2007; Bernardinelli et al. 2016)
Torsemide	NKCC2 (Tamargo et al. 2014a)	No effect (Bernardinelli et al. 2016)
Triamterene	ENaC (Tamargo et al. 2014a)	No effect (Pedemonte et al. 2007; Bernardinelli et al. 2016)
Spironolactone	Aldosterone antagonist (Tamargo et al. 2014a; Frishman and Stier 2004)	No effect (Pedemonte et al. 2007; Bernardinelli et al. 2016)
CFTR-inhibitor 172	CFTR (Taddei et al. 2004)	No effect (Pedemonte et al. 2007) No effect (1.6 μM) (Bernardinelli et al. 2016)

Table 11.3 (continued)

Test compound	Conventional target	Effect on pendrin
DIDS	Anion transport (Jessen et al. 1986; Zhang et al. 2015; Shen et al. 2015)	No effect (Pedemonte et al. 2007; Bernardinelli et al. 2016) No effect (0.5 mM) (Reimold et al. 2011) No effect (0.5 mM) (Dossena et al. 2006b) 88 % inhibition (0.5 mM) (Soleimani et al. 2001) Full inhibition (0.5 mM) (Azroyan et al. 2011) 62 % inhibition (1 mM) (Scott et al. 1999) 48 % inhibition (1 mM) (Bernardinelli et al. 2016)
DNDS	Anion transport (Barzilay and Cabantchik 1979)	No effect (Bernardinelli et al. 2016)
NPPB	Chloride channels (Brown and Dudley 1996; Zhang et al. 2015)	No effect (Pedemonte et al. 2007) 59 % inhibition (Dossena et al. 2006b) 33 % inhibition (Bernardinelli et al. 2016)
Probenecid	OAT1 (Takeda et al. 2001)	No effect (Pedemonte et al. 2007; Bernardinelli et al. 2016) 37 % inhibition (1 mM) (Scott et al. 1999) No effect (1 mM) (Bernardinelli et al. 2016)
Glybenclamide	KCNJ1 (Gribble and Reimann 2003)	No effect (Pedemonte et al. 2007; Bernardinelli et al. 2016)
Hydroxycinnamate		No effect (Pedemonte et al. 2007; Bernardinelli et al. 2016) No effect (1 mM) (Reimold et al. 2011)
Niflumic acid	COX, PLA2 (Jabeen et al. 2005; Shen et al. 2015)	Inhibition (0.25 mM) (Nakao et al. 2008) Inhibition (Pedemonte et al. 2007; Dossena et al. 2006b) 64 % inhibition (Bernardinelli et al. 2016)
Tenidap	COX, LOX (?) (Moore et al. 1996) (Blackburn et al. 1991)	49 % inhibition (Reimold et al. 2011) 30 % inhibition (Bernardinelli et al. 2016)

Modified from Bernardinelli et al. (2016), with permission

The transporter or enzyme known to be blocked by each compound ("conventional target") is indicated together with the effect observed on pendrin transport activity. Compounds were tested at 0.1 mM, unless otherwise specified

CFTR cystic fibrosis transmembrane conductance regulator, *COX* cyclooxygenase, *DIDS* 4,4′-diisothiocyano-2,2′-stilbene-disulfonic acid, *DNDS* 4,4'-dinitrostilbene-2,2'-disulfonate, *DRA* down-regulated in adenoma, *ENaC* epithelial sodium channel, *KCNJ1* potassium channel, inwardly rectifying subfamily J, member 1, *LOX* lipooxygenase, *NCC* sodium chloride cotransporter, *NKCC2* sodium potassium chloride cotransporter 2, *NPPB* 5-nitro-2-[(3-phenylpropyl)amino]benzoic acid, *OAT1* organic anion transporter 1, *PLA2* phospholipase A2, *SLC26A3* solute carrier family 26, member A3

The search for potent and specific pendrin ligands has seen a breakthrough with the recently published studies of Verkman and colleagues (Haggie et al. 2016). These investigators used a semi-automated, high-throughput method to screen 36,000 commercially available small compounds from multiple chemical classes for the ability to inhibit pendrin-mediated Cl^-/SCN^- exchange in Fischer rat thyroid cells. Compounds identified from two chemical classes – tetrahydropyrazolopyridines and pyrazolothiophenesulfonamides – exhibited 50% inhibition of pendrin transport activity in the low micromolar concentration range. Further characterization of two of these candidates (PDSinh-A01 and PDSinh-C01) revealed similar inhibitory potency and efficacy against pendrin-dependent Cl^-/SCN^-, Cl^-/I^-, and Cl^-/NO_3^- exchange activities. PDSinh-A01 showed slightly greater inhibitory potency for pendrin's Cl^-/HCO_3^- exchange activity, with 50% inhibition at 2.5 μM, while exhibiting no inhibition of the pendrin homolog SLC26A3/DRA, or of NKCC1, ENaC, CFTR, CaCC, or epithelial K^+ channels. The 5 minutes lag time for onset of inhibition and the kinetics of reversibility upon compound washout suggested an intracellular site of action and excluded possible effects on transcript and/or protein levels. PDSinh-A01 was also found to increase airway surface liquid (ASL) depth in IL-13-treated human bronchial epithelial cells from healthy subjects and also from patients with cystic fibrosis, thus suggesting that pendrin inhibition can be applied therapeutically to increase ASL hydration in cystic fibrosis and other pulmonary diseases (Haggie et al. 2016).

A more recent study by the same group showed that *in vivo* administration of PDSinh-C01 in the setting of short- or long-term furosemide treatment increased urine output by 30 % and 60%, respectively. The increased urine output was associated with increased natriuresis and chloruresis. Interestingly, treatment of mice with similar concentrations of PDSinh-C01 as a sole agent had no effect on urine output, salt excretion, or acid-base balance. These results are consistent with earlier mouse knockout studies (Soleimani 2015; Wall 2015), and strengthen the hypothesis that addition of pendrin inhibitors to a regimen of multiple diuretics may offer substantial benefit in the treatment of hypertension, edema, and other fluid overload states resistant to combination therapy with currently available diuretics (Cil et al. 2016).

11.3.2 Rescuing Pendrin Function: Future Perspectives

As previously mentioned (see Sect. 11.2.3), studies in heterologous expression systems showed that polypeptide retention within intracellular compartments, a hallmark of protein misfolding, represents a major mechanism leading to loss of anion transport function of pathogenic pendrin protein variants. Exposure to physical or chemical chaperones was therefore envisioned as a promising approach to assist folding of pendrin mutants, allowing possible functional recovery. This strategy was successfully applied to pathogenic variants of the CFTR chloride channel (Nieddu et al. 2013). Incubation at low temperature (27 °C) restored complex-glycosylation of several pathogenic pendrin variants (Table 11.2), a finding reproduced for pendrin variant p.H723R by treatment with the non-specific histone deacetylase inhibitor, Na butyrate. Furthermore, the combination of Na butyrate treatment and reduced temperature rescued the Cl^-/HCO_3^- exchange activity of this pendrin mutant. However, the trafficking and activity of pendrin p.L236P and some other mutants were unmodified by these treatments, demonstrating heterogeneity of conformational defects among pendrin loss-of-function variants (Yoon et al. 2008). Incubation with 10 mM salicylate, a pharmacological chaperone used to stimulate plasmalemmal trafficking of inactive prestin/SLC26A5 mutants (Kumano et al. 2010), also restored plasma membrane targeting and iodide transport activity in four of eight tested pendrin loss-of-function variants (Ishihara et al. 2010) (Table 11.2), confirming that a single strategy to assist folding of all mutant pendrin polypeptides will not likely be successful. However, these studies did document that at least some pendrin variants *per se* retain a certain level of ion transport ability, but their function results impaired by premature degradation and insufficient amounts of protein reaching the cell surface.

We envision notable future effort devoted to the search of strategies to stimulate plasmalemmal trafficking of pathogenic pendrin variants. Jung and collaborators recently showed that blockade of ER-to-Golgi transport or imposition of ER stress could induce a Golgi-independent cell surface expression of pendrin p.H723R *via* the unconventional protein secretion pathway. Importantly, restoration of cell surface expression was accompanied by recovery of ion transport activity through a mechanism dependent upon heat shock cognate protein 70 (Hsc70) and HSP70 co-chaperone DNAJC14 activities (Jung et al. 2016).

As mentioned above, reduced total cell expression level of pendrin variant polypeptides is a common feature of hypomorphic or loss-of-function variants, likely reflecting increased degradation of the misfolded polypeptides. The mechanisms controlling turnover, and hence, expression levels of wild-type pendrin and its variants remain largely unexplored. An early report suggested that polyubiquitination may play a role in controlling degradation of wild-type and p.L236P pendrin (Shepshelovich et al. 2005). Lee and collaborators later identified a specific ER-resident E3 ubiquitin ligase (namely, Rma1) involved in degradation of wild-type and mutated pendrin (Lee et al. 2012). Thus, current data suggest that an enhanced degradation by the ubiquitin proteasome system may explain the reduced expression levels of hypofunctional pendrin variants. We therefore suggest that inhibition of pendrin variant degradation, combined with chaperonin-mediated assistance to achieve proper protein folding, could together constitute a therapeutic approach to restore or increase the activity of pathogenic pendrin variants.

The irreversibility of inner ear malformations resulting from pendrin malfunction poses a great challenge to any therapeutic approach targeting the pendrin polypeptide or its interactors. Studies on mouse models lacking pendrin expression selectively in the endolymphatic sac (Hulander et al. 2003) or in the cochlea and the vestibular labyrinth (Li et al. 2013) have shown that pendrin expression in the endolymphatic sac is both necessary and sufficient to rescue hearing and balance. In addition, studies with bi-transgenic pendrin null mice in which pendrin transgene expression can be induced by doxycycline under temporal control (Choi et al. 2011) showed that pendrin activity is not required for hearing maintenance in the mature inner ear, but is essential during a critical window of embryonic development. Importantly, a recent study on the latter mouse model, in which hearing fluctuations were induced by reversal of doxycycline-induced pendrin expression at embryonic day 17.5, showed that post-natal re-induction of pendrin expression could reduce or prevent later hearing fluctuations (Nishio et al. 2016) (for a detailed description of mice models used to investigate pendrin function, see Chap. 2). Collectively, these findings suggest that a spatially and temporally limited therapy directed to the endolymphatic sac and focused on the prenatal or early post-natal phase could restore hearing or prevent hearing fluctuations in patients with hearing loss linked to pendrin mutations (Wangemann 2013).

Acknowledgments The authors gratefully acknowledge the valuable help of Roberta Costa.

References

Akyuz N, Georgieva ER, Zhou Z, Stolzenberg S, Cuendet MA, Khelashvili G, Altman RB, Terry DS, Freed JH, Weinstein H, Boudker O, Blanchard SC (2015) Transport domain unlocking sets the uptake rate of an aspartate transporter. Nature 518(7537):68–73

Alper SL, Sharma AK (2013) The *SLC26* gene family of anion transporters and channels. Mol Aspects Med 34(2–3):494–515

Arakawa T, Kobayashi-Yurugi T, Alguel Y, Iwanari H, Hatae H, Iwata M, Abe Y, Hino T, Ikeda-Suno C, Kuma H, Kang D, Murata T, Hamakubo T, Cameron AD, Kobayashi T, Hamasaki N, Iwata S (2015) Crystal structure of the anion exchanger domain of human erythrocyte band 3. Science 350(6261):680–684

Aravind L, Koonin EV (2000) The STAS domain – a link between anion transporters and antisigma-factor antagonists. Curr Biol 10(2):R53–R55

Azroyan A, Laghmani K, Crambert G, Mordasini D, Doucet A, Edwards A (2011) Regulation of pendrin by pH: dependence on glycosylation. Biochem J 434(1):61–72

Babu M, Greenblatt JF, Emili A, Strynadka NC, Reithmeier RA, Moraes TF (2010) Structure of a SLC26 anion transporter STAS domain in complex with acyl carrier protein: implications for *E. coli* YchM in fatty acid metabolism. Structure 18(11):1450–1462

Barneaud-Rocca D, Etchebest C, Guizouarn H (2013) Structural model of the anion exchanger 1 (SLC4A1) and identification of transmembrane segments forming the transport site. J Biol Chem 288(37):26372–26384

Barzilay M, Cabantchik ZI (1979) Anion transport in red blood cells II. Kinetics of reversible inhibition by nitroaromatic sulfonic acids. Membr Biochem 2(2):255–281

Belsky H (1953) Use of a new oral diuretic, diamox, in congestive heart failure. N Engl J Med 249(4):140–143

Bernardinelli E, Costa R, Nofziger C, Paulmichl M, Dossena S (2016) Effect of known inhibitors of ion transport on pendrin (SLC26A4) activity in a human kidney cell line. Cell Physiol Biochem 38(5):1984–1998

Blackburn WD Jr, Heck LW, Loose LD, Eskra JD, Carty TJ (1991) Inhibition of 5-lipoxygenase product formation and polymorphonuclear cell degranulation by tenidap sodium in patients with rheumatoid arthritis. Arthritis Rheum 34(2):204–210

Bogazzi F, Bartalena L, Raggi F, Ultimieri F, Martino E (2000) Pendrin does not increase sulfate uptake in mammalian COS-7 cells. J Endocrinol Invest 23(3):170–172

Brown CD, Dudley AJ (1996) Chloride channel blockers decrease intracellular pH in cultured renal epithelial LLC-PK1 cells. Br J Pharmacol 118(3):443–444

Brownstein ZN, Dror AA, Gilony D, Migirov L, Hirschberg K, Avraham KB (2008) A novel *SLC26A4 (PDS)* deafness mutation retained in the endoplasmic reticulum. Arch Otolaryngol Head Neck Surg 134(4):403–407

Chernova MN, Jiang L, Shmukler BE, Schweinfest CW, Blanco P, Freedman SD, Stewart AK, Alper SL (2003) Acute regulation of the SLC26A3 congenital chloride diarrhoea anion exchanger (DRA) expressed in *Xenopus* oocytes. J Physiol 549(Pt 1):3–19

Choi BY et al (2009a) Hypo-functional *SLC26A4* variants associated with nonsyndromic hearing loss and enlargement of the vestibular aqueduct: genotype-phenotype correlation or coincidental polymorphisms? Hum Mutat 30(4):599–608

Choi BY, Stewart AK, Nishimura KK, Cha WJ, Seong MW, Park SS, Kim SW, Chun YS, Chung JW, Park SN, Chang SO, Kim CS, Alper SL, Griffith AJ, Oh SH (2009b) Efficient molecular genetic diagnosis of enlarged vestibular aqueducts in East Asians. Genet Test Mol Biomarkers 13(5):679–687

Choi BY, Kim HM, Ito T, Lee KY, Li X, Monahan K, Wen Y, Wilson E, Kurima K, Saunders TL, Petralia RS, Wangemann P, Friedman TB, Griffith AJ (2011) Mouse model of enlarged vestibular aqueducts defines temporal requirement of *Slc26a4* expression for hearing acquisition. J Clin Invest 121(11):4516–4525

Cil O, Haggie PM, Phuan PW, Tan JA, Verkman AS (2016) Small-molecule inhibitors of pendrin potentiate the diuretic action of furosemide. J Am Soc Nephrol. pii:ASN.2015121312

Cirello V, Bazzini C, Vezzoli V, Muzza M, Rodighiero S, Castorina P, Maffini A, Botta G, Persani L, Beck-Peccoz P, Meyer G, Fugazzola L (2012) Molecular and functional studies of 4 candidate loci in Pendred syndrome and nonsyndromic hearing loss. Mol Cell Endocrinol 351(2):342–350

Compton EL, Karinou E, Naismith JH, Gabel F, Javelle A (2011) Low resolution structure of a bacterial Slc26 transporter reveals dimeric stoichiometry and mobile intracellular domains. J Biol Chem 286(30):27058–27067

Compton EL, Page K, Findlay HE, Haertlein M, Moulin M, Zachariae U, Norman DG, Gabel F, Javelle A (2014) Conserved structure and domain organization among bacterial Slc26 transporters. Biochem J 463(2):297–307

Dai P et al (2009) Distinct and novel *SLC26A4/Pendrin* mutations in Chinese and U.S. patients with nonsyndromic hearing loss. Physiol Genomics 38(3):281–290

Davis N, Lunardi C, Shield JP (2006) Sensori-neural deafness and hypothyroidism: autoimmunity causing 'pseudo-Pendred syndrome'. Horm Res 65(6):267–268

de Moraes V, Bernardinelli E, Zocal N, Fernandez J, Nofziger C, Castilho AM, Sartorato EL, Paulmichl M, Dossena S (2016) Reduction of cellular expression levels is a common feature of functionally affected pendrin (SLC26A4) protein variants. Mol Med. doi:10.2119/molmed.2015.00226

Detro-Dassen S, Schanzler M, Lauks H, Martin I, zu Berstenhorst SM, Nothmann D, Torres-Salazar D, Hidalgo P, Schmalzing G, Fahlke C (2008) Conserved dimeric subunit stoichiometry of SLC26 multifunctional anion exchangers. J Biol Chem 283(7):4177–4188

Dorwart MR, Shcheynikov N, Baker JM, Forman-Kay JD, Muallem S, Thomas PJ (2008a) Congenital chloride-losing diarrhea causing mutations in the STAS domain result in misfolding and mistrafficking of SLC26A3. J Biol Chem 283(13):8711–8722

Dorwart MR, Shcheynikov N, Yang D, Muallem S (2008b) The solute carrier 26 family of proteins in epithelial ion transport. Physiology (Bethesda) 23:104–114

Dossena S, Maccagni A, Vezzoli V, Bazzini C, Garavaglia ML, Meyer G, Furst J, Ritter M, Fugazzola L, Persani L, Zorowka P, Storelli C, Beck-Peccoz P, Botta G, Paulmichl M (2005) The expression of wild-type pendrin (SLC26A4) in human embryonic kidney (HEK 293 Phoenix) cells leads to the activation of cationic currents. Eur J Endocrinol 153(5):693–699

Dossena S, Rodighiero S, Vezzoli V, Bazzini C, Sironi C, Meyer G, Furst J, Ritter M, Garavaglia ML, Fugazzola L, Persani L, Zorowka P, Storelli C, Beck-Peccoz P, Botta G, Paulmichl M (2006a) Fast fluorometric method for measuring pendrin (SLC26A4) Cl^-/I^- transport activity. Cell Physiol Biochem 18(1–3):67–74

Dossena S, Vezzoli V, Cerutti N, Bazzini C, Tosco M, Sironi C, Rodighiero S, Meyer G, Fascio U, Furst J, Ritter M, Fugazzola L, Persani L, Zorowka P, Storelli C, Beck-Peccoz P, Botta G, Paulmichl M (2006b) Functional characterization of wild-type and a mutated form of SLC26A4 identified in a patient with Pendred syndrome. Cell Physiol Biochem 17(5–6):245–256

Dossena S, Bizhanova A, Nofziger C, Bernardinelli E, Ramsauer J, Kopp P, Paulmichl M (2011a) Identification of allelic variants of pendrin (SLC26A4) with loss and gain of function. Cell Physiol Biochem 28(3):467–476

Dossena S, Nofziger C, Brownstein Z, Kanaan M, Avraham KB, Paulmichl M (2011b) Functional characterization of pendrin mutations found in the Israeli and Palestinian populations. Cell Physiol Biochem 28(3):477–484

Dossena S, Nofziger C, Tamma G, Bernardinelli E, Vanoni S, Nowak C, Grabmayer E, Kossler S, Stephan S, Patsch W, Paulmichl M (2011c) Molecular and functional characterization of human pendrin and its allelic variants. Cell Physiol Biochem 28(3):451–466

Everett LA, Glaser B, Beck JC, Idol JR, Buchs A, Heyman M, Adawi F, Hazani E, Nassir E, Baxevanis AD, Sheffield VC, Green ED (1997) Pendred syndrome is caused by mutations in a putative sulphate transporter gene (*PDS*). Nat Genet 17(4):411–422

Felce J, Saier MH Jr (2004) Carbonic anhydrases fused to anion transporters of the *SulP* family: evidence for a novel type of bicarbonate transporter. J Mol Microbiol Biotechnol 8(3):169–176

Frishman WH, Stier CT Jr (2004) Aldosterone and aldosterone antagonism in systemic hypertension. Curr Hypertens Rep 6(3):195–200

Fugazzola L, Mannavola D, Cerutti N, Maghnie M, Pagella F, Bianchi P, Weber G, Persani L, Beck-Peccoz P (2000) Molecular analysis of the Pendred's syndrome gene and magnetic resonance imaging studies of the inner ear are essential for the diagnosis of true Pendred's syndrome. J Clin Endocrinol Metab 85(7):2469–2475

Fugazzola L, Cerutti N, Mannavola D, Crino A, Cassio A, Gasparoni P, Vannucchi G, Beck-Peccoz P (2002) Differential diagnosis between Pendred and pseudo-Pendred syndromes: clinical, radiologic, and molecular studies. Pediatr Res 51(4):479–484

Fugazzola L, Cirello V, Dossena S, Rodighiero S, Muzza M, Castorina P, Lalatta F, Ambrosetti U, Beck-Peccoz P, Botta G, Paulmichl M (2007) High phenotypic intrafamilial variability in patients with Pendred syndrome and a novel duplication in the *SLC26A4* gene: clinical characterization and functional studies of the mutated SLC26A4 protein. Eur J Endocrinol 157(3):331–338

Garay RP, Alvarez-Guerra M, Alda JO, Nazaret C, Soler A, Vargas F (1998) Regulation of renal Na-K-Cl cotransporter NKCC2 by humoral natriuretic factors: relevance in hypertension. Clin Exp Hypertens 20(5–6):675–682

Garnett JP, Hickman E, Burrows R, Hegyi P, Tiszlavicz L, Cuthbert AW, Fong P, Gray MA (2011) Novel role for pendrin in orchestrating bicarbonate secretion in cystic fibrosis transmembrane conductance regulator (CFTR)-expressing airway serous cells. J Biol Chem 286(47):41069–41082

Geertsma ER, Chang YN, Shaik FR, Neldner Y, Pardon E, Steyaert J, Dutzler R (2015) Structure of a prokaryotic fumarate transporter reveals the architecture of the SLC26 family. Nat Struct Mol Biol 22(10):803–808

Gigolashvili T, Kopriva S (2014) Transporters in plant sulfur metabolism. Front Plant Sci 5:442

Gillam MP, Sidhaye AR, Lee EJ, Rutishauser J, Stephan CW, Kopp P (2004) Functional characterization of pendrin in a polarized cell system. Evidence for pendrin-mediated apical iodide efflux. J Biol Chem 279(13):13004–13010

Gillam MP, Bartolone L, Kopp P, Benvenga S (2005) Molecular analysis of the *PDS* gene in a nonconsanguineous Sicilian family with Pendred's syndrome. Thyroid 15(7):734–741

Gorbunov D, Sturlese M, Nies F, Kluge M, Bellanda M, Battistutta R, Oliver D (2014) Molecular architecture and the structural basis for anion interaction in prestin and SLC26 transporters. Nat Commun 5:3622

Gribble FM, Reimann F (2003) Sulphonylurea action revisited: the post-cloning era. Diabetologia 46(7):875–891

Haggie PM, Phuan PW, Tan JA, Zlock L, Finkbeiner WE, Verkman AS (2016) Inhibitors of pendrin anion exchange identified in a small molecule screen increase airway surface liquid volume in cystic fibrosis. FASEB J 30(6):2187–2197

Hastbacka J, De la Chapelle A, Mahtani MM, Clines G, Reeve-Daly MP, Daly M, Hamilton BA, Kusumi K, Trivedi B, Weaver A (1994) The diastrophic dysplasia gene encodes a novel sulfate transporter: positional cloning by fine-structure linkage disequilibrium mapping. Cell 78(6):1073–1087

Hong JH, Yang D, Shcheynikov N, Ohana E, Shin DM, Muallem S (2013) Convergence of IRBIT, phosphatidylinositol (4,5) bisphosphate, and WNK/SPAK kinases in regulation of the Na^+-HCO_3^- cotransporters family. Proc Natl Acad Sci U S A 110(10):4105–4110

Hulander M, Kiernan AE, Blomqvist SR, Carlsson P, Samuelsson EJ, Johansson BR, Steel KP, Enerback S (2003) Lack of pendrin expression leads to deafness and expansion of the endolymphatic compartment in inner ears of *Foxi1* null mutant mice. Development 130(9):2013–2025

Ishihara K, Okuyama S, Kumano S, Iida K, Hamana H, Murakoshi M, Kobayashi T, Usami S, Ikeda K, Haga Y, Tsumoto K, Nakamura H, Hirasawa N, Wada H (2010) Salicylate restores transport function and anion exchanger activity of missense pendrin mutations. Hear Res 270(1–2):110–118

Isogai S, Deupi X, Opitz C, Heydenreich FM, Tsai CJ, Brueckner F, Schertler GF, Veprintsev DB, Grzesiek S (2016) Backbone NMR reveals allosteric signal transduction networks in the beta1-adrenergic receptor. Nature 530(7589):237–241

Jabeen T, Singh N, Singh RK, Sharma S, Somvanshi RK, Dey S, Singh TP (2005) Non-steroidal anti-inflammatory drugs as potent inhibitors of phospholipase A2: structure of the complex of phospholipase A2 with niflumic acid at 2.5 Angstroms resolution. Acta Crystallogr D Biol Crystallogr 61(Pt 12):1579–1586

Jang JH, Jung J, Kim AR, Cho YM, Kim MY, Lee SY, Choi JY, Lee JH, Choi BY (2014) Identification of novel functional null allele of *SLC26A4* associated with enlarged vestibular aqueduct and its possible implication. Audiol Neurootol 19(5):319–326

Jessen F, Sjoholm C, Hoffmann EK (1986) Identification of the anion exchange protein of Ehrlich cells: a kinetic analysis of the inhibitory effects of 4,4'-diisothiocyano-2,2'-stilbene-disulfonic acid (DIDS) and labeling of membrane proteins with 3H-DIDS. J Membr Biol 92(3):195–205

Jung J, Kim J, Roh SH, Jun I, Sampson RD, Gee HY, Choi JY, Lee MG (2016) The HSP70 co-chaperone DNAJC14 targets misfolded pendrin for unconventional protein secretion. Nat Commun 7:11386

Kara C, Kilic M, Ucakturk A, Aydin M (2010) Congenital goitrous hypothyroidism, deafness and iodide organification defect in four siblings: Pendred or pseudo-Pendred syndrome? J Clin Res Pediatr Endocrinol 2(2):81–84

Karinou E, Compton EL, Morel M, Javelle A (2013) The *Escherichia coli* Slc26 homologue YchM (DauA) is a C(4)-dicarboxylic acid transporter. Mol Microbiol 87(3):623–640

Keller JP, Homma K, Duan C, Zheng J, Cheatham MA, Dallos P (2014) Functional regulation of the SLC26-family protein prestin by calcium/calmodulin. J Neurosci 34(4):1325–1332

Kopp P, Arseven OK, Sabacan L, Kotlar T, Dupuis J, Cavaliere H, Santos CL, Jameson JL, Medeiros-Neto G (1999) Phenocopies for deafness and goiter development in a large inbred Brazilian kindred with Pendred's syndrome associated with a novel mutation in the *PDS* gene. J Clin Endocrinol Metab 84(1):336–341

Kovacs H, Comfort D, Lord M, Campbell ID, Yudkin MD (1998) Solution structure of SpoIIAA, a phosphorylatable component of the system that regulates transcription factor sigmaF of *Bacillus subtilis*. Proc Natl Acad Sci U S A 95(9):5067–5071

Kraiem Z, Heinrich R, Sadeh O, Shiloni E, Nassir E, Hazani E, Glaser B (1999) Sulfate transport is not impaired in Pendred syndrome thyrocytes. J Clin Endocrinol Metab 84(7):2574–2576

Kumano S, Iida K, Ishihara K, Murakoshi M, Tsumoto K, Ikeda K, Kumagai I, Kobayashi T, Wada H (2010) Salicylate-induced translocation of prestin having mutation in the GTSRH sequence to the plasma membrane. FEBS Lett 584(11):2327–2332

Lee K, Hong TJ, Hahn JS (2012) Roles of 17-AAG-induced molecular chaperones and Rma1 E3 ubiquitin ligase in folding and degradation of Pendrin. FEBS Lett 586(16):2535–2541

Li X, Sanneman JD, Harbidge DG, Zhou F, Ito T, Nelson R, Picard N, Chambrey R, Eladari D, Miesner T, Griffith AJ, Marcus DC, Wangemann P (2013) SLC26A4 targeted to the endolymphatic sac rescues hearing and balance in *Slc26a4* mutant mice. PLoS Genet 9(7), e1003641

Lindskog S (1997) Structure and mechanism of carbonic anhydrase. Pharmacol Ther 74(1):1–20

Lu F, Li S, Jiang Y, Jiang J, Fan H, Lu G, Deng D, Dang S, Zhang X, Wang J, Yan N (2011) Structure and mechanism of the uracil transporter UraA. Nature 472(7342):243–246

Masuda S, Murakami KS, Wang S, Anders OC, Donigian J, Leon F, Darst SA, Campbell EA (2004) Crystal structures of the ADP and ATP bound forms of the *Bacillus* anti-sigma factor SpoIIAB in complex with the anti-anti-sigma SpoIIAA. J Mol Biol 340(5):941–956

McDonald FJ, Snyder PM, McCray PB Jr, Welsh MJ (1994) Cloning, expression, and tissue distribution of a human amiloride-sensitive Na^+ channel. Am J Physiol 266(6 Pt 1):L728–L734

Monroy A, Plata C, Hebert SC, Gamba G (2000) Characterization of the thiazide-sensitive Na(+)-Cl(−) cotransporter: a new model for ions and diuretics interaction. Am J Physiol Renal Physiol 279(1):F161–F169

Moore PF, Larson DL, Otterness IG, Weissman A, Kadin SB, Sweeney FJ, Eskra JD, Nagahisa A, Sakakibara M, Carty TJ (1996) Tenidap, a structurally novel drug for the treatment of arthritis: antiinflammatory and analgesic properties. Inflamm Res 45(2):54–61

Moradi M, Enkavi G, Tajkhorshid E (2015) Atomic-level characterization of transport cycle thermodynamics in the glycerol-3-phosphate:phosphate antiporter. Nat Commun 6:8393

Nakao I et al (2008) Identification of pendrin as a common mediator for mucus production in bronchial asthma and chronic obstructive pulmonary disease. J Immunol 180(9):6262–6269

Nieddu E, Pollarolo B, Merello L, Schenone S, Mazzei M (2013) F508del-CFTR rescue: a matter of cell stress response. Curr Pharm Des 19(19):3476–3496

Nishio A, Ito T, Cheng H, Fitzgerald TS, Wangemann P, Griffith AJ (2016) *Slc26a4* expression prevents fluctuation of hearing in a mouse model of large vestibular aqueduct syndrome. Neuroscience 329:74–82

Ohana E, Shcheynikov N, Moe OW, Muallem S (2013) SLC26A6 and NaDC-1 transporters interact to regulate oxalate and citrate homeostasis. J Am Soc Nephrol 24(10):1617–1626

Palos F, Garcia-Rendueles ME, raujo-Vilar D, Obregon MJ, Calvo RM, Cameselle-Teijeiro J, Bravo SB, Perez-Guerra O, Loidi L, Czarnocka B, Alvarez P, Refetoff S, Dominguez-Gerpe L, Alvarez CV, Lado-Abeal J (2008) Pendred syndrome in two Galician families: insights into clinical phenotypes through cellular, genetic, and molecular studies. J Clin Endocrinol Metab 93(1):267–277

Pasqualetto E, Aiello R, Gesiot L, Bonetto G, Bellanda M, Battistutta R (2010) Structure of the cytosolic portion of the motor protein prestin and functional role of the STAS domain in SLC26/SulP anion transporters. J Mol Biol 400(3):448–462

Pedemonte N, Caci E, Sondo E, Caputo A, Rhoden K, Pfeffer U, Di CM, Bandettini R, Ravazzolo R, Zegarra-Moran O, Galietta LJ (2007) Thiocyanate transport in resting and IL-4-stimulated human bronchial epithelial cells: role of pendrin and anion channels. J Immunol 178(8):5144–5153

Pera A, Dossena S, Rodighiero S, Gandia M, Botta G, Meyer G, Moreno F, Nofziger C, Hernandez-Chico C, Paulmichl M (2008) Functional assessment of allelic variants in the *SLC26A4* gene involved in Pendred syndrome and nonsyndromic EVA. Proc Natl Acad Sci U S A 105(47):18608–18613

Pfarr N, Borck G, Turk A, Napiontek U, Keilmann A, Muller-Forell W, Kopp P, Pohlenz J (2006) Goitrous congenital hypothyroidism and hearing impairment associated with mutations in the *TPO* and *SLC26A4/PDS* genes. J Clin Endocrinol Metab 91(7):2678–2681

Pryor SP, Madeo AC, Reynolds JC, Sarlis NJ, Arnos KS, Nance WE, Yang Y, Zalewski CK, Brewer CC, Butman JA, Griffith AJ (2005) *SLC26A4/PDS* genotype-phenotype correlation in hearing loss with enlargement of the vestibular aqueduct (EVA): evidence that Pendred syndrome and non-syndromic EVA are distinct clinical and genetic entities. J Med Genet 42(2):159–165

Reardon W, OMahoney CF, Trembath R, Jan H, Phelps PD (2000) Enlarged vestibular aqueduct: a radiological marker of Pendred syndrome, and mutation of the *PDS* gene. QJM 93(2):99–104

Rebeh IB, Yoshimi N, Hadj-Kacem H, Yanohco S, Hammami B, Mnif M, Araki M, Ghorbel A, Ayadi H, Masmoudi S, Miyazaki H (2010) Two missense mutations in *SLC26A4* gene: a molecular and functional study. Clin Genet 78(1):74–80

Reimold FR, Heneghan JF, Steward AK, Zelikovic I, Vandorpe DH, Shmukler BE, Alper SL (2011) Pendrin function and regulation in *Xenopus* oocytes. Cell Physiol Biochem 28(3):435–450

Rotman-Pikielny P, Hirschberg K, Maruvada P, Suzuki K, Royaux IE, Green ED, Kohn LD, Lippincott-Schwartz J, Yen PM (2002) Retention of pendrin in the endoplasmic reticulum is a major mechanism for Pendred syndrome. Hum Mol Genet 11(21):2625–2633

Scott DA, Karniski LP (2000) Human pendrin expressed in *Xenopus laevis* oocytes mediates chloride/formate exchange. Am J Physiol Cell Physiol 278(1):C207–C211

Scott DA, Wang R, Kreman TM, Sheffield VC, Karniski LP (1999) The Pendred syndrome gene encodes a chloride-iodide transport protein. Nat Genet 21(4):440–443

Scott DA, Wang R, Kreman TM, Andrews M, McDonald JM, Bishop JR, Smith RJ, Karniski LP, Sheffield VC (2000) Functional differences of the *PDS* gene product are associated with phenotypic variation in patients with Pendred syndrome and non-syndromic hearing loss (DFNB4). Hum Mol Genet 9(11):1709–1715

Sharma AK, Rigby AC, Alper SL (2011a) STAS domain structure and function. Cell Physiol Biochem 28(3):407–422

Sharma AK, Ye L, Baer CE, Shanmugasundaram K, Alber T, Alper SL, Rigby AC (2011b) Solution structure of the guanine nucleotide-binding STAS domain of SLC26-related SulP protein Rv1739c from *Mycobacterium tuberculosis*. J Biol Chem 286(10):8534–8544

Sharma AK, Zelikovic I, Alper SL (2014) Molecular dynamics simulations of the STAS domains of rat prestin and human pendrin reveal conformational motions in conserved flexible regions. Cell Physiol Biochem 33(3):605–620

Shcheynikov N, Yang D, Wang Y, Zeng W, Karniski LP, So I, Wall SM, Muallem S (2008) The Slc26a4 transporter functions as an electroneutral $Cl^-/I^-/HCO_3^-$ exchanger: role of Slc26a4 and Slc26a6 in I^- and $HCO3^-$ secretion and in regulation of CFTR in the parotid duct. J Physiol 586(16):3813–3824

Shen B, Fu J, Guo J, Zhang J, Wang X, Pan X, Chen M, Zhou Y, Zhu M, Du J (2015) Role of Na^+-K^+-$2Cl^-$ Cotransporter 1 in Phenylephrine-Induced Rhythmic Contraction in the Mouse Aorta: Regulation of Na^+-K^+-$2Cl^-$ Cotransporter 1 by Ca^{2+} Sparks and KCa Channels. Cell Physiol Biochem 37(2):747–758

Shepshelovich J, Goldstein-Magal L, Globerson A, Yen PM, Rotman-Pikielny P, Hirschberg K (2005) Protein synthesis inhibitors and the chemical chaperone TMAO reverse endoplasmic reticulum perturbation induced by overexpression of the iodide transporter pendrin. J Cell Sci 118(Pt 8):1577–1586

Sherman T, Chernova MN, Clark JS, Jiang L, Alper SL, Nehrke K (2005) The abts and sulp families of anion transporters from *Caenorhabditis elegans*. Am J Physiol Cell Physiol 289(2):C341–C351

Shibagaki N, Grossman AR (2004) Probing the function of STAS domains of the *Arabidopsis* sulfate transporters. J Biol Chem 279(29):30791–30799

Shibagaki N, Grossman AR (2010) Binding of cysteine synthase to the STAS domain of sulfate transporter and its regulatory consequences. J Biol Chem 285(32):25094–25102

Soleimani M (2012) A novel target for diuretic therapy. Iran J Kidney Dis 6(6):419–425

Soleimani M (2015) The multiple roles of pendrin in the kidney. Nephrol Dial Transplant 30(8):1257–1266

Soleimani M, Greeley T, Petrovic S, Wang Z, Amlal H, Kopp P, Burnham CE (2001) Pendrin: an apical $Cl^-/OH^-/HCO_3^-$ exchanger in the kidney cortex. Am J Physiol Renal Physiol 280(2):F356–F364

Srinivasan L, Baars TL, Fendler K, Michel H (2016) Functional characterization of solute carrier (SLC) 26/sulfate permease (SulP) proteins in membrane mimetic systems. Biochim Biophys Acta 1858(4):698–705

Stenson PD, Mort M, Ball EV, Shaw K, Phillips A, Cooper DN (2014) The Human Gene Mutation Database: building a comprehensive mutation repository for clinical and molecular genetics, diagnostic testing and personalized genomic medicine. Hum Genet 133(1):1–9

Taddei A, Folli C, Zegarra-Moran O, Fanen P, Verkman AS, Galietta LJ (2004) Altered channel gating mechanism for CFTR inhibition by a high-affinity thiazolidinone blocker. FEBS Lett 558(1–3):52–56

Takeda M, Narikawa S, Hosoyamada M, Cha SH, Sekine T, Endou H (2001) Characterization of organic anion transport inhibitors using cells stably expressing human organic anion transporters. Eur J Pharmacol 419(2–3):113–120

Tamargo J, Segura J, Ruilope LM (2014a) Diuretics in the treatment of hypertension. Part 1: thiazide and thiazide-like diuretics. Expert Opin Pharmacother 15(4):527–547

Tamargo J, Segura J, Ruilope LM (2014b) Diuretics in the treatment of hypertension. Part 2: loop diuretics and potassium-sparing agents. Expert Opin Pharmacother 15(5):605–621

Taylor JP, Metcalfe RA, Watson PF, Weetman AP, Trembath RC (2002) Mutations of the *PDS* gene, encoding pendrin, are associated with protein mislocalization and loss of iodide efflux: implications for thyroid dysfunction in Pendred syndrome. J Clin Endocrinol Metab 87(4):1778–1784

Toure A, Morin L, Pineau C, Becq F, Dorseuil O, Gacon G (2001) Tat1, a novel sulfate transporter specifically expressed in human male germ cells and potentially linked to rhogtpase signaling. J Biol Chem 276(23):20309–20315

Van Hauwe P, Everett LA, Coucke P, Scott DA, Kraft ML, Ris-Stalpers C, Bolder C, Otten B, de Vijlder JJ, Dietrich NL, Ramesh A, Srisailapathy SC, Parving A, Cremers CW, Willems PJ, Smith RJ, Green ED, Van CG (1998) Two frequent missense mutations in Pendred syndrome. Hum Mol Genet 7(7):1099–1104

Wall SM (2015) The role of pendrin in blood pressure regulation. Am J Physiol Renal Physiol:2015 Nov 4:ajprenal.00400.2015

Walsh T, Abu RA, Abu SJ, Shahin H, Shepshelovich J, Lee MK, Hirschberg K, Tekin M, Salhab W, Avraham KB, King MC, Kanaan M (2006) Genomic analysis of a heterogeneous Mendelian phenotype: multiple novel alleles for inherited hearing loss in the Palestinian population. Hum Genomics 2(4):203–211

Wang Y, Liu Y, Deberg HA, Nomura T, Hoffman MT, Rohde PR, Schulten K, Martinac B, Selvin PR (2014) Single molecule FRET reveals pore size and opening mechanism of a mechanosensitive ion channel. Elife 3, e01834

Wangemann P (2013) Mouse models for pendrin-associated loss of cochlear and vestibular function. Cell Physiol Biochem 32:157–165

Xu JBS, Zahedi K, Soleimani M (2015) Identification of IQGAP-1 as a pendrin-binding protein in the kidney. J Am Soc Nephrol 26:SA-PO944

Yoon JS, Park HJ, Yoo SY, Namkung W, Jo MJ, Koo SK, Park HY, Lee WS, Kim KH, Lee MG (2008) Heterogeneity in the processing defect of *SLC26A4* mutants. J Med Genet 45(7):411–419

Yuan Y, Guo W, Tang J, Zhang G, Wang G, Han M, Zhang X, Yang S, He DZ, Dai P (2012) Molecular epidemiology and functional assessment of novel allelic variants of *SLC26A4* in non-syndromic hearing loss patients with enlarged vestibular aqueduct in China. PLoS One 7(11), e49984

Zhang JY, Wu F, Gu XM, Jin ZX, Kong LH, Zhang Y, Zhou JJ, Gao F (2015) The Blockade of Transmembrane Cl(−) Flux Mitigates I/R-Induced Heart Injury via the Inhibition of Calpain Activity. Cell Physiol Biochem 35(6):2121–2134

Zheng ZL, Zhang B, Leustek T (2014) Transceptors at the boundary of nutrient transporters and receptors: a new role for *Arabidopsis* SULTR1;2 in sulfur sensing. Front Plant Sci 5:710

Zolotarev AS, Unnikrishnan M, Shmukler BE, Clark JS, Vandorpe DH, Grigorieff N, Rubin EJ, Alper SL (2008) Increased sulfate uptake by *E. coli* overexpressing the SLC26-related SulP protein Rv1739c from *Mycobacterium tuberculosis*. Comp Biochem Physiol A Mol Integr Physiol 149(3):255–266

Index

S. Dossena, M. Paulmichl (eds.), *The Role of Pendrin in Health and Disease*,
DOI 10.1007/978-3-319-43287-8